Design of RF and Microwave Amplifiers and Oscillators

Second Edition

DISCLAIMER OF WARRANTY

For a listing of recent titles in the
Artech House Microwave Library,
turn to the back of this book.

Design of RF and Microwave Amplifiers and Oscillators

Second Edition

Pieter L. D. Abrie

ARTECH
HOUSE

BOSTON | LONDON
artechhouse.com

Library of Congress Cataloging-in-Publication Data
A catalog record for this book is available from the U.S. Library of Congress.

British Library Cataloguing in Publication Data
A catalogue record for this book is available from the British Library.

ISBN-13 978-1-59693-098-8

Cover design by Yekaterina Ratner

© 2009 ARTECH HOUSE
685 Canton Street
Norwood, MA 02062

10 9 8 7 6 5 4 3 2 1

To my wife, Hilda,

and

our sons, Albert, Dewald, and Willem

Contents

Preface   xv

Chapter 1   **Characterization and Analysis of Linear Circuits at RF and Microwave Frequencies**   1

1.1   Introduction   1

1.2   *Y*-Parameters   1

    1.2.1   The Indefinite Admittance Matrix   7

1.3   *Z*-Parameters   8

1.4   Transmission Parameters   10

1.5   Scattering Parameters   11

    1.5.1   *S*-Parameter Definitions   12
    1.5.2   The Physical Meanings of the Normalized Incident and Reflected Components of an *N*-Port   18
    1.5.3   The Physical Interpretations of the Scattering Parameters   20
    1.5.4   Constraints Imposed on the Normalized Incident and Reflected Components by the Terminations of an *N*-Port   23
    1.5.5   Derivation of Expressions for the Gain Ratios and Reflection Parameters of a Two-Port   26
    1.5.6   Signal Flow Graphs   31
    1.5.7   The Indefinite *S*-Matrix   35
    1.5.8   Extension of the Single-Frequency *S*-Parameter Definitions to the Complex Frequency Plane   37
    1.5.9   Constraints on the Scattering Matrix of a Lossless *N*-Port   40
    1.5.10   Conversion of *S*-Parameters to Other Parameters   44

Questions and Problems   46

References   48

Selected Bibliography   48

**Chapter 2   Characterization and Analysis of Active Circuits
              at RF and Microwave Frequencies**   49

2.1   Introduction   49

2.2   Noise Parameters   52

       2.2.1   Modeling the Noise Contribution of a Two-Port with
               Equivalent Circuits   52
       2.2.2   Noise Correlation Matrices   58
       2.2.3   Calculating the Noise Figure of a Cascade Network   62

2.3   The Output Power of Linear Amplifiers   64

       2.3.1   Load-Line Considerations in Class-A and Class-B Amplifiers   64
       2.3.2   Class-AB and Class-F Load Line   70
       2.3.3   Class-E Load Line   73
       2.3.4   Doherty Amplifiers   74
       2.3.5   Distortion in Linear Amplifiers   77
       2.3.6   The Cripps Approach to Estimating the Maximum
               Output Power Obtainable from a Transistor   83
       2.3.7   Estimation of the Maximum Output Power of a Linear Network
               by Using the Power Parameters   85

Questions and Problems   100

References   102

Chapter 3   Radio-Frequency Components   105

3.1   Introduction   105

3.2   Capacitors   106

3.3   Inductors   109

       3.3.1   The Influence of Parasitic Capacitance on an Inductor   110
       3.3.2   Low-Frequency Losses in Inductors   112
       3.3.3   The Skin Effect   112
       3.3.4   The Proximity Effect   115
       3.3.5   Magnetic Materials   115
       3.3.6   The Design of Single-Layer Solenoidal Coils   118
       3.3.7   The Design of Inductors with Magnetic Cores   124

3.4   Transmission Lines   128

3.4.1   Coaxial Cables   128
3.4.2   Microstrip Transmission Lines   129
3.4.3   Twisted Pairs   133

Questions and Problems   134

References   136

Selected Bibliography   137

Chapter 4   Narrowband Impedance-Matching with LC Networks   139

4.1   Introduction   139

4.2   Parallel Resonance   140

4.3   Series Resonance   144

4.4   L-Sections   146

4.5   PI-Sections and T-Sections   151

4.5.1   The PI-Section   152
4.5.2   The T-Section   155

4.6   The Design of PI-Sections and T-Sections with Complex Terminations   156

4.7   Four-Element Matching Networks   159

4.8   Calculation of the Insertion Loss of an LC Impedance-Matching Network   160

4.9   Calculation of the Bandwidth of Cascaded LC Networks   162

Questions and Problems   162

Selected Bibliography   164

Chapter 5   Coupled Coils and Transformers   165

5.1   Introduction   165

5.2   The Ideal Transformer   165

5.3   Equivalent Circuits for Practical Transformers   167

5.4   Wideband Impedance Matching with Transformers   170

5.5   Single-Tuned Transformers   172

5.6   Tapped Coils   173

5.7   Parallel Double-Tuned Transformers   179

5.8   Series Double-Tuned Transformers   185

5.9   Measurement of the Coupling Factor of a Transformer   188

        5.9.1   Measurement of the Coupling Factor by Short-Circuiting the
                Secondary Winding   188
        5.9.2   Measurement of the Coupling Factor by Measuring the Open-
                Circuit Voltage Gain   189
        5.9.3   Deriving the Coupling Factor from S-Parameter Measurements   189

Questions and Problems   190

References   192

Chapter 6   Transmission-Line Transformers   193

6.1   Introduction   193

6.2   Transmission-Line Transformer Configurations   195

6.3   Analysis of Transmission-Line Transformers   201

6.4   Design of Transmission Line Transformers   209

        6.4.1   Determining the Optimum Characteristic Impedance and
                Diameter of the Transmission Line to Be Used   210
        6.4.2   Determining the Minimum Value of the Magnetizing
                Inductance of the Transformer   211
        6.4.3   Determining the Type and Size of the Magnetic Core to
                Be Used   213
        6.4.4   Compensation of Transmission-Line Transformers for
                Nonoptimum Characteristic Impedances   216
        6.4.5   The Design of Highpass LC Networks to Extend the
                Bandwidth of a Transmission-Line Transformer   220

6.5   Considerations Applying to RF Power Amplifiers   223

Questions and Problems   228

References   230

Selected Bibliography   230

**Chapter 7   Film Resistors, Parallel-Plate Capacitors, Inductors,
                and Microstrip Discontinuities**   231

7.1   Introduction   231

7.2   Film Resistors   233

7.3   Single-Layer Parallel-Plate Capacitors   234

        7.3.1   Parallel-Plate Capacitors on a Ground Plane   237
        7.3.2   Parallel-Plate Capacitors Used as Series Stubs   238
        7.3.3   Series Connected Parallel-Plate Capacitors   240

7.4   Inductors   251

        7.4.1   Strip Inductors   252
        7.4.2   Single-Turn Circular Loop   252
        7.4.3   Bond Wire Inductors   252
        7.4.4   Single-Layer Solenoidal Air-Cored Inductors   255
        7.4.5   Spiral Inductors   255

7.5   Microstrip Discontinuity Effects at the Lower Microwave Frequencies   259

        7.5.1   Open-Ended Stubs   259
        7.5.2   Steps in Width   260
        7.5.3   Microstrip Bends and Curves   261
        7.5.4   T-Junctions and Crosses   264
        7.5.5   Via Holes   265

Questions and Problems   266

References   267

Selected Bibliography   269

Chapter 8   The Design of Wideband Impedance-Matching Networks   271

8.1   Introduction   271

8.2   Fitting an Impedance or Admittance Function to a Set of Impedance
        Versus Frequency Coordinates   272

8.3   The Analytical Approach to Impedance Matching   279

        8.3.1   Darlington Synthesis of Impedance-Matching Networks   281
        8.3.2   LC Transformers   286
        8.3.3   The Gain-Bandwidth Constraints Imposed by Simple RC and
                RL Loads   289
        8.3.4   Direct Synthesis of Impedance-Matching Networks When the
                Load (or Source) Is Reactive   291
        8.3.5   Synthesis of Networks for Matching a Reactive Load to a
                Purely Resistive or a Reactive Source by Using the Principle
                of Parasitic Absorption   295
        8.3.6   The Analytical Approach to Designing Commensurate
                Distributed Impedance-Matching Networks   299

8.4   The Iterative Design of Impedance-Matching Networks   305

        8.4.1   The Line-Segment Approach to Matching a Complex Load to
                a Resistive Source   307
        8.4.2   The Reflection Coefficient Approach to Solving Double-
                Matching Problems   320
        8.4.3   The Transformation-Q Approach to the Design of Impedance-
                Matching Networks   331

8.5   The Design of RLC Impedance-Matching Networks   354

Questions and Problems   360

References   365

Selected Bibliography   367

**Chapter 9   The Design of Radio-Frequency and Microwave
               Amplifiers and Oscillators**   369

9.1   Introduction   369

9.2   Stability   370

    9.2.1   Stability Circles on the Admittance Plane   372
    9.2.2   Stability Circles on the Smith Chart and Associated Stability
            Factors and Figures of Merit   376
    9.2.3   The Reflection Gain Approach   382
    9.2.4   The Loop Gain Approach   383
    9.2.5   Stabilization of a Two-Port with Shunt or Series Resistance   386

9.3   Tunability   388

9.4   Controlling the Gain of an Amplifier   389

    9.4.1   Circles of Constant Mismatch for a Passive Problem   392
    9.4.2   Constant Operating Power Gain Circles   394
    9.4.3   Constant Available Power Gain Circles   397
    9.4.4   Constant Transducer Power Gain Circles   398

9.5   Controlling the Noise Figure of an Amplifier   400

9.6   Controlling the Output Power or the Effective Output Power of a Transistor   404

9.7   The Equivalent Passive Impedance-Matching Problem
                    409
    9.7.1   Constant Operating Power Gain Case   410
    9.7.2   Constant Available Power Gain Case   412
    9.7.3   Constant Noise Figure Case   413

9.8   Resistive Feedback and Loading   414

9.9   Designing Cascade Amplifiers   421

9.10   Lossless Feedback Amplifiers   425

9.11   Reflection Amplifiers   433

9.12   Balanced Amplifiers   436

9.13   Oscillator Design   438

    9.13.1 Estimation of the Compression Associated with the Maximum
             Effective Output Power   446
    9.13.2 Derivation of the Equations for the T- and PI-Section Feedback
             Components Required   448
    9.13.3 High-Q Resonator Circuits   452
    9.13.4 Transforming the Impedance Presented by a Resonator Network
             to That Required in the T- and PI-Section Feedback Network   455

9.13.5 Designing Varactor Circuits to Realize the Varactor-Type
         Reactance Required   457
9.13.6 Considerations Applying to Oscillators with Low Phase Noise   459

Questions and Problems   460

References   463

Selected Bibliography   465

Appendix   467

About the Author   471

Index   473

PREFACE

In this book the design of radio-frequency and microwave amplifiers and oscillators is addressed. The focus will be on synthesis techniques (iterative, as well as analytical) for designing linear amplifiers. Class A, class B, class AB, class E (not linear), class F, and Doherty amplifiers will be considered. The design of small-signal amplifiers, including low-noise amplifiers, will also be considered in detail.

Most of the material covered previously in [1, 2] is also included in this book. Where necessary, the material has been revised and extended. The executable of a new integrated version (LSM) of the main computer programs provided previously is also licensed with this book. The new program is a Visual C++ 2008 program with a menu driven interface. This program will run under Windows Vista and Windows XP. The Fortran source code of the previous versions are also supplied with the book. Wideband single-, as well as double-matching (complex source and complex load) problems can be solved with LSM. A wizard for fitting a resistance (conductance) function to a set of resistance versus frequency data points is also provided in LSM.

To be of real practical use at microwave and millimeter-wave frequencies, it is essential to design a matching network with the pads required for any lumped components in place. The transformation-Q technique described can easily be extended to provide this capability. The same approach can be used to design mixed lumped/distributed matching networks in which the lumped components are used to reduce the line lengths required. Matching networks with spiral inductors and parallel-plate (MIM) capacitors can also be designed directly by following this approach. In order to do this, accurate modeling of spiral inductors and parallel-plate capacitors is required. Modeling of these components will also be considered in this book.

A main feature of this book remains the power parameter approach introduced in [2] to estimate the output power (1-dB compression point) of a linear amplifier without resorting to nonlinear analysis techniques. Apart from the advantage that close to optimum designs can be created in many cases without load-pull information or an accurate large-signal model, the power parameter approach also serves to generate solutions that can be refined in nonlinear circuit simulators.

The basic principle in the power parameter approach is that the output power of a linear amplifier is limited by the maximum amplitudes of the current and voltage associated with the intrinsic current-source in the transistor model [3]. The output power extracted from a transistor can be current- and/or voltage-limited. Harmonics can complicate or improve the situation. In a class-F amplifier, the square (ideal case) voltage waveform allows the fundamental tone voltage-swing to be larger than the supply voltage (minus the knee voltage), with a corresponding increase in power. The square waveform also increases the efficiency.

A small-signal model, the dc operating point (at full power), and four boundary lines on the dc or pulsed I/V-curves of each transistor used are required in this approach. The power parameters of the transistor can be derived from the small-signal model. All the normal operations associated with the S-parameters and the noise parameters of a linear

xv

circuit (feedback, loading, cascading, changes in configuration) are also allowed with the power parameters.

The external load line associated with any intrinsic load line can be found easily by using the power parameter approach. With this capability in place, it is a simple matter to generate load-pull contours for any linear amplifier stage, and to find the external terminations associated with the required intrinsic harmonic terminations.

The idea that amplifier design essentially reduces to the design of specialized impedance-matching networks is still prevalent. However, it was found that when more demanding amplifiers, especially wideband amplifiers, are designed, the performance could only be obtained by modifying the characteristics of the transistors used with frequency-selective resistive networks (feedback and/or loading). In doing this, the excess in capability (noise figure, gain) at the lower frequencies is exchanged for more desirable characteristics in the passband of interest (stability, gain leveled in the passband, reduced gain-bandwidth constraints, and correspondingly, improved VSWRs (voltage standing wave ratios), and optimum noise/power and optimum match points closer to each other). This preconditioning step will be referred to as device modification.

Because of the desensitizing effect of the resistive networks added, it was found that amplifiers based on device modification and impedance matching are frequently first-time right. It was also found that choosing the right transistors for an amplifier can be critical in this respect. While transistors may seem to be equivalent on superficial inspection, the performance obtainable during the device-modification stage may differ greatly.

The normal approach to deciding the stability of an amplifier was also found to be inadequate in some cases. Small changes in some circuits can easily change them from inherently stable to potentially (or actually) unstable. Satisfactory results were obtained in some cases when the stability analysis was extended to include calculation of the well-known gain and phase margins used in feedback theory. Because the actual cause of the oscillations is often the feedback loops introduced, it makes sense to investigate these loops in addition to calculating the usual "black-box" stability factors. Knowing that a loop is 1 dB away from oscillation is also useful.

When narrowband high power amplifiers are designed it may not be realistic to add modification networks to a transistor. Measures and techniques for designing conditionally stable amplifiers are then required. Useful concepts like the maximum single-sided-matched stable gain, and the maximum mismatched (double-sided) stable gain were introduced in [4, 5]. Stability factors, introduced in [4–6], will indicate the range of reflection coefficients (VSWRs) that can be tolerated before the potential instability of a selected transistor results in oscillations. Note that these reflection coefficients are defined in terms of the default normalization resistance used (50Ω). These stability factors can be generalized by using the actual terminations of interest (generally, complex impedances) as normalization impedances, as was done in [7].

The power parameters approach combined with loop gain calculations also lead directly to the design of RF and microwave oscillators. A major shortcoming in the regular approach to the design of oscillators is that only the negative resistance is considered during synthesis. Clearly, clipping of the voltage and current is as important as in the case of amplifiers. In the design approach proposed here the loop gain is controlled with the load line presented to the transistor. An immediate advantage of a well-behaved load line is that the main nonlinear effect in the oscillator will be g_m compression. Assuming an exponential saturation curve for the output power, the loop gain can then be controlled to maximize the output power of the oscillator. If low phase noise is required, the loop gain must be kept low in order to minimize upconversion of the flicker noise, and the loaded Q of the circuit must be maximized.

The material in this book is organized as follows.

Analysis and characterization of linear RF and microwave circuits with Y-, Z-, T-, and S-parameters are considered in Chapter 1. Analysis by using flow diagrams is also considered. Characterization and analysis of the noise and the power performance of active linear circuits are considered in Chapter 2. Noise correlation matrices and the power parameters are also covered, with the load-line considerations applying to the different classes of linear power amplifiers. Doherty amplifiers are also considered.

Radio-frequency components are considered in Chapter 3. Basic inductor, capacitor, and resistor models are considered, with the skin effect and the proximity effect. The design of single-layer air-cored inductors, and inductors with magnetic cores, is also investigated in detail. Coaxial cables and microstrip transmission lines are also considered in this chapter.

Resonant circuits and the design of narrowband impedance-matching networks (L-, T-, and PI-sections) are investigated in Chapter 4. Coupled coils and conventional transformers are covered in Chapter 5. Transmission-line transformers are widely used in RF and UHF circuits and are covered in Chapter 6. The design of RF power amplifiers is also considered in this chapter. Film resistors, single-layer parallel-plate capacitors (including MIM capacitors), spiral inductors, and microstrip discontinuities are considered in Chapter 7. Chapter 8 is devoted to the design of wideband impedance-matching networks. The design of RF and microwave amplifiers and oscillators is considered in Chapter 10. Cascade amplifiers, lossless feedback amplifiers, reflection amplifiers, and balanced amplifiers are considered in this chapter.

This book was improved significantly by the feedback provided by the reviewer, and I would like to acknowledge his efforts and contributions.

REFERENCES

[1]   Abrie, P.L.D., *The Design of Impedance-Matching Networks for Radio-Frequency and Microwave Amplifiers*, Dedham, MA: Artech House, 1985.

[2]   Abrie, P.L.D., *The Design of RF and Microwave Amplifiers and Oscillators*, Norwood, MA: Artech House, 1999.

[3]   Cripps, S.C., "GaAs Power Amplifier Design," Technical Notes 3.2, Palo Alto, CA: Matcom Inc.

[4]   Babak, L.I., "Comments on 'A Deterministic Approach for Designing Conditionally Stable Amplifiers'," *IEEE Trans. Microwave Theory Tech.*, Vol. MTT-47, No. 2, February 1999.

[5]   Edwards, M.L., and Cheng, S., "A Deterministic Approach for Designing Conditionally Stable Amplifiers," *IEEE Trans. Microwave Theory Tech.*, Vol. MTT-43, No. 7, July 1995.

[6]   Edwards, M.L., and Cheng, S., "Conditionally Stable Amplifier Design Using Constant μ-Contours," *IEEE Trans. Microwave Theory Tech.*, Vol. MTT-44, No. 12, December 1996.

[7]   *Multimatch RF and Microwave Impedance-Matching, Amplifier and Oscillator Synthesis Software*, Stellenbosch, South Africa, Ampsa (Pty) Ltd, http://www.ampsa.com, 2009.

CHAPTER 1

CHARACTERIZATION AND ANALYSIS OF LINEAR CIRCUITS AT RF AND MICROWAVE FREQUENCIES

1.1 INTRODUCTION

Low-frequency circuits are usually analyzed in terms of transfer functions. This approach is seldom used at RF and microwave frequencies. Analysis at these frequencies is usually in terms of one of the many sets of single-frequency parameters.

The parameters most frequently used are the *Y-, Z-, T-,* and *S*-parameters. The first three sets of parameters relate the terminal voltages and currents in different ways, while the *S*-parameters are closely related to the power incident to, and reflected from, a network.

Because of the relative ease with which *S*-parameters can be measured and the useful information directly obtained from them, components are usually characterized by measuring their *S*-parameters, and circuits are analyzed by calculating their *S*-parameters. The other parameters are often used to simplify the computations necessary for circuit analysis and synthesis.

Each of these sets of parameters will be considered in detail in the following sections.

Any of the voltages, currents, or power levels in a linear *N*-port network can be calculated in terms of the external signals (independent variables) when one of these sets of parameters is known at the frequency of interest. Conversion between the different parameters is straightforward.

1.2 *Y*-PARAMETERS

The *Y*-parameters of an *N*-port network are defined by the expression

$$I = Y V \tag{1.1}$$

where

$$I = \begin{bmatrix} I_1 \\ I_2 \\ \cdot \\ \cdot \\ I_N \end{bmatrix} \hspace{8cm} (1.2)$$

$$V = \begin{bmatrix} V_1 \\ V_2 \\ \cdot \\ \cdot \\ V_N \end{bmatrix} \hspace{8cm} (1.3)$$

$$Y = \begin{bmatrix} y_{11} & y_{12} & \cdots & y_{1N} \\ y_{21} & y_{22} & \cdots & y_{2N} \\ \cdots & \cdots & \cdots & \cdots \\ y_{N1} & y_{N2} & \cdots & y_{NN} \end{bmatrix} \hspace{5cm} (1.4)$$

I_i is the current flowing into the ith terminal, and V_i is the voltage across the ith port of the network.

Each element of the Y-parameter matrix can be calculated or measured by using the relationship

$$y_{ij} = \frac{I_i}{V_j}\bigg|_{V_h = 0} \hspace{1cm} h \in [1, 2, 3,..., N] \hspace{0.5cm} h \neq j \hspace{3cm} (1.5)$$

That is, y_{ij} is given by the ratio of the current flowing into the ith terminal (output signal) and the voltage across the jth port (input signal), with all the other voltages set to zero.

By using (1.1), the terminal currents corresponding to any given set of terminal voltages can be determined. The linear response of the network is, therefore, completely characterized when the N^2 elements of the Y-parameter matrix are known.

As with any other set of parameters, the Y-parameters can be used to calculate the impedances and gain ratios corresponding to any set of terminations. By using the equivalent circuit in Figure 1.1, it can be easily shown that the following expressions apply to a two-port network terminated as shown:

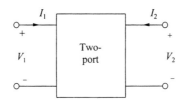

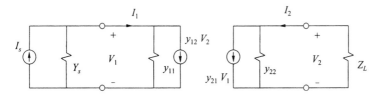

Figure 1.1   An equivalent circuit for a two-port in terms of its Y-parameters.

$$Y_{in} = \frac{I_1}{V_1} = y_{11} - \frac{y_{12}y_{21}}{y_{22} + Y_L} \tag{1.6}$$

$$Y_{out} = \frac{I_2}{V_2} = y_{22} - \frac{y_{12}y_{21}}{y_{11} + Y_s} \tag{1.7}$$

$$A_V = \frac{V_2}{V_1} = -\frac{y_{21}}{y_{22} + Y_L} \tag{1.8}$$

$$A_I = \frac{I_0}{I_1} = -\frac{I_2}{I_1} = A_V Y_L / Y_{in} \tag{1.9}$$

$$G_\omega = \frac{P_L}{P_{in}} = \frac{|V_2|^2 G_L}{|V_1|^2 G_{in}} = \left| \frac{y_{21}}{y_{22} + Y_L} \right|^2 \frac{G_L}{Re(Y_{in})} \tag{1.10}$$

$$G_T = \frac{P_L}{P_{av-E}} = \frac{|V_2|^2 G_L}{|I_s|^2 / (4G_s)} = \left| \frac{y_{21}}{(y_{11} + Y_s)(y_{22} + Y_L) - y_{12}y_{21}} \right|^2 4 G_L G_s \tag{1.11}$$

$$G_A = \frac{P_{av-o}}{P_{av-E}} = \frac{|y_{21} I_s / (y_{11} + Y_s)|^2 / [4 Re(Y_{out})]}{I_s / (4G_s)} = \left| \frac{y_{21}}{y_{11} + Y_s} \right|^2 \frac{G_s}{Re(Y_{out})} \tag{1.12}$$

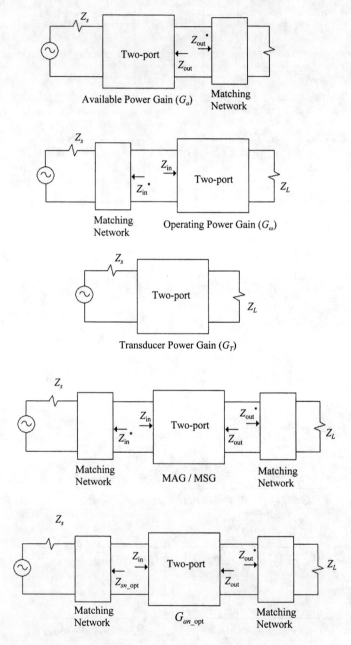

Figure 1.2   The equivalent circuits relevant to the different power gain definitions.

In these equations, $Y_L = G_L + jB_L$ is the load admittance, $Y_s = G_s + jB_s$ is the source admittance, P_L is the power dissipated in the load, P_{in} is the power entering the input port of the two-port, P_{av-E} is the power available from the source, P_{av-o} is the available power at the output terminals of the two-port, A_V and A_I are the voltage gain and current gain of the two-port, Y_{in} is the input admittance of the two-port, and Y_{out} is the output admittance.

The available power of a source is defined as the power dissipated in a load that conjugately matches the source, and is given by the expression

$$P_{av-E} = \frac{|E|^2}{4\,R_s} = \frac{|I_s|^2}{4\,G_s} \tag{1.13}$$

where E is the source voltage, I_s is the equivalent (Norton) source current, and R_s and G_s are defined by

$$Y_s = G_s + jB_s \tag{1.14}$$

$$Z_s = R_s + jX_s \tag{1.15}$$

where Z_s is the source impedance, and Y_s is its inverse. To calculate the available power at the output port of a two-port [as required in (1.12)], a Thevenin or Norton equivalent source must be set up for that port first.

Note that the operating power gain (G_ω) will be equal to the transducer power gain if the input is conjugately matched (see Figure 1.2). Similarly, the available power gain (G_a) will be equal to the transducer power gain (G_T) when the output is conjugately matched.

The maximum available gain (MAG) of a two-port is defined as the transducer power gain when both sides are conjugately matched (if possible). If the MAG cannot be calculated (negative resistance), the maximum stable gain (MSG) is of interest. The maximum stable gain (MSG) is the MAG associated with the device after adding the minimum shunt conductance or series resistance required for the MAG to exist.

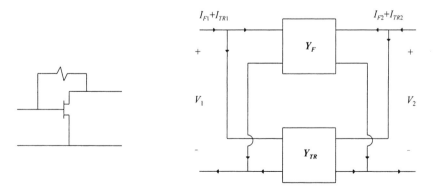

Figure 1.3   Two networks connected in parallel.

G_{an_opt} is the available power gain associated with an optimum noise match on the input side (i.e., Z_s is chosen to minimize the noise figure of the two-port).

When a circuit is analyzed, the Y-parameters are frequently used to find a single set of parameters characterizing two networks connected in parallel. This is illustrated in Figure 1.3. Note that the terminal voltages for the two networks are the same.

The Y-parameters of two networks connected in parallel simply equal the sum of the Y-parameters of each individual network:

$$Y_T = Y_A + Y_B \tag{1.16}$$

EXAMPLE 1.1   Derivation of the equation for the input admittance of a two-port terminated in a load with admittance Y_L.

The input admittance is defined by (1.6):

$$Y_{in} = I_1 / V_1$$

To find the input admittance it is therefore necessary to find an expression for V_1 in terms of I_1. Ohm's law and Kirchhoff's current law applied to the input port yield

$$V_1 = [I_1 - y_{12} V_2] / y_{11} \tag{1.17}$$

The output voltage is given by

$$V_2 = -I_2 / Y_L = -[y_{21} V_1 + y_{22} V_2] / Y_L \tag{1.18}$$

that is,

$$V_2 = -\frac{y_{21}}{y_{22} + Y_L} V_1 \tag{1.19}$$

After some manipulation, substitution of (1.19) into (1.17) yields (1.6):

$$Y_{in} = y_{11} - \frac{y_{12} y_{21}}{y_{22} Y_L}$$

1.2.1 The Indefinite Admittance Matrix

The indefinite admittance matrix is a useful tool by which the Y-parameters of a network can be determined if they are known for the same network connected differently. For example, if the common-emitter parameters of a bipolar transistor are known, this matrix can be used to determine the common-base or common-collector parameters.

An admittance matrix is indefinite when none of the network terminals have been connected to ground, and the total current flowing into it is therefore equal to the sum of the currents flowing into each terminal.

It can easily be shown that the sum of the elements in each row or each column of an indefinite admittance matrix is equal to zero. Considering a three-port network, this implies that if four of the nine parameters are known, then all the parameters are known.

The proof that the sum of the elements of each row must equal zero follows by choosing the terminal voltages to be equal. Each of the currents will then be zero and extraction of each individual equation from (1.1) yields the desired result.

That the sum of the elements in each column should also equal zero follows by setting two of the voltages equal to zero and adding the three currents, the sum of which must be equal to zero.

EXAMPLE 1.2   Calculation of the common-base parameters in terms of the common-emitter parameters.

The common-base parameters of a transistor will be determined in terms of its common-emitter parameters, as an example of using the indefinite admittance matrix.

The indefinite admittance parameters, which correspond to the common-emitter parameters, can be identified by setting V_2 in Figure 1.4 and in (1.20) equal to zero.

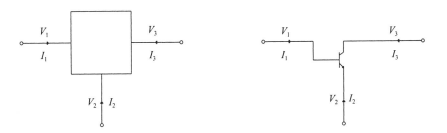

Figure 1.4   An indefinite three-port.

$$\begin{bmatrix} I_1 \\ I_2 \\ I_3 \end{bmatrix} = \begin{bmatrix} y_{11} & y_{12} & y_{13} \\ y_{21} & y_{22} & y_{23} \\ y_{31} & y_{32} & y_{33} \end{bmatrix} \begin{bmatrix} V_1 \\ V_2 \\ V_3 \end{bmatrix} \tag{1.20}$$

Because the current in the emitter (I_2) is not of interest when the common-emitter configuration is considered, (1.20) then reduces to

$$\begin{bmatrix} I_1 \\ I_3 \end{bmatrix} = \begin{bmatrix} y_{11} & y_{13} \\ y_{31} & y_{33} \end{bmatrix} \begin{bmatrix} V_1 \\ V_3 \end{bmatrix} = \begin{bmatrix} y_{11e} & y_{12e} \\ y_{21e} & y_{22e} \end{bmatrix} \begin{bmatrix} V_1 \\ V_3 \end{bmatrix} \tag{1.21}$$

With the common-emitter parameters known, y_{11}, y_{13}, y_{31}, and y_{33} are also known, and the rule for the zero column and row can now be applied to determine the other parameters. The only remaining step is to identify the common-base parameters in (1.20). Similar to the common-emitter parameters, this is done by setting V_1 in (1.20) equal to zero and eliminating the equation giving the base current (I_1) as a function of the voltages. It follows that

$$\begin{bmatrix} y_{11b} & y_{12b} \\ y_{21b} & y_{22b} \end{bmatrix} = \begin{bmatrix} y_{22} & y_{23} \\ y_{32} & y_{33} \end{bmatrix} \tag{1.22}$$

The common-collector parameters are given by

$$\begin{bmatrix} y_{11c} & y_{12c} \\ y_{21c} & y_{22c} \end{bmatrix} = \begin{bmatrix} y_{11} & y_{12} \\ y_{21} & y_{22} \end{bmatrix} \tag{1.23}$$

1.3   Z-PARAMETERS

The Z-parameters of an N-port network are defined by the expression

$$V = ZI \tag{1.24}$$

where

$$Z = \begin{bmatrix} z_{11} & z_{12} & \cdots & z_{1N} \\ z_{21} & z_{22} & \cdots & z_{2N} \\ \cdots & \cdots & \cdots & \cdots \\ z_{N1} & z_{N1} & \cdots & z_{NN} \end{bmatrix} \tag{1.25}$$

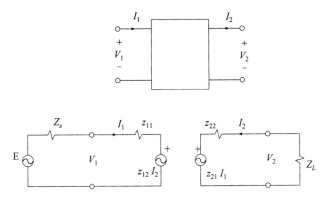

Figure 1.5   An equivalent circuit for a two-port network in terms of its Z-parameters.

and V and I are defined by (1.3) and (1.2), respectively. The equivalent circuit associated with the two-port case is shown in Figure 1.5.

Each element in (1.25) can be computed or measured by using the relationship

$$z_{ij} = \frac{V_i}{I_j}\bigg|_{I_h=0} \quad h \in [1,2,3,...,N] \quad h \neq j \tag{1.26}$$

that is, z_{ij} is the ratio of the voltages across the jth port (output signal) and the current at the ith port (input signal) with all the other ports idle (open-circuited).

Equation (1.24) can be used to find the terminal voltages corresponding to any given set of terminal currents.

Comparison of (1.24) and (1.1) reveals that the Z-parameters of a network are related to its Y-parameters in the following way:

$$\boldsymbol{Z} = \boldsymbol{Y}^{-1} \tag{1.27}$$

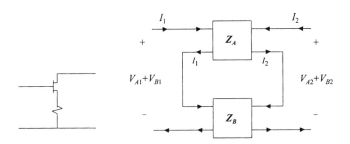

Figure 1.6   Two networks connected in series.

Z-parameters are frequently used to find an equivalent set of parameters for two networks connected in series, as illustrated in Figure 1.6. Note that when networks are connected in series, the terminal currents are the same, while the voltages add.

The Z-parameters of two networks connected in series are given in terms of the individual Z-parameters by

$$Z_T = Z_A + Z_B \tag{1.28}$$

1.4   TRANSMISSION PARAMETERS

The transmission parameters (T-parameters or ABCD parameters) of a two-port are defined by the equation

$$\begin{bmatrix} V_1 \\ I_1 \end{bmatrix} = \begin{bmatrix} A & B \\ C & D \end{bmatrix} \begin{bmatrix} V_2 \\ I_2 \end{bmatrix} \tag{1.29}$$

with the voltage and current as defined in Figure 1.7. Note that I_2 is the output current and not the current entering the output terminal as in the case of the Y- and Z-parameters.

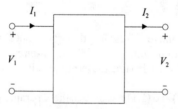

Figure 1.7   The voltage and current relevant to the definition of the transmission parameters.

The expressions for the individual elements of the transmission matrix can be obtained by setting either V_2 or I_2 in (1.28) equal to zero after extracting the individual equations from the matrix equation.

T-parameters can be converted to Y-parameters by using the following set of equations:

$$y_{11} = D / B \tag{1.30}$$

$$y_{12} = C - AD / B \tag{1.31}$$

$$y_{21} = -1/B \tag{1.32}$$

$$y_{22} = A/B \tag{1.33}$$

The inverse expressions are

$$A = -y_{22}/y_{21} \tag{1.34}$$

$$B = -1/y_{21} \tag{1.35}$$

$$C = y_{12} - y_{11}y_{22}/y_{21} \tag{1.36}$$

$$D = -y_{11}/y_{21} \tag{1.37}$$

Transmission parameters are frequently used to find an equivalent set of parameters for two cascaded networks. The transmission matrix for the equivalent network is given in terms of the matrices for the individual networks by

$$T_T = T_A T_B \tag{1.38}$$

This is illustrated in Figure 1.8.

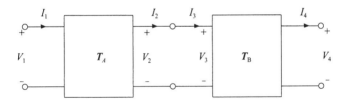

Figure 1.8   Two cascaded two-port networks.

1.5   SCATTERING PARAMETERS

Because of the ease with which scattering parameters (*S*-parameters) can be measured, as well as stability considerations and the physical meanings attached to them, *S*-parameters are used extensively to characterize components and also to analyze circuits.

The definitions relevant to these parameters, their physical meanings, and their application in analyzing circuits will be considered in the following sections. Both single-frequency *S*-parameters and those in the complex frequency plane will be considered. Because lossless networks are of considerable interest in this text, the constraints on the *S*-

matrix of a lossless network will also be examined.

1.5.1   S-Parameter Definitions

Similar to the reflection coefficients in transmission-line theory, S-parameters are defined in terms of incident and reflected components. In S-parameter theory, however, an incident component is defined as that component that would exist if the port under consideration were conjugately matched to the normalizing impedance at that port. The normalizing impedances are the equivalents of the short-circuit and open-circuit terminations used to characterize a network in terms of its Y-, Z-, or T-parameters. They can be defined to have any arbitrary value (as long as the resistive part is positive and not equal to zero), but 50-Ω impedances are used in most cases.

In terms of the current and voltage at each terminal, the incident components (V_i, I_i, a) and the reflected components (V_r, I_r, b) of an N-port are defined by the following set of matrix equations:

$$E_0 = V + Z_0 I \tag{1.39}$$

$$I_i = [Z_0 + Z_0^*]^{-1} E_0 \tag{1.40}$$

$$I = I_i - I_r \tag{1.41}$$

$$V_i = Z_0^* I_i \tag{1.42}$$

$$V = V_i + V_r \tag{1.43}$$

$$a = \frac{1}{\sqrt{2}} [Z_0 + Z_0^*]^{1/2} I_i \tag{1.44}$$

$$b = \frac{1}{\sqrt{2}} [Z_0 + Z_0^*]^{1/2} I_r \tag{1.45}$$

$$Z_0 = \begin{bmatrix} Z_{01} & 0 & 0 & \cdots & 0 \\ 0 & Z_{02} & 0 & \cdots & 0 \\ 0 & 0 & Z_{03} & \cdots & 0 \\ \cdots & \cdots & \cdots & \cdots & \cdots \\ 0 & 0 & 0 & \cdots & Z_{0N} \end{bmatrix} \tag{1.46}$$

$$\frac{1}{\sqrt{2}}[Z_0 + Z_0^*]^{1/2} = \begin{bmatrix} \sqrt{R}_{01} & 0 & 0 & \dots & 0 \\ 0 & \sqrt{R}_{02} & 0 & \dots & 0 \\ 0 & 0 & \sqrt{R}_{03} & \dots & 0 \\ \dots & \dots & \dots & \dots & \dots \\ 0 & 0 & 0 & \dots & \sqrt{R}_{0N} \end{bmatrix} \tag{1.47}$$

with Z_{0j} the normalizing impedance at port j, Z_0 the corresponding matrix, Z_0^* the matrix with conjugate elements of those of Z_0, I_{ji} and V_{ji} the incident current and voltage at port j, I_{jr} and V_{jr} the reflected current and voltage at port j, a_j the normalized incident component at port j, and b_j the normalized reflected component at port j.

The voltage and current relationships are illustrated in Figure 1.9 for a two-port network. Note that (1.39) follows directly from Figure 1.9(c) and application of Kirchhoff's voltage law at the different ports. Equations (1.40) and (1.42) follow from the definition of the incidents components [see the first paragraph of this section and Figure 1.9(d)]. The reflected current is defined as the difference between the incident current and the actual current [see (1.41)], while the reflected voltage is defined as the difference between the actual voltage and the incident voltage [see (1.43)].

Note that the incident voltage is equal to the product of the conjugate of the normalizing impedance and the incident current; that is,

$$V_i = Z_0^* I_i$$

The equivalent relationship in transmission-line theory is

$$V_i = Z_0 I_i$$

By using (1.40) to eliminate E_0 in (1.39) and substituting (1.41) and (1.43) in the resulting equation, it can be shown easily that, similar to transmission-line theory, the relationship between the reflected currents and voltages is

$$V_r = Z_0 I_r \tag{1.48}$$

There are three different types of S-parameters, which are defined in the following way:

$$I_r = S^I I_i \tag{1.49}$$

$$V_r = S^V V_i \tag{1.50}$$

$$b = Sa \tag{1.51}$$

These parameter sets are the current, voltage, and normalized S-parameters, respectively. For a two-port network, (1.51) reduces to

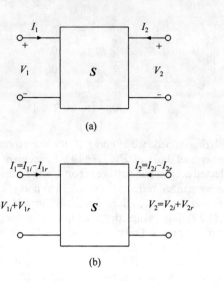

(a)

(b)

(c)

(d)

Figure 1.9   (a, b) The voltage and current relevant to the S-parameter definitions; (c) the two-port of (a, b) augmented by the normalizing impedances; and (d) the equivalent circuit for calculating the incident current and voltage.

$$\begin{bmatrix} b_1 \\ b_2 \end{bmatrix} = \begin{bmatrix} s_{11} & s_{12} \\ s_{21} & s_{22} \end{bmatrix} \begin{bmatrix} a_1 \\ a_2 \end{bmatrix} \tag{1.52}$$

The definitions given above are summarized together with other useful relationships in Figure 1.10.

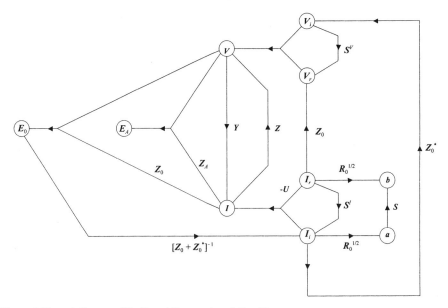

Figure 1.10   A diagram of S-, Y-, and Z-parameter relationships.

The impedance matrix $\boldsymbol{Z}_A$ in Figure 1.10 is defined by

$$\boldsymbol{Z}_A = \begin{bmatrix} Z_{A1} & 0 & ... & 0 \\ 0 & Z_{A2} & ... & 0 \\ ... & ... & ... & ... \\ 0 & 0 & ... & Z_{AN} \end{bmatrix} \tag{1.53}$$

and the matrix $\boldsymbol{E}_A$ by

$$\boldsymbol{E}_A = \begin{bmatrix} E_{A1} \\ E_{A2} \\ . \\ . \\ E_{AN} \end{bmatrix} \tag{1.54}$$

where E_{Aj} refers to the source voltage at the jth port of the N-port augmented by the actual

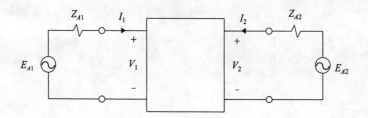

Figure 1.11   The two-port augmented by the actual load and source terminations (E_{A2} is usually equal to zero).

source and load impedances of interest ($\boldsymbol{Z}_A$). These definitions are illustrated in Figure 1.11 for a two-port network.

Note that the vectors in Figure 1.10 flow into the dependent variables and emanate from the independent variables. The branch multipliers are shown next to each branch. If no multiplier is shown, the unit matrix ($\boldsymbol{U}$) should be used.

It can be shown that $\boldsymbol{E}_0$ (the source voltages of the N-port augmented by its normalizing impedances as illustrated in Figure 1.9) is given in terms of $\boldsymbol{E}_A$ (the source voltages of the N-port augmented by the actual impedances and source voltage of interest) by the expression

$$E_0 = [I_N - (Z_0 - Z_A)(I_N - S^I)(Z_0 + Z_0^*)^{-1}]^{-1} E_A \qquad (1.55)$$

EXAMPLE 1.3   Derivation of the relationship between the reflected current and voltage.

To use the diagram in Figure 1.10, consider the derivation of the equality (1.48):

$$V_r = Z_0 I_r$$

It follows by inspection of the diagram that in order to find a relationship between V_r and I_r it is necessary to relate V to I. The easiest possible way would be to use the expression

$$E_0 = V + Z_0 I$$

$\boldsymbol{E}_0$ can then be replaced in terms of $\boldsymbol{I}_i$, $\boldsymbol{V}$ in terms of $\boldsymbol{V}_i$ and $\boldsymbol{V}_r$, $\boldsymbol{V}_i$ in terms of $\boldsymbol{Z}_0^*$ and $\boldsymbol{I}_i$, and $\boldsymbol{I}$ in terms of $\boldsymbol{I}_r$ and $\boldsymbol{I}_i$.

After a few manipulations on the equation thus obtained, (1.48) follows.

EXAMPLE 1.4   Calculation of the incident and reflected components for a two-port.

In order to make the definitions given above more real, consider finding the incident and reflected components when the terminal voltage and current of a two-port are given by

$$V_1 = 1.0\text{V}$$
$$V_2 = 0.5\text{V}$$
$$I_1 = 0.1\text{A}$$
$$I_2 = -0.2\text{A}$$

and the normalizing impedances are chosen to be

$$Z_{01} = 5\Omega$$
$$Z_{02} = 10\Omega$$

The first step is to find the source voltage in the equivalent circuit shown in Figure 1.9(d) in order to find the incident current and voltage. Inspection of the diagram yields (1.39):

$$E_0 = V + Z_0 I$$

$$= \begin{bmatrix} 1.0 \\ 0.5 \end{bmatrix} + \begin{bmatrix} 5 & 0 \\ 0 & 10 \end{bmatrix} \begin{bmatrix} 0.1 \\ -0.2 \end{bmatrix}$$

$$= \begin{bmatrix} 1.5 \\ -1.5 \end{bmatrix}$$

The incident components can now be obtained by using the equivalent circuit in Figure 1.9(d):

$$\begin{bmatrix} I_{1i} \\ I_{2i} \end{bmatrix} = \begin{bmatrix} 1/(2R_{01}) & 0 \\ 0 & 1/(2R_{02}) \end{bmatrix} \begin{bmatrix} E_{01} \\ E_{02} \end{bmatrix} = \begin{bmatrix} 0.150 \\ -0.075 \end{bmatrix}$$

$$\begin{bmatrix} V_{1i} \\ V_{2i} \end{bmatrix} = \begin{bmatrix} Z_{01}^* I_{1i} \\ Z_{02}^* I_{2i} \end{bmatrix} = \begin{bmatrix} 0.75 \\ -0.75 \end{bmatrix}$$

The normalized incident components follow by the application of (1.44):

$$\begin{bmatrix} a_1 \\ a_2 \end{bmatrix} = \begin{bmatrix} \sqrt{R_{01}}\, I_{1i} \\ \sqrt{R_{02}}\, I_{2i} \end{bmatrix} = \begin{bmatrix} 0.3354 \\ -0.2372 \end{bmatrix}$$

The reflected components can be obtained by applying (1.41), (1.48), and (1.45):

$$\begin{bmatrix} I_{1r} \\ I_{2r} \end{bmatrix} = \begin{bmatrix} I_{1i} - I_1 \\ I_{2i} - I_2 \end{bmatrix} = \begin{bmatrix} 0.050 \\ 0.125 \end{bmatrix}$$

$$\begin{bmatrix} V_{1r} \\ V_{2r} \end{bmatrix} = \begin{bmatrix} Z_{01}\, I_{1r} \\ Z_{02}\, I_{2r} \end{bmatrix} = \begin{bmatrix} 0.25 \\ 1.25 \end{bmatrix}$$

$$\begin{bmatrix} b_1 \\ b_2 \end{bmatrix} = \begin{bmatrix} \sqrt{R_{01}}\, I_{1r} \\ \sqrt{R_{02}}\, I_{2r} \end{bmatrix} = \begin{bmatrix} 0.1118 \\ 0.3953 \end{bmatrix}$$

1.5.2 The Physical Meanings of the Normalized Incident and Reflected Components of an *N*-Port

The normalized incident and reflected components are defined in (1.44) and (1.45) in terms of the incident and reflected components of the terminal current. It is useful to have expressions for these components in terms of the terminal voltage and current. The inverse relationships are also of interest.

The required expression for a_j can be obtained easily by using the relationship between the incident current and E_0:

$$a_j = \sqrt{R_{0j}}\, I_{ji} \tag{1.56}$$

$$= \sqrt{R_{0j}}\, E_{0j} / [R_{0j} + R_{0j}]$$

$$= \frac{V_j + Z_{0j} I_j}{2\sqrt{R_{0j}}} \tag{1.57}$$

The expression for the normalized reflected component can be derived by using this result in the following way:

$$b_j = \sqrt{R_{0j}}\, I_{jr} \tag{1.58}$$

$$= \sqrt{R_{0j}}\, I_{ji} - \sqrt{R_{0j}}\, I_j$$

$$= \frac{V_j + Z_{0j}I_j}{2\sqrt{R_{0j}}} - \sqrt{R_{0j}}\, I_j$$

$$= \frac{V_j - Z_{0j}^* I_j}{2\sqrt{R_{0j}}} \tag{1.59}$$

The inverse relationships follow easily by manipulating (1.57) and (1.59):

$$I_j = \frac{a_j - b_j}{\sqrt{R_{0j}}} \tag{1.60}$$

$$V_j = \frac{Z_{0j}^* a_j + Z_{0j} b_j}{\sqrt{R_{0j}}} \tag{1.61}$$

It follows from (1.60) that the normalized current at any point in the circuit can be obtained as the difference between the normalized incident and reflected components at that point.

Note that, if squared, the units of the normalized current would be that of power. When

$$Z_{0j} = Z_{0j}^* = R_{0j}$$

(1.61) simplifies to

$$V_j = \sqrt{R_{0j}}\,[a_j + b_j] \tag{1.62}$$

In this case, the normalized voltage at any point can be obtained as the sum of the normalized incident and reflected components. The units of the normalized voltage are again that of power if it is squared.

An expression for the power entering any port can be derived in terms of the normalized components by using (1.60) and (1.61) in conjunction with the expression for the input power:

$$P_{in,j} = 0.5(V_j I_j^* + V_j^* I_j) \tag{1.63}$$

$$= 0.5\{\frac{Z_{0j}a_j^* + Z_{0j}^* b_j^*}{2\sqrt{R_{0j}}} \frac{a_j - b_j}{\sqrt{R_{0j}}} + \frac{Z_{0j}^* a_j + Z_{0j} b_j}{2\sqrt{R_{0j}}} \frac{a_j^* - b_j^*}{\sqrt{R_{0j}}}\}$$

$$= |a_j|^2 - |b_j|^2 \qquad\qquad (1.64)$$

The power entering any port is, therefore, simply equal to the difference between the square of the normalized incident and reflected components at that port.

The last statement can be taken a step further. It can be shown easily that $|a_j|^2$ is the available power at the jth port of the N-port augmented by its normalizing impedances [see Figures 1.9(c, d)]. From this and from (1.64), it follows that $|b_j|^2$ is the reflected power at the jth port of the augmented N-port, and, consequently, the power entering any port of a network is equal to the difference between the available and reflected power at the jth port of the N-port augmented by its reference impedances.

It is important to realize that the available power in the N-port augmented by the normalizing impedances is not equal to the available power in the N-port augmented by the actual source and load impedances, unless the two sets of impedances are identical.

The simple expressions for the voltage (1.61), current (1.60), and power (1.64) in terms of the normalized incident and reflected components are summarized below.

$$I_j = (a_j - b_j) / \sqrt{R_{0j}}$$

$$V_j = (Z_{0j}^* a_j + Z_{0j} b_j) / \sqrt{R_{0j}}$$

$$= \sqrt{R_{0j}}(a_j + b_j) \quad \text{if} \quad Z_{0j} = Z_{0j}^*$$

$$P_j = |a_j|^2 - |b_j|^2$$

1.5.3   The Physical Interpretations of the Scattering Parameters

Consider the definitions of the elements of a two-port scattering matrix. The input reflection parameter s_{11} is defined by

$$s_{11} = \frac{b_1}{a_1}\bigg|_{a_2 = 0} \qquad\qquad (1.65)$$

and the forward transmission parameter s_{21} by

$$S_{21} = \frac{b_2}{a_1}\bigg|_{a_2 = 0} \tag{1.66}$$

The constraints on the current and voltage at the output terminals, when $a_2 = 0$, can be determined by using (1.57):

$$0 = a_2 = \frac{V_2 + Z_{02}I_2}{2\sqrt{R_{02}}}$$

leading to

$$V_2 = Z_{02}[-I_2] \tag{1.67}$$

In order for a_2 to be equal to zero, the load impedance across the output port must therefore be equal to the normalizing impedance at that port, and the electromotive force (EMF) must be equal to zero. This is illustrated in Figure 1.12(a).

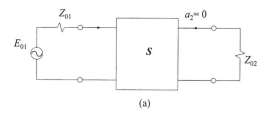

(a)

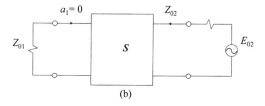

(b)

Figure 1.12   The conditions under which (a) $a_2 = 0$ and (b) $a_1 = 0$.

At this stage (1.57) and (1.59) can be substituted into (1.65) and (1.66) to find an expression for the parameters in terms of the terminal current and voltage:

$$s_{11} = \frac{V_1 - Z_{01}^* I_1}{V_1 + Z_{01} I_1}\bigg|_{a_2=0} = \frac{Z_{in} - Z_{01}^*}{Z_{in} + Z_{01}}\bigg|_{a_2=0} \tag{1.68}$$

$$s_{21} = \sqrt{\frac{R_{01}}{R_{02}}}\,\frac{V_2 - Z_{02}^* I_2}{V_1 + Z_{01} I_1}\bigg|_{a_2=0} = \sqrt{\frac{R_{01}}{R_{02}}}\,\frac{Z_{02}(-I_2) - Z_{02}^* I_2}{E_{01}}\bigg|_{a_2=0}$$

$$= -2\sqrt{R_{01} R_{02}}\,\frac{I_2}{E_{01}}\bigg|_{a_2=0} \tag{1.69}$$

where Z_{in} is the input impedance of the two-port terminated, as shown in Figure 1.12(a). The equivalence between (1.68) and the expression

$$\Gamma_{in} = \frac{Z_{in} - Z_{01}}{Z_{in} + Z_{01}} \tag{1.70}$$

for a reflection coefficient in transmission-line theory is obvious. When

$$Z_{01} = R_{01}$$

as is often the case, the two expressions will be identical.

When the normalizing resistance is equal, the forward transmission parameter s_{21} is simply the voltage gain $V_L/(E_{01}/2)$ of the two-port augmented with its normalizing impedances and with E_{02} set equal to zero.

Because the S-parameters are defined in terms of the normalized incident and reflected components, and the square of these components was shown to be the incident and reflected power at the relevant port of the two-port augmented with its normalizing impedances, respectively, it follows that

$$|s_{11}|^2 = \left|\frac{b_1}{a_1}\right|^2\bigg|_{a_2=0} = \frac{P_{1r}}{P_{av-E_{01}}}\bigg|_{a_2=0} \tag{1.71}$$

and

$$|s_{21}|^2 = \left|\frac{b_2}{a_1}\right|^2\bigg|_{a_2=0}$$

$$= \frac{|b_2|^2 - |a_2|^2}{|a_1|^2}\bigg|_{a_2=0}$$

$$= \frac{P_L}{P_{av-E}}\bigg|_{a_2=0} \tag{1.72}$$

where P_{av-E} is the power available from the source when the two-port is augmented by the normalizing impedances, and P_{1r} is the power reflected from the input port when it is augmented by the normalizing impedances and E_{02} is set equal to zero.

The meanings of $|s_{11}|^2$ and $|s_{21}|^2$ are illustrated in Figure 1.13.

Similar expressions apply to the output reflection parameter s_{22} and the reverse transmission parameter s_{12}.

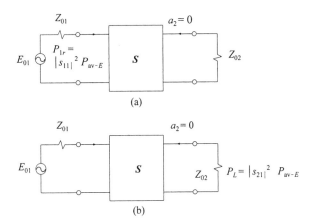

Figure 1.13   (a, b) The physical meanings of the scattering parameters (s_{11}; s_{21}) illustrated.

When the normalizing impedances are also the impedances in the actual network of interest, the transducer power gain and the ratio of the reflected power at the input to the available power from the source are given directly by s_{21} and s_{11}, respectively.

When the normalizing impedances are purely resistive, and s_{11} and s_{22} are displayed on a Smith chart, the input and output impedances of the network can be read directly.

1.5.4   Constraints Imposed on the Normalized Incident and Reflected Components by the Terminations of an N-Port

In order to derive expressions for the gains and impedances of an N-port with arbitrary terminations, it is necessary to derive expressions for the constraints imposed by the terminations on the normalized incident and reflected components.

Consider port n of the N-port terminated in an impedance Z_{An} in series with a voltage source E_{An} as shown in Figure 1.14.

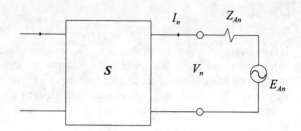

Figure 1.14   The N-port under consideration.

The termination forces the following relationship between the terminal voltage and current:

$$E_{An} = V_n + Z_{An}I_n \tag{1.73}$$

By using this relationship in conjunction with (1.57) and (1.59), it follows that

$$2\sqrt{R_{0n}}\,a_n = V_n + Z_{0n}I_n = E_{An} - (Z_{An} - Z_{0n})I_n$$

leading to

$$2\sqrt{R_{0n}}\,a_n - E_{An} = -(Z_{An} - Z_{0n})I_n \tag{1.74}$$

and

$$2\sqrt{R_{0n}}\,b_n = V_n - Z_{0n}^{*}I_n = E_{An} - (Z_{An} - Z_{0n}^{*})I_n$$

which leads to

$$2\sqrt{R_{0n}}\,b_n - E_{An} = -(Z_{An} + Z_{0n}^{*})I_n \tag{1.75}$$

Dividing (1.74) by (1.75) yields

$$\frac{2\sqrt{R_{0n}}\,a_n - E_{An}}{2\sqrt{R_{0n}}\,b_n - E_{An}} = \frac{-[Z_{An} - Z_{0n}]I_n}{-[Z_{An} + Z_{0n}^{*}]I_n}$$

which leads to

$$a_n = \frac{Z_{AN} - (Z_{0n}^{**})}{Z_{AN} + (Z_{0n}^{*})}\,b_n + \frac{\sqrt{R_{0n}}}{Z_{AN} + (Z_{0n}^{*})}\,E_{An} \tag{1.76}$$

With

$$E_{An} = 0$$

the second term in (1.76) is equal to zero, and the following relationship applies:

$$S_n = a_n / b_n = \frac{Z_{An} - (Z_{0n}^{**})}{Z_{An} + (Z_{0n}^{*})} \tag{1.77}$$

This expression clearly has the form of a reflection parameter with normalizing impedance Z_{0n}^{*}. The termination can therefore be considered to be the interconnection of a one-port network with a port of the two-port. The normalizing impedance of the one-port must be the conjugate of that at the corresponding port of the two-port. This is illustrated in Figure 1.15.

One would expect that the normalized component incident on the one-port (a_L) should be equal to the component reflected from the two-port (b_2) and that the component reflected from the one-port (b_L) should be equal to that incident on the two-port (a_2), that is,

$$a_L = b_2 \text{ and } b_L = a_2$$

The proof follows easily from the fact that the voltage across the one-port is the same as that at the corresponding port of the two-port ($V_L = V_2$) and that the currents are also identical except for a difference in sign ($I_L = -I_2$). It follows from (1.57) and (1.59) that

$$a_2 = \frac{V_2 + Z_{02}\,I_2}{2\sqrt{R_{02}}} = \frac{V_L - (Z_{02}^{*})^{*}I_L}{2\sqrt{R_{02}}} = b_L \tag{1.78}$$

$$b_2 = \frac{V_2 - Z_{02}^{*}\,I_2}{2\sqrt{R_{02}}} = \frac{V_L + (Z_{02}^{*})I_L}{2\sqrt{R_{02}}} = a_L \tag{1.79}$$

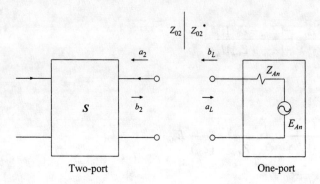

Figure 1.15   Cascading a one-port with a two-port network.

The component incident on the N-port (a_n) is, therefore, reflected from the one-port, and the component reflected from the N-port (b_n) is incident on the one-port.

The normalizing impedance for the single-port is the conjugate of that for the N-port.

1.5.5   Derivation of Expressions for the Gain Ratios and Reflection Parameters of a Two-port

Consider the two-port with terminations as shown in Figure 1.16 and the associated S-parameter expression:

$$\begin{bmatrix} b_1 \\ b_2 \end{bmatrix} = \begin{bmatrix} s_{11} & s_{12} \\ s_{21} & s_{22} \end{bmatrix} \begin{bmatrix} a_1 \\ a_2 \end{bmatrix}$$

(1.80)

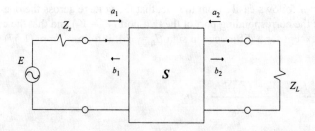

Figure 1.16   The two-port under consideration.

In (1.80) a_1 is an independent variable, the magnitude and phase of which are determined by the source voltage E and the fixed normalizing impedance Z_{01}. According to (1.77), b_2 is constrained to

$$b_2 = a_2 \, / \, \frac{Z_L - (Z_{02}^{**})}{Z_L + (Z_{02}^{*})}$$

$$= a_2 \, / \, S_L \qquad (1.81)$$

With a_1 the independent variable and b_2 known in terms of a_2, (1.80) amounts to two equations with two unknowns and values for a_2, b_1, and b_2 can be determined in terms of the scattering parameters and a_1. The results are as follows:

$$a_1 = 1 \qquad (1.82)$$

$$b_1 = s_{11} + \frac{s_{12} s_{21} S_L}{1 - s_{22} S_L} \qquad (1.83)$$

$$a_2 = \frac{s_{21} S_L}{1 - s_{22} S_L} \qquad (1.84)$$

$$b_2 = a_2 \, / \, S_L \qquad (1.85)$$

At this stage, the reflection parameters and the gain ratios of interest can be determined. The expressions most frequently used are repeated here.

$$s_{11\omega} = \frac{Z_{in} - Z_{01}^{*}}{Z_{in} + Z_{01}} = \frac{V_1 - Z_{01}^{*} I_1}{V_1 + Z_{01} I_1} = \frac{b_1}{a_1} = s_{11}' = s_{11} + \frac{s_{12} s_{21} S_L}{1 - s_{22} S_L} \qquad (1.86)$$

$$s_{22a} = \frac{Z_{out} - Z_{02}^{*}}{Z_{out} + Z_{02}} = \frac{V_2 - Z_{02}^{*} I_2}{V_2 + Z_{02} I_2} = \frac{b_2}{a_2} = s_{22}' = s_{22} + \frac{s_{12} s_{21} S_s}{1 - s_{11} S_s} \qquad (1.87)$$

$$S_S = \frac{Z_s - (Z_{01}^{**})}{Z_s + (Z_{01}^{*})} \qquad (1.88)$$

$$G_\omega = \frac{|b_2|^2 - |a_2|^2}{|a_1|^2 - |b_1|^2}$$

$$= \frac{|s_{21}|^2 [1 - |S_L|^2]}{|1 - s_{22}S_L|^2 - |s_{22}(1 - s_{22}S_L) + s_{12}s_{21}S_L|^2} \tag{1.89}$$

$$G_T = \frac{|b_2|^2 - |a_2|^2}{P_{av-E}}$$

$$= \frac{|s_{21}|^2 [1 - |S_L|^2][1 - |S_S|^2]}{|[1 - s_{11}S_S][1 - s_{22}S_L] - s_{12}s_{21}S_S S_L|^2} \tag{1.90}$$

$$G_{T s_{12}=0} = G_{T,u}$$

$$= \frac{1 - |S_S|^2}{|1 - s_{11}S_S|^2} |s_{21}|^2 \frac{1 - |S_L|^2}{|1 - s_{22}S_L|^2} \tag{1.91}$$

where $G_{T,u}$ is the unilateral transducer power gain

$$G_A = \frac{P_{av-O}}{P_{av-E}} \tag{1.92}$$

$$= \frac{|s_{21}|^2 [1 - |S_S|^2]}{|1 - |s_{22}|^2 + |S_S|^2 [|s_{11}|^2 - |\Delta|^2] - 2\operatorname{Re}(C_1 S_S)|} \tag{1.93}$$

where P_{av-O} is the maximum available power at the output terminals of the transistor

$$\Delta = s_{11}s_{22} - s_{12}s_{21} \tag{1.94}$$

and

$$C_1 = s_{11} - \Delta s_{22}^* \tag{1.95}$$

$$A_V = \frac{\sqrt{R_{02}}}{\sqrt{R_{01}}} \frac{a_2 + b_2}{a_1 + b_1} = \sqrt{\frac{R_{02}}{R_{01}}} \frac{s_{21}[1 + S_L]}{1 + s_{11} - s_{22}S_L - s_{11}s_{22}S_L + s_{12}s_{21}S_L} \tag{1.96}$$

In order for (1.96) to apply, the normalizing impedances must be purely resistive.

In (1.86) $s_{11\omega}$ is defined to be the input reflection parameter with the two-port terminated in the actual load of interest (normalizing impedance on the input side: Z_{01}), while s_{22a} is defined in (1.87) as the output reflection parameter with the two-port terminated in the source impedance of interest (normalizing impedance on the output side, Z_{02}).

Similarly, $s_{21\omega}$ is defined here as s_{21} when the output normalizing impedance is the actual load impedance of interest ($Z_{02} = Z_L$) and the input normalizing impedance is taken to be the conjugate of the input impedance of the two-port ($Z_{01} = Z_{in}^*$). It follows from this definition that

$$|s_{21\omega}|^2 = G_\omega \tag{1.97}$$

Similarly, s_{21a} is defined as s_{21} when the input normalizing impedance is the actual source impedance of interest ($Z_{01} = Z_s$) and the output normalizing impedance is taken to be the conjugate of the output impedance of the two-port ($Z_{02} = Z_{out}^*$). It follows that

$$|s_{21a}|^2 = G_A \tag{1.98}$$

s_{21T} is defined as s_{21} when the normalizing impedance on the load side is the actual load of interest ($Z_{02} = Z_L$) and that on the input side the actual source impedance of interest ($Z_{01} = Z_s$). This implies that

$$|s_{21T}|^2 = G_T \tag{1.99}$$

These definitions are relevant during circuit synthesis.

EXAMPLE 1.5   Derivation of the expression for the transducer power gain.

As an example of the application of (1.82) to (1.85), consider the derivation of (1.90).

An expression for the power dissipated in the load follows directly from (1.84) and (1.85):

$$P_L = |b_2|^2 - |a_2|^2$$

$$= \left|\frac{s_{21}}{1 - s_{22}S_L}\right|^2 [1 - |S_L|^2]|a_1|^2 \tag{1.100}$$

In order to derive an expression for P_{av-E} it is necessary to use (1.76). Application of (1.76) to port 1 yields

$$a_1 = \frac{Z_s - (Z_{01}^{**})}{Z_s + (Z_{01}^{*})} b_1 + \frac{\sqrt{R_{01}}}{Z_s + Z_{01}} E_1$$

from which it follows that

$$E_1 = \frac{a_1 - S_S b_1}{\sqrt{R_{01}}} (Z_s + Z_{01}) \tag{1.101}$$

Substitution of (1.101) in the expression for the available power gain yields

$$P_{av-E} = E_1^2 / [4R_s] = \frac{\left|Z_s + Z_{01}^*\right|^2}{4R_{01}R_s} \left|a_1 - S_S b_1\right|^2 = \frac{\left|a_1 - S_S b_1\right|^2}{1 - \left|S_S\right|^2} \tag{1.102}$$

After substitution of b_1 in terms of a_1 [see (1.83)] in this equation, it follows that

$$P_{av-E} = \frac{\left|[1 - s_{11}S_S][1 - s_{22}S_L] - s_{21}s_{12}S_S S_L\right|^2}{[1 - \left|S_S\right|^2][1 - s_{22}S_L]^2} \tag{1.103}$$

Combination of (1.103) and (1.100) yields the desired expression.

The S-parameters (50-Ω normalization) for a typical microwave transistor are displayed in Figure 1.17. The performance with different terminations can be obtained by using the equations provided in this section.

1.5.6   Signal Flow Graphs

The S-parameter equations shown above can also be derived by using signal flow graphs [1]. Additional insight into the different relationships are also gained from the flow graphs. The following rules apply when a signal flow graph is created:

1.   Each variable is designated with a node (in the case of the two-port, nodes will be used for a_1, a_2, b_1, b_2, and b_s).

2.   A multiplier is associated with each branch.

3.   Branches emanate from independent variable nodes and terminate on dependent variable nodes (dependence and independence are established by the associated equation). The direction of the flow is indicated with an

arrow on each branch. The branch multipliers are applied to the independent variable nodes.

4.   The value of each dependent variable is determined by the multipliers and independent variables associated with the branches entering the relevant node.

These rules are illustrated by building a signal flow graph for the normalized incident and reflected components of a two-port in Figure 1.18.

Apart from representing the relationships of interest graphically, flow graphs can also be used to calculate the value of any of the dependent variables in the graph in terms of the independent variable of the graph (b_s in this case). This is done by applying Mason's rule to the graph. The following terms are required before the rule can be formulated:

1.   A first-order loop product is defined as the product of the branch multipliers encountered in a journey starting from any specific node and moving back to the same node in the direction of the arrows. The first-order loop products in Figure 1.18 are $s_{11} \Gamma_s$, $s_{22} \Gamma_L$, and $s_{21} \Gamma_L s_{12} \Gamma_s$.

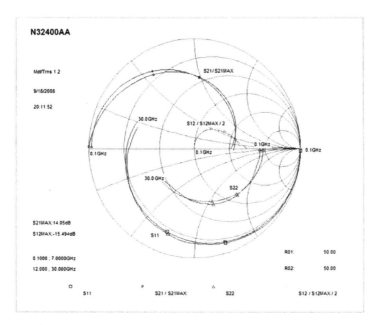

Figure 1.17   The *S*-parameters (50 Ω normalization) of a transistor displayed on a polar plot (the constant resistance and constant reactance circles only apply to s_{11} and s_{22}; s_{12} and s_{21} were normalized as shown). The one set of traces is used for the parameters as supplied by the manufacturer (mixed color traces), while the other set is used for the *S*-parameters of the small-signal model fitted [2]. Note that the highest frequency point on each curve is not marked.

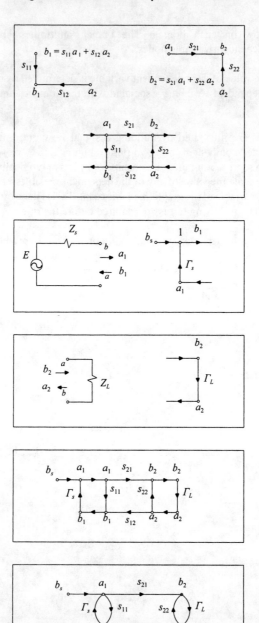

Figure 1.18   A flow graph for the incident and reflected components of a two-port.

2.   Loops are nontouching when they have no nodes or branches in common.

3.   A second-order loop product is the product formed by combining the loop products of any two nontouching first-order loops.

4.   A third-order loop product is the product associated with any three nontouching first-order loops.

5.   An nth-order loop product is the product associated with any n nontouching first-order loops.

6.   A path is any forward route (route in the direction of the arrows) emanating from the independent variable of the graph and terminating on the dependent variable of interest.

Mason's rule can be formulated at this point:

$$T = \frac{P_1[1 - \sum L^1_{NT\,P1} + \sum L^2_{NT\,P1} -] + P_2[1 - \sum L^1_{NT\,P2} + ...}{1 - \sum L^1 + \sum L^2 - \sum L^3 + ...} \qquad (1.104)$$

where

$\sum L^n$ is the sum of all the nth order loop products, $\sum L^n_{NT\,P_m}$ is the sum of all the nth order products associated with the loops not touching path m, and P_m is the product of the branch terms along the path m.

Note that the denominator of (1.104) is only a function of the graph topology and is the same for all the dependent variables. It follows that this term will be cancelled if the ratio of any of the dependent variables is taken.

EXAMPLE 1.6   Calculation of a_1 in terms of b_s, and b_1, b_2, and a_2 in terms of a_1.

To demonstrate application of (1.104), a_1 in Figure 1.18 will be calculated as a function of b_s.

The sum of all the first-order loop products (see Figure 1.19) is

$$s_{11}\,\Gamma_s + s_{22}\,\Gamma_L + s_{21}\,\Gamma_s\,s_{12}\,\Gamma_L$$

There is only one second-order loop (loop factor $s_{11}\,s_{22}\,\Gamma_s\,\Gamma_L$).
The only loop that does not touch the path leading to a_1 is the loop on the right-hand side of the flow graph (loop factor $s_{22}\,\Gamma_L$).
This leads to

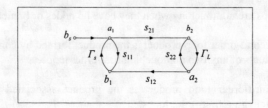

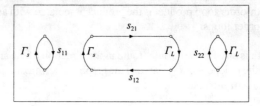

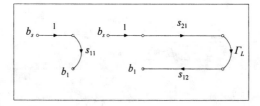

Figure 1.19   The first-order loops and the forward paths relevant to calculation of the ratio b_1/b_s.

$$\frac{a_1}{b_s} = \frac{1 - s_{22}\,\Gamma_L}{1 - [s_{11}\Gamma_s + s_{22}\Gamma_L + s_{21}\Gamma_s\,s_{12}\Gamma_L] + s_{11}s_{22}\Gamma_s\,\Gamma_L} \qquad (1.105)$$

In the previous section a_1 was taken to be unity, which leads to

$$b_s = \frac{1 - [s_{11}\Gamma_s + s_{22}\Gamma_L + s_{21}\Gamma_s\,s_{12}\Gamma_L] + s_{11}s_{22}\Gamma_s\,\Gamma_L}{1 - s_{22}\,\Gamma_L} \qquad (1.106)$$

b_1, b_2, and a_2 can now be derived in terms of a_1 by applying Mason's rule in each case. The results obtained will be same as those in the previous section. To illustrate this, consider the derivation for b_1:

$$\frac{b_1}{b_s} = \frac{s_{11}\,(1 - s_{22}\,\Gamma_L) + s_{21}\Gamma_L\,s_{12}\,(1)}{1 - [s_{11}\Gamma_s + s_{22}\Gamma_L + s_{21}\Gamma_s\,s_{12}\Gamma_L] + s_{11}s_{22}\Gamma_s\,\Gamma_L} \qquad (1.107)$$

Note that there are no nontouching loops associated with the second path term in the numerator of (1.105) $[s_{21} \Gamma_L s_{12}(1)]$.

Substituting b_s in this equation produces the same result as (1.83).

1.5.7   The Indefinite S-Matrix

Similar to the indefinite admittance matrix, the sum of the elements in each row or column of the indefinite S-matrix is equal to a constant. In this case the constant is unity.

In order to prove that the sum of the elements in each row must equal 1, consider the three-port shown in Figure 1.20.

Under the conditions shown, all the incident components are equal, and

$$b_j = s_{j1}a_1 + s_{j2}a_2 + s_{j3}a_3$$

simplifies to

$$b_j = [s_{j1} + s_{j2} + s_{j3}]a_1$$

Substitution of b_j and a_1 in terms of the reflected and incident currents yields

$$I_{jr} = [s_{j1} + s_{j2} + s_{j3}]I_{1i}$$

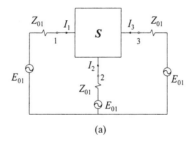

(a)

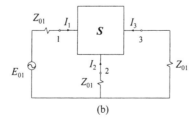

(b)

Figure 1.20   Circuits used to prove that the sum of the elements in (a) any row or (b) any column of an indefinite S-matrix is equal to 1.

and because the terminal currents (I_j) must equal zero when all the source voltages are equal, I_{jr} must equal I_{1i}. It follows that

$$s_{j1} + s_{j2} + s_{j3} = 1 \qquad (1.108)$$

The circuit in Figure 1.20(b) can be used to prove that the sum of the elements of the first column of the indefinite matrix is equal to 1. Because the incident components at terminals two and three are equal to zero, the necessary condition

$$I_1 + I_2 + I_3 = 0$$

simplifies to

$$I_{1i} = I_{1r} + I_{2r} + I_{3r} \qquad (1.109)$$

with

$$a_2 = 0 = a_3$$

$$b_1 = s_{11}a_1 + s_{12}a_2 + s_{13}a_3$$
$$b_2 = s_{21}a_1 + s_{22}a_2 + s_{23}a_3$$
$$b_3 = s_{31}a_1 + s_{32}a_2 + s_{33}a_3$$

simplifies to

$$b_1 = s_{11}a_1$$
$$b_2 = s_{21}a_1$$
$$b_3 = s_{31}a_1$$

and, therefore,

$$I_{1r} = s_{11}I_{1i} \qquad (1.110a)$$
$$I_{2r} = s_{21}I_{1i} \qquad (1.110b)$$
$$I_{3r} = s_{31}I_{1i} \qquad (1.110c)$$

Equation (1.109) combined with (1.110) yields

$$s_{11} + s_{21} + s_{32} = 1 \qquad (1.111)$$

By moving the voltage source in Figure 1.20(b) to the other two-ports and following the same procedure, it can also be shown that the sum of the elements in each of the other two columns of the indefinite S-matrix is equal to 1.

1.5.8   Extension of the Single-Frequency S-Parameter Definitions to the Complex Frequency Plane

A necessary condition for a matrix to be the S-parameter matrix of a linear, lumped, passive network normalized to N minimum reactance functions (i.e., impedance functions with no poles on the real-frequency axis) is that none of its elements may have any poles in the closed right-hand side (RHS) of the complex frequency ($s=\sigma+j\omega$) plane [3]. (Note that the imaginary axis complex frequencies ($j\omega$) are the real-world frequencies, or abbreviated, real frequencies. The imaginary axis complex frequencies are also referred to as the radian frequencies. The real axis complex frequencies (σ) are referred to as neper frequencies [3].)

The definitions given for a and b in Section 1.5.1 are adequate for any single-frequency application, as well as in the complex plane when the normalizing impedance functions [$Z_{0j}(s)$] do not have any finite poles (i.e., purely resistive normalizing impedances, or impedances of the form $R_{0j} + sL_{0j}$). However, when these impedance functions are more complex, it is necessary to extend the definitions of the normalized incident and reflected components. The following definitions [3] are relevant to the more general case:

$$\mathbf{Z}_0(s) = \begin{bmatrix} Z_{01}(s) & 0 & \dots & 0 \\ 0 & Z_{02}(s) & \dots & 0 \\ \dots & \dots & \dots & 0 \\ 0 & 0 & \dots & Z_{0N}(s) \end{bmatrix} \tag{1.112}$$

where $Z_{0j}(s)$ is the normalizing impedance at port j,

$$\mathbf{r}(s) = \begin{bmatrix} r_{01}(s) & 0 & \dots & 0 \\ 0 & r_{02}(s) & \dots & 0 \\ \dots & \dots & \dots & 0 \\ 0 & 0 & \dots & r_{0N}(s) \end{bmatrix} \tag{1.113}$$

$$= \mathbf{h}(s)\mathbf{h}(-s) \tag{1.114}$$

where

$$r_{0j}(s) = 0.5\,[Z_{0j}(s) + Z_{0j}(-s)] \tag{1.115}$$

and

$$h(s) = \begin{bmatrix} m_1(s)/n_1(s) & 0 & \cdots & 0 \\ 0 & m_2(s)/n_2(s) & \cdots & 0 \\ \cdots & \cdots & \cdots & \cdots \\ 0 & 0 & \cdots & m_N(s)/n_N(s) \end{bmatrix} \tag{1.116}$$

where $m_j(s)$ and $n_j(s)$ are polynomials and the zeros of $n_j(s)$ [poles of $h_j(s)$] are constrained to the open left-hand plane (LHP) and the zeros of $m_j(s)$ [zeros of $h_j(s)$] are constrained to the closed right-hand plane (RHP).

$$a(s) = h(-s)I_i(s) \tag{1.117}$$

$$b(s) = h(s)I_r(s) \tag{1.118}$$

where $a(s)$ is the matrix of normalized incident components, $b(s)$ the normalized reflected components, and with I_i and I_r as defined in Section 1.5.1.

Note that the elements of $r(s)$ are even functions (i.e., $r_{0i}(s) = r_{0i}(-s)$) and are the effective series resistance parts of the corresponding normalizing impedances.

With these definitions for the normalized and reflected components, it follows that

$$a_j(s) = \frac{V_j(s) + Z_{0j}(s)I_j(s)}{2h_j(s)} \tag{1.119}$$

$$b_j(s) = \frac{V_j(s) - Z_{0j}(-s)I_j(s)}{2h_j(-s)} \tag{1.120}$$

$$S_{jj}(s) = \frac{h_j(s)}{h_j(-s)} \frac{Z_{\text{in}\,j}(s) - Z_{0j}(-s)}{Z_{\text{in}\,j}(s) + Z_{0j}(s)} \tag{1.121}$$

$$s_{jk}(s) = -2h_j(s)h_k(s)\frac{I_j(s)}{E_{0k}} \tag{1.122}$$

These relationships are identical to those derived previously for single-frequency applications as long as

$$h_j(s) = \sqrt{R_{0j}} = h_j(-s) \tag{1.123}$$

This relationship will apply in all cases where the normalizing impedances are purely resistive or of the form $R_{0j} + sL_{0j}$.

Independent of the complexity of the normalizing impedances, the incident and reflected power are still given by $|a_j|^2$ and $|b_j|^2$, respectively.

EXAMPLE 1.7   Calculating $h(s)$ for a two-port.

As an example, $h(s)$ will be calculated for the normalizing impedances shown in Figure 1.21.

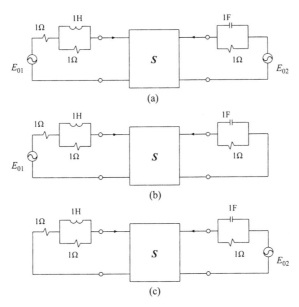

(a)

(b)

(c)

Figure 1.21   (a) The normalizing impedances under consideration; (b) the equivalent circuit used to determine $s_{11}(s)$ and $s_{21}(s)$; and (c) the equivalent circuit used to determine $s_{22}(s)$ and $s_{12}(s)$.

Because

$$Z_{01}(s) = 1 + 1/[1 + 1/s] = 1 + s/[1 + s]$$

it follows that

$$r_{01}(s) = 0.5\,[Z_{01}(s) + Z_{01}(-s)]$$

$$= \frac{1 - 2s^2}{1 - s^2}$$

$$= \frac{1 - \sqrt{2}s}{1 + s}\frac{1 + \sqrt{2}s}{1 - s}$$

and, therefore,

$$h_1(s) = (1 - \sqrt{2}s) / (1 + s)$$

Similarly,

$$h_2(s) = s / (1 + s)$$

1.5.9   Constraints on the Scattering Matrix of a Lossless N-Port

The average power entering a passive lossless device must be equal to zero. This imposes the following constraints on the scattering matrix:

$$0 = P_{ave} = 0.5 \, [V^*(j\omega) \, I(j\omega) + I^*(j\omega) \, V(j\omega)]$$

$$= [a^{*'}(j\omega) \, a(j\omega) - b^{*'}(j\omega) \, b(j\omega)]$$

$$= a^{*'}(j\omega) \, [I_n - S^{*'}(j\omega) \, S(j\omega)] \, a(j\omega) \tag{1.124}$$

leading to

$$S^{*'}(j\omega) \, S(j\omega) = I_n \tag{1.125}$$

In these equations the superscript *' indicates the transposed conjugate of the relevant matrix.

It is clear from (1.125) that the inverse of the scattering matrix of a lossless network is constrained to be equal to its transposed conjugate, that is,

$$S^{-1}(j\omega) = S^{*'}(j\omega) \tag{1.126}$$

A matrix whose inverse is equal to its transposed conjugate is called a unitary matrix. A necessary and sufficient condition for a matrix to be unitary is that its columns (or rows) should be mutually orthogonal unit vectors [3]. In terms of the elements of the scattering matrix, this implies that the following equations must be satisfied:

$$\sum_{j=1}^{N} s_{ij}(j\omega) s_{kj}^*(j\omega) = \delta_{ik} \tag{1.127}$$

$$\sum_{j=1}^{N} s_{ji}(j\omega)s_{jk}^{*}(j\omega) = \delta_{ik} \qquad (1.128)$$

where δ_{ik} is the Kronecker delta ($\delta_{ik} = 0$, if $i \neq k$; $\delta_{ik} = 1$, if $i = k$).

The unitary constraint on the magnitude of each row or column vector of the scattering matrix forces the following two relationships on the elements of each row and column, respectively:

$$\sum_{j=1}^{N} s_{ij}(j\omega)s_{ij}^{*}(j\omega) = \sum_{j=i}^{N} \left| s_{ij}(j\omega) \right|^{2} = 1 \qquad (1.129)$$

$$\sum_{j=1}^{N} s_{ji}(j\omega)s_{ji}^{*}(j\omega) = \sum_{j=i}^{N} \left| s_{ji}(j\omega) \right|^{2} = 1 \qquad (1.130)$$

The magnitude of each element of the S-matrix of a lossless (and also passive) network is therefore bounded by unity; that is,

$$\left| s_{ij}(j\omega) \right| \leq 1 \qquad (1.131)$$

By applying (1.129) and (1.130) to a two-port network, it follows that

$$\left| s_{11}(j\omega) \right|^{2} = 1 - \left| s_{12}(j\omega) \right|^{2} \qquad (1.132)$$

$$\left| s_{11}(j\omega) \right|^{2} = 1 - \left| s_{21}(j\omega) \right|^{2} \qquad (1.133)$$

$$\left| s_{22}(j\omega) \right|^{2} = 1 - \left| s_{21}(j\omega) \right|^{2} \qquad (1.134)$$

$$\left| s_{22}(j\omega) \right|^{2} = 1 - \left| s_{12}(j\omega) \right|^{2} \qquad (1.135)$$

$$s_{11}(j\omega)s_{12}^{*}(j\omega) = -s_{21}(j\omega)s_{22}^{*}(j\omega) \qquad (1.136)$$

$$s_{11}(j\omega)s_{21}^{*}(j\omega) = -s_{12}(j\omega)s_{22}^{*}(j\omega) \qquad (1.137)$$

Combining (1.133) and (1.134) yields

$$|s_{11}(j\omega)| = |s_{22}(j\omega)|$$ (1.138)

while (1.134) and (1.135) can be combined to show that

$$|s_{12}(j\omega)| = |s_{21}(j\omega)|$$ (1.139)

Equation (1.138) can be extended to

$$s_{12}(j\omega) = s_{21}(j\omega)$$ (1.140)

whenever the network considered is passive and reciprocal. This can be proved easily by using the reciprocity theorem.

By combining (1.137) and (1.140), it can be shown that

$$s_{22}(j\omega) = -\frac{s_{21}(j\omega)}{s_{21}^*(j\omega)} \, s_{11}^*(j\omega)$$ (1.141)

These relationships will prove useful in later chapters.

EXAMPLE 1.8   Calculation of the S-parameters of a lossless two-port.

As an example, the S-parameters of the lossless two-port in Figure 1.22 will be derived, and some of the relationships given above will be illustrated.

Because the normalizing impedances in Figure 1.22 are purely resistive, it follows that

$$h_j(s) = \sqrt{R_{0j}} = h_j(-s)$$

and the input reflection and forward transmission parameters can, therefore, be determined by using (1.68) and (1.69), respectively.

The equivalent circuit corresponding to $a_2 = 0$ and the chosen normalizing impedances are shown in Figure 1.22.

The input impedance necessary for determining $s_{11}(j\omega)$ is given by

$$Z_{in}(s) = \frac{1}{G_{02} + sC} + sL$$

and $s_{11}(j\omega)$ is therefore given by

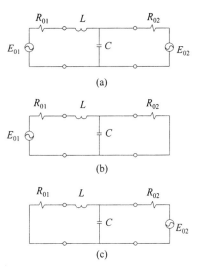

(a)

(b)

(c)

Figure 1.22 (a) The lossless two-port under consideration; (b) the equivalent circuit used to determine $s_{11}(j\omega)$ and $s_{21}(j\omega)$; and (c) the equivalent circuit used to determine $s_{22}(j\omega)$ and $s_{12}(j\omega)$.

$$s_{11}(j\omega) = \frac{Z_{in}(j\omega) - Z_{01}^*}{Z_{in}(j\omega) - Z_{01}}\bigg|_{a_2 = 0}$$

$$= \frac{\dfrac{1}{G_{02} + j\omega C} + j\omega L - R_{01}}{\dfrac{1}{G_{02} + j\omega C} + j\omega L + R_{01}} \qquad (1.142)$$

Similarly, the output reflection parameter is given by

$$s_{22}(j\omega) = \frac{\dfrac{R_{01} + j\omega L}{1 + (R_{01} + j\omega L)j\omega C} - R_{02}}{\dfrac{R_{01} + j\omega L}{1 + (R_{01} + j\omega L)j\omega C} - R_{02}}$$

$$= \frac{(R_{02} - R_{01} - \omega^2 LCR_{02}) + j\omega(L - R_{01}R_{02}C)}{R_{02} + R_{01} - \omega^2 LCR_{02} + j\omega(L + R_{01}R_{02}C)} \qquad (1.143)$$

Comparison of (1.142) and (1.143) yields that

$$|s_{11}(j\omega)| = |s_{22}(j\omega)|$$

as expected.

The forward transmission parameter is given by

$$s_{21}(j\omega) = -2\sqrt{R_{01}R_{02}}\,\frac{I_{02}}{E_{01}}\bigg|_{a_2=0}$$

$$= 2\frac{\sqrt{R_{01}R_{02}}}{R_{01}+R_{02}}\frac{R_{02}}{R_{02}-\omega^2 LCR_{02}+j\omega(L+CR_{01}R_{02})} \tag{1.144}$$

$$s_{12}(j\omega) = -2\sqrt{R_{01}R_{02}}\,\frac{I_{01}}{E_{02}}\bigg|_{a_1=0}$$

$$= 2\frac{\sqrt{R_{01}R_{02}}}{R_{01}+R_{02}}\frac{R_{02}}{R_{02}-\omega^2 LCR_{02}+j\omega(L+CR_{01}R_{02})} \tag{1.145}$$

Equations (1.144) and (1.145) are clearly identical and, therefore,

$$s_{21}(j\omega) = s_{12}(j\omega)$$

as expected. The same result can be obtained directly by application of the reciprocity theorem. It is a simple matter to show that

$$|s_{21}|^2 = 1 - |s_{11}|^2$$

and that

$$|s_{12}|^2 = 1 - |s_{22}|^2$$

1.5.10   Conversion of S-Parameters to Other Parameters

The schematic representation of the S-parameter and related relationships in Figure 1.10 can be used to derive expressions for the conversion of the normalized S-parameters to the other S-parameters as well as to Z- or Y-parameters. The results are

$$S = R_0^{1/2} S' R_0^{-1/2} \tag{1.146}$$

$$S^I = Z_0^{-1} S^V Z_0^* \qquad (1.147)$$

$$S^I = [Z + Z_0]^{-1}[Z - Z_0^*] \qquad (1.148)$$

$$S^V = -[Y + Y_0]^{-1}[Y - Y_0^*] \qquad (1.149)$$

$$Y = Y_0[I_n - S^V][I_n + S^V]^{-1} \qquad (1.150)$$

with

$$Y_0 = Z_0^{-1} \qquad (1.151)$$

When the normalizing impedances are all purely resistive and equal

$$S = S^I = S^V \qquad (1.152)$$

EXAMPLE 1.9   Derivation of the expression for the reflected voltages of an N-port in terms of the incident voltages.

It follows by inspection of the diagram in Figure 1.10 that

$$I = YV$$

is equivalent to

$$I_i - I_r = Y(V_i + V_r)$$

which is equivalent to

$$Z_0^{*-1} V_i - Z_0^{-1} V_r = YV_i + YV_r$$

This equation can be manipulated to

$$V_r = -(Y + Y_0)^{-1}(Y - Y_0^*)V_i$$

which yields the required expression.

QUESTIONS AND PROBLEMS

1.1   Derive (1.6)–(1.12) by using the equivalent circuit in Figure 1.1.

1.2   Determine the common-base parameters of a bipolar transistor with the following common-emitter parameters:

$y_{11e} = (4.0+j2.0)$ mS
$y_{12e} = (0.0-j0.05)$ mS
$y_{21e} = (40.0+j10)$ mS
$y_{22e} = (1.0+j1.0)$ mS

1.3   Derive (1.30) through (1.37).

1.4   Use the diagram in Figure 1.10 to prove (1.55).

1.5   What are the physical meanings of $|a_i|^2$, $|b_i|^2$, $|s_{11}|^2$, and $|s_{22}|^2$?

1.6   Prove that $|a_1|^2$ is the available power in an two-port augmented by its normalizing impedances.

1.7   Prove that $|b_1|^2$ is the power reflected from the input port of a two-port network augmented by its normalizing impedances.

1.8   Under what condition will the available power in a two-port network augmented by the actual load and source impedances be equal to that of the two-port augmented by the normalizing impedances?

1.9   The terminal voltages and currents of a two-port are

$V_1 = 0.2$V
$V_2 = 1.0$V
$I_1 = 0.05$A
$I_2 = -0.2$A

If the normalizing impedances are $Z_{01} = 20\Omega$ and $Z_{02} = 10\Omega$, determine the incident and reflected current and voltage, and the normalized components of the two-port.

1.10   Use the normalized components to calculate the power entering the network, as well as that dissipated in the load. Use the normalized components to calculate the terminal current and voltage.

1.11   Derive (1.89) through (1.93).

1.12   The S-parameters of a transistor at 2 GHz are (20 log $|s_{ij}|$; normalizing impedances of 50Ω):

$s_{11} = -0.355$ dB, $-23°$
$s_{12} = -30.5$ dB, $83°$
$s_{21} = 7.20$ dB, $159°$
$s_{22} = -2.715$ dB, $-6°$

Calculate the input and output impedances and the transducer power gain, if a 100-Ω load is connected to it and the internal impedance of the source is equal to 50Ω.

1.13   Convert the *S*-parameters of the transistor in Problem 1.12 to *Y*-parameters and find the *S*-parameters for the transistor if the normalizing impedances are changed to Z_{01}=50Ω and Z_{02}=100Ω. Compare the results with those obtained in the previous problem.

1.14   Use (1.102) to prove that the transducer power gain of a one-port network is given by

$$G_T = \frac{[1-|S_L|^2]\,[1-|S_S|^2]}{|1-S_S\,\Gamma_L|^2} \tag{1.153}$$

where S_s is the reflection parameter of the source and S_L that of the load.

1.15   Prove (1.146) through (1.150).

1.16   Assuming that the *S*-parameters given in Problem 1.12 are the parameters of a common-source GaAs transistor, determine the *S*-parameters corresponding to the common-gate configuration.

1.17   Is the following matrix unitary?

$$\frac{1}{2-\omega^2 + j2\omega}\begin{bmatrix} -\omega^2 & 0.5 \\ 0.5 & \omega^2 \end{bmatrix}$$

1.18   Use the reciprocity theorem to prove that the following equalities apply to any passive reciprocal network:

$s_{21}(s) = s_{12}(s)$

$y_{21}(s) = y_{12}(s)$

$z_{21}(s) = z_{12}(s)$

1.19   Prove that $|s_{21}|^2 = 1 - |s_{11}|^2$ for any lossless passive network.

1.20   Instead of using the T-parameters to find an equivalent set of parameters for two cascaded networks, the transmission matrix defined by the equation:

$$\begin{bmatrix} a_1 \\ b_1 \end{bmatrix} = P \begin{bmatrix} a_2 \\ b_2 \end{bmatrix} \qquad (1.154)$$

and its inverse can be used. Derive the equations relating the elements of P and the corresponding S-matrix.

REFERENCES

[1]   *S-Parameter Design*, Application Note 154, Palo Alto, CA: Hewlett Packard, April 1972.

[2]   *MultiMatch RF and Microwave Impedance-Matching, Amplifier and Oscillator Synthesis Software*, Stellenbosch, South Africa: Ampsa (Pty) Ltd; http://www.ampsa.com, 2009.

[3]   Chen, W. K., *Theory and Design of Broadband Matching Networks*, Oxford, U.K.: Pergamon Press, 1976.

SELECTED BIBLIOGRAPHY

Carson, R. S., *High Frequency Amplifiers*, New York: John Wiley and Sons, 1975.

CHAPTER 2

CHARACTERIZATION AND ANALYSIS OF ACTIVE CIRCUITS AT RF AND MICROWAVE FREQUENCIES

2.1 INTRODUCTION

Characterization and analysis of linear circuits in terms of Y-, Z-, T-, and S-parameters were considered in Chapter 1. When active circuits are designed, the noise performance, the output power, the distortion, and the efficiency are also of interest.

Noise parameters are used to characterize the noise behavior of linear circuits at RF and microwave frequencies. These parameters will be considered in Section 2.2. The noise figure of a linear circuit follows easily from the noise parameters. The relevant equations will also be derived in Section 2.2.

Noise characterization and analysis in terms of equivalent circuits and correlation matrices will also be considered in this section. The effect of feedback and loading on the noise parameters of a transistor can be established easily by using noise correlation matrices. Calculation of these effects will be considered in Section 2.2.2.

The focus in this book is on synthesizing linear circuits. Ways of controlling and predicting the output power without directly using nonlinear models will be explored where possible. This does not imply that nonlinear models should not be used. In fact, the S-parameters required for a design are frequently obtained from a large-signal model. Measured parameters are, however, preferable. This follows because many large-signal models are not accurate at all bias points.

Large-signal modeling and simulation is of course essential to fine-tune and verify the performance of the circuit designed, but this step can usually be performed adequately with a good circuit simulator. Large-signal modeling and simulation, therefore, will not be covered in this book.

The power obtainable from a linear amplifier (class A, class B, class AB, and class F) is a strong function of the bias point and the intrinsic voltage and current associated with each of the transistors used. The power level at which the intrinsic output current and/or voltage starts to clip usually provides a close estimate of the 1-dB compression point of a linear amplifier [1]. The Cripps approach to calculating the maximum output power will be considered in Section 2.3, and a new set of parameters (power parameters) will be introduced [2]; these can be used to simplify calculation of the expected output power.

The power parameters map the intrinsic voltages in each transistor to the external voltages and also map the intrinsic output current to the intrinsic voltages. The relationship between the intrinsic load and the external load of each transistor follows easily from the power parameters. The derivation will be considered in Section 2.3. The need for this mapping arises from the fact that the output power is limited by intrinsic constraints.

Boundary lines on the I/V curves (preferable pulsed) of the transistor are used to establish the intrinsic constraints on the transistor current and voltage. When these curves are not available, the appropriate boundaries can be estimated from the data sheet information available.

The power parameters of a transistor can be adjusted easily to incorporate the effect of feedback and/or loading or any change in the configuration (common-source, common-gate, and so forth). All of these aspects will also be considered in Section 2.3.

Multistage (linear) power amplifiers can also be designed effectively by using the power parameters. A set of power parameters is associated with each transistor used in the circuit. The influence of each transistor is considered with the other transistors in the circuit assumed to be ideal. The output power is mainly determined by the stage in which voltage and/or current clipping first occurs. The general case can follow an approach similar to when the output power of a cascade of power amplifiers is calculated. The intercept point and the 1-dB compression point of a cascade are considered in Section 2.3.2.

A model is required for each transistor used in the circuit in order to calculate the power parameters. The model used should provide a good fit over the complete frequency range over which data are available, and should accurately represent the intrinsic part and the parasitics of the actual transistor. As long as the transistors used operate in a linear mode, conventional small-signal models are found to be adequate for this purpose.

When this approach is followed, adequate control over the output power is usually possible, but control over the distortion generated is largely indirect. There is usually a tight relationship between the 1-dB compression point and the third-order intercept point of a transistor (this can be improved by predistortion), which is useful in CW (continuous wave) applications. When the average signal level is significantly below the peak level (complex modulation schemes), this relationship is, however, not of much use. A design for maximum CW power will, therefore, not be identical to one for the best linearity.

Load-pull information, and/or large-signal simulation results for synthesized circuits, can be used to establish synthesis rules for such cases. As an example of this, examination of the load-pull data provided for at least one gallium nitride transistor [3] which showed that the distortion (OFDM signal) will be close to optimum if the load line is selected to maximize the power gain with the maximum power level (class-B operation) set to around 6 dBm above the average power level. The source-pull data (fundamental frequency) indicates resonance around the average power level, with the optimum source resistance 3 to 4 times higher than that required for a conjugate match at the average power level (class-A operation assumed). Harmonic load-pull data was not provided. It may very well be that with the second harmonic controlled, the optimum fundamental frequency input match will be closer to a conjugate match.

The efficiency is largely controlled by the class of operation selected and the load line provided. In addition to controlling the load line at the fundamental frequency, short circuits and open circuits must be provided for the different harmonics of interest in order to establish the required mode of operation [4, 5]. This can obviously only be done in narrowband cases.

In a narrowband single-ended class-B amplifier the harmonics should ideally be short-circuited to ensure that the intrinsic output voltage will be sinusoidal with the half-sine intrinsic output current. Open circuits at the odd harmonics and short circuits at the even harmonics are required in the class-F case in order to force the output voltage towards a square wave with the half-sine output current. In a class-E amplifier [6] high impedances are targeted at the harmonic frequencies [4]. The high external impedance requirement ensures that the harmonic power in the actual load will be insignificant. Note that the transistor in a class-E amplifier is used as a switch and not a voltage-controlled current source. Also note that the current and voltage at the transistor terminals consist of all harmonics.

The short and open circuit terminations required are intrinsic in all cases, except the class-E case. The power parameters can be used to calculate the external loads associated with the required harmonic terminations when the transistor operates in a linear mode.

Efficient operation can also be obtained with Doherty power amplifiers [7, 8]. An extra transistor is used in this configuration to modulate the load line of the main transistor. The main transistor usually operates as a class-AB amplifier at low signal levels with the extra transistor inactive. The load line has a value higher than what is possible with a single transistor, which allows for higher gain and efficiency at lower signal levels. When the main transistor saturates (low voltage across the transistor), the extra transistor kicks in and allows the current of the main transistor to increase towards its maximum value with the voltage across the main transistor remaining constant at its saturated level. The class-AB amplifier can be designed by using the power parameters, but a large-signal approach is required to design the class-C auxiliary amplifier and to fine-tune the final circuit.

In addition to controlling load terminations at harmonic frequencies, improvement in the performance can also be obtained by controlling the source impedance presented to the transistor at harmonic frequencies. Such an improvement is demonstrated in [9] for a pHemt. The second harmonic was short-circuited at the input in this demonstration. Without the short-circuited harmonic, the input voltage is distorted by C_{gs}, which is strongly voltage dependent. This distortion increases the conduction angle of the output current, which reduces the efficiency because of the increased overlap between the voltage and current. The short circuit restores the input voltage to a sinusoid and corrects the conduction angle of the output current, which brings about the observed improvement in the efficiency. Because of the reduced second harmonic content on the input side, one would also expect an improvement in the third-order distortion.

Alternative compensation methods for the strong variation in C_{gs} is demonstrated in [10, 11]. In both cases the variation in the input capacitance is reduced by effectively connecting a junction capacitor with opposite voltage dependence across the input capacitor. The compensation on [10] was demonstrated with a MESFET. The compensation in [11] is illustrated for a BJT and is more extensive. The power parameters can be used in these cases too if S-parameters are generated for the composite device with a harmonic-balance simulator.

An interesting concept is introduced in [12]. The input signal of a class-B amplifier is usually a sinusoid, but half a sinusoid would be perfectly adequate to drive the transistor. When this is done, the amplitude of the associated fundamental tone is half of that of the original sinusoid, which leads to an improvement of up to 6 dB in the gain. The same half sinusoid will also be adequate for the class-F case. A technique for generating the required half sinusoid is introduced in [12]. The increased harmonic content may, however, increase the distortion.

It should be mentioned that distortion cancellation can sometimes be obtained by balancing distortion contributions with opposite signs [13]. This can only be done with large-signal or Volterra models and is outside the scope of this book.

2.2   NOISE PARAMETERS

Instead of considering the noise contribution of each physical noise source in a linear network, its noise contribution can be modeled in terms of equivalent noise sources at its input and/or output ports or by using correlation matrices. Both approaches will be considered here. The relationship between the equivalent noise sources or the correlation matrices and the noise parameters typically supplied for a transistor (F_{min}, Γ_{n_opt}, and R_n) will also be established.

The noise figure of a transistor is a function of its noise parameters at the bias point of interest and the source impedance presented to its input terminals by the circuit. The dependence of the noise figure on the source impedance will also be considered.

The equivalent noise sources or the correlation matrices can be used to find the noise parameters for parallel networks, series networks, or cascaded networks in terms of the noise parameters of the individual networks. The influence of matching networks or filters or of adding series and parallel feedback on the noise parameters of a transistor can be established easily by using the results.

The noise figure of cascaded networks will also be considered.

2.2.1   Modeling the Noise Contribution of a Two-Port with Equivalent Circuits

The noise generated by any two-port device can be modeled with two equivalent partially correlated noise sources [14]. The cascade, current, and voltage representations are shown in Figure 2.1. These representations are equivalent to the T-parameter, Y-parameter, and Z-parameter approaches, respectively.

The two sources used in each representation are partially correlated. The correlated and uncorrelated parts in each case are defined by the following set of equations:

$$I_n(t) = I_{nu}(t) + Y_{cor}V_n(t) \tag{2.1}$$

$$I_{2n}(t) = I_{2nu}(t) + X_i I_{1n}(t) \tag{2.2}$$

$$V_{2n}(t) = V_{2nu}(t) + X_v V_{1n}(t) \tag{2.3}$$

By using the definition of the noise figure and these equivalent circuits, expressions for the noise figure in terms of the equivalent noise sources and the correlation factors can be derived. This will be done here for the cascade representation.

The noise figure of a device is defined as the ratio of the total noise power at the load to that which would have been delivered to the load if the device was noiseless, that is,

$$F = \frac{P_{no}}{P_{no-\text{ide}}} \qquad (2.4)$$

If the spot noise figure (narrowband noise figure at a particular frequency) is considered, the noise power at the output in the ideal case is given by

$$P_{no-\text{ide}}(f) = kTBG_T(f) \qquad (2.5)$$

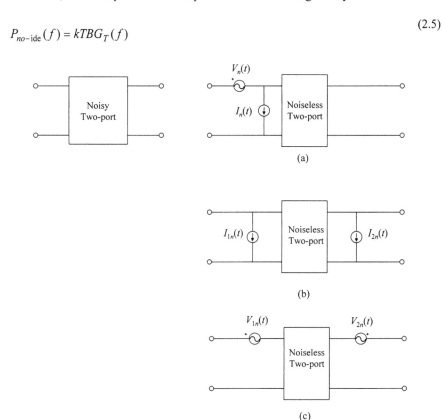

Figure 2.1   Equivalent circuits for the noise contributed by a two-port device: (a) cascade, (b) current, and (c) voltage representation.

where k is Boltzmann's constant (1.38×10^{-23} J/K), T is the absolute temperature (in kelvin), B is the bandwidth (in hertz), and $G_T(f)$ is the transducer power gain of the two-port at frequency f. A noise reference temperature of 290K (room temperature) is typically used.

When the spot noise figure is considered, the output power can be referenced easily to the input side, in which case (2.5) becomes

$$F = \frac{P_{ni-av}}{P_{ni-av\_ide}} \tag{2.6}$$

where P_{ni-av} is the effective noise power available at the input side, and P_{ni-av_ide} is the noise power that would have been available at the input side if the device was noiseless.

The available power at the input terminals can be obtained by terminating the noise source(s) in the conjugate of the source admittance, as is illustrated in Figure 2.2.

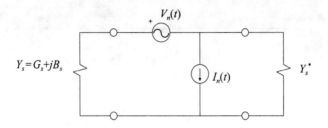

Figure 2.2   The equivalent circuit used to calculate the available noise power at the input of a two-port (cascade representation).

It should be noted that superposition can be applied to the uncorrelated components of the noise power. In general, superposition only applies to the voltage and current in a circuit.

Because of the above, the noise power resulting from $I_u(t)$ can simply be added to that resulting from $V_n(t)$ and the correlated part of $I_n(t)$ $[Y_{cor} V_n(t)]$.

The correlated fraction of the available power can be calculated in the following way:

$$V_{oc} = V_n \, Y_s \, / \, (Y_s + Y_s^*) + V_n \, Y_{cor} \, / \, (2G_s)$$

$$= V_n \, (Y_s + Y_{cor}) \, / \, (2G_s) \tag{2.7}$$

$$\frac{P_{o-cor}}{G_s} = \frac{1}{T} \int_0^T V_{oc}(t) \, V_{oc}^*(t) \, \partial t$$

$$= \overline{V_{oc} \, V_{oc}^*}$$

$$= \overline{V_n \, V_n^*} \; (Y_s + Y_{cor})(Y_s + Y_{cor})^* \, / \, (4G_s^2) \tag{2.8}$$

where T is the period over which the noise power is averaged.

The uncorrelated fraction of the output power can be calculated in terms of I_n and V_n by using (2.1):

$$\overline{I_n\,I_n^*} = \overline{I_{nu}I_{nu}^*} + Y_{cor}Y_{cor}^* \,\overline{V_n\,V_n^*} \tag{2.9}$$

$$I_{nu} = I_n - Y_{cor}\,V_n \tag{2.10}$$

$$V_{ou} = I_{nu}/(2\,G_s) \tag{2.11}$$

Therefore,

$$\frac{P_{ou}}{G_s} = \overline{V_{ou}\,V_{ou}^*} = \overline{I_{nu}I_{nu}^*}\,/\,(4G_s^{\,2})$$

$$= \overline{I_n I_n^*}\,/\,(4G_s^{\,2}) - \overline{V_n V_n^*}\;Y_{cor}Y_{cor}^*\,/\,(4G_s^{\,2}) \tag{2.12}$$

Equation (2.9) follows from (2.1) because I_{nu} and V_n are uncorrelated.
The total available noise power at the input terminals can now be calculated:

$$\frac{P_o}{G_s} = \frac{\overline{I_n I_n^*}}{4G_s^{\,2}} - \frac{Y_{cor}Y_{cor}^*\,\overline{V_n V_n^*}}{4G_s^{\,2}} + \frac{(Y_s+Y_{cor})(Y_s+Y_{cor})^*\,\overline{V_n V_n^*}}{4G_s^{\,2}}$$

$$= \frac{\overline{I_n I_n^*}}{4G_s^{\,2}} + \frac{Y_s Y_s^* + 2\,\Re(Y_{cor}Y_s^*)}{4G_s^{\,2}}\,\overline{V_n V_n^*} \tag{2.13}$$

Because the available noise power in the ideal case is simply

$$\frac{P_{o\text{-ide}}}{G_s} = V_{o\text{-thermal}}^2 = kTB\,/\,G_s \tag{2.14}$$

it follows that

$$F = \frac{\dfrac{kTB}{G_s} + \dfrac{\overline{I_n\,I_n^*}}{4G_s^2} + \dfrac{Y_s\,Y_s^* + 2\,\Re\,(Y_{cor}\,Y_s^*)}{4G_s^2}\,\overline{V_n\,V_n^*}}{kTB\,/\,G_s}$$

$$= 1+G_{ni}/G_s+R_{nv}\,[(G_s+G_{cor})^2 + (B_s+B_{cor})^2 - (G_{cor}^2+B_{cor}^2)]\,/\,G_s \tag{2.15}$$

where

$$\overline{I_n\,I_n^{\,*}} \;=\; 4kTBG_{ni} \tag{2.16}$$

$$\overline{V_n\,V_n^{\,*}} \;=\; 4kTBR_{nv} \tag{2.17}$$

$$Y_{\text{cor}} = G_{\text{cor}} + jB_{\text{cor}} \tag{2.18}$$

and

$$Y_s \;=\; G_s \,+\, jB_s \tag{2.19}$$

For any given value of G_s, there exists an optimum value for B_s that will minimize the noise figure. Taking the derivative of (2.15) and setting it equal to zero yields

$$\partial F \,/\, \partial B_s \;=\; 0 \;=\; 0 \,+\, 0 \,+\, 0 \,+\, 2R_{nv}\,(B_s \,+\, B_{\text{cor}}) \,/\, G_s \tag{2.20}$$

from which it follows that

$$B_{s\text{-opt}} \;=\; -B_{\text{cor}} \tag{2.21}$$

Note that the optimum value of B_s is the same for all values of G_s.

The value of G_s that will minimize the noise figure can now be obtained from (2.15), after replacing B_s with its optimum value. Taking the derivative yields

$$\partial F/\partial G_s \;=\; 0 \;=\; 0 \,-\, G_{ni}\,/\,G_s^{\,2} \,+\, 2\,R_{nv}\,(G_s \,+\, G_{\text{cor}}) \,/\, G_s$$

$$-R_{nv}\,[(G_s \,+\, G_{\text{cor}})^2 \,+\, (B_{s\text{-opt}} \,+\, B_{\text{cor}})^2 \,-\, (G_{\text{cor}}^{\,2} \,+\, B_{\text{cor}}^{\,2})] \,/\, G_s^{\,2} \tag{2.22}$$

that is,

$$G_{s\text{-opt}} \;=\; \sqrt{G_{ni}\,/\,R_{nv} \,-\, B_{\text{cor}}^{\,2}} \tag{2.23}$$

Substituting (2.21) and (2.23) into (2.15) yields

$$F_{\min} \;=\; 1 \,+\, 2R_{nv}\,(G_{s\text{-opt}} \,+\, G_{\text{cor}}) \tag{2.24}$$

The inverse relationships and an expression for the noise figure in terms of $F_{\min}$, $G_{s\text{-opt}}$, $B_{s\text{-opt}}$, and R_n can be derived easily by using (2.21), (2.23), and (2.24) and making the necessary substitutions in (2.15):

$$G_{cor} = (F_{min} - 1) / (2R_{nv}) - G_{s-opt} \tag{2.25}$$

$$B_{cor} = -B_{s-opt} \tag{2.26}$$

$$F = F_{min} + R_{nv} [(G_s - G_{s-opt})^2 + (B_s - B_{s-opt})^2] / G_s \tag{2.27}$$

$$= F_{min} + \frac{R_{nv}}{G_s} | Y_s - Y_{s-opt} |^2 \tag{2.28}$$

By inspecting (2.27) it is clear that the loci of constant noise figures are circles in the linear admittance plane, as is illustrated in Figure 2.3.

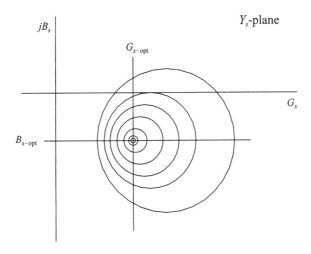

Figure 2.3   An example of the constant noise figure circles associated with a linear two-port device. The best noise performance is obtained when the source admittance (Y_s) is equal to $G_{s-opt} + jB_{s-opt}$.

It will be shown in Chapter 9 that the constant noise figure contours are also circles on the Smith chart. Constant noise figure circles for a low-noise transistor are displayed in Figure 2.4. Note that the optimum noise impedances rotate counterclockwise on the Smith chart. The output reflection coefficients associated with the selected noise terminations are also displayed (lowest trace).

The highest trace shown in Figure 2.4 is the conjugate of the input reflection coefficient associated with a conjugate match on the output side of the transistor. If this trace was on top of the optimum noise reflection coefficient trace, maximum gain would have been obtained with the same impedances as the optimum noise figure. The trace on the right (s_{21a}) represents the available power gain associated with the selected noise terminations.

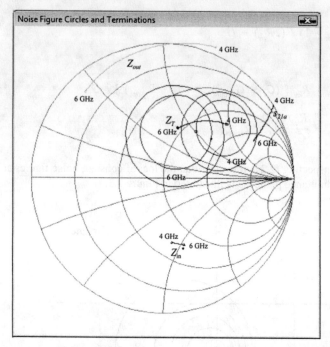

Figure 2.4   An example of constant noise figure circles displayed on a Smith chart. The output reflection
coefficients (Z_{out} trace) and the conjugate of the input reflection coefficients (Z_{in}^{*} trace)
associated with the optimum noise terminations (Z_T trace) are also displayed. The magnitude
of s_{21a} is equal to the available power gain associated with the optimum noise termination at the
frequency of interest.

2.2.2   Noise Correlation Matrices

A correlation matrix is defined for each of the three representations considered in the
previous section [15]:

Cascade representation

$$C_a = \overline{\begin{bmatrix} V_n(t) \\ I_n(t) \end{bmatrix} \begin{bmatrix} V_n^*(t) & I_n^*(t) \end{bmatrix}} \Big/ (2B)$$

$$= \begin{bmatrix} \overline{V_n(t)\ V_n^*(t)} & \overline{V_n(t)\ I_n^*(t)} \\ \overline{V_n^*(t)\ I_n(t)} & \overline{I_n(t)\ I_n^*(t)} \end{bmatrix} \Big/ (2B) \qquad (2.29)$$

Current representation

$$C_y = \begin{bmatrix} I_{n1}(t) \\ I_{n2}(t) \end{bmatrix} [I_{n1}^*(t) \quad I_{n2}^*(t)] / (2B)$$

$$= \begin{bmatrix} \overline{I_{n1}(t) I_{n1}^*(t)} & \overline{I_{n1}(t) I_{n2}^*(t)} \\ \overline{I_{n1}^*(t) I_{n2}(t)} & \overline{I_{n2}(t) I_{n2}^*(t)} \end{bmatrix} / (2B) \tag{2.30}$$

Voltage representation

$$C_z = \begin{bmatrix} V_{n1}(t) \\ V_{n2}(t) \end{bmatrix} [V_{n1}^*(t) \quad V_{n2}^*(t)] / (2B)$$

$$= \begin{bmatrix} \overline{V_{n1}(t) V_{n1}^*(t)} & \overline{V_{n1}(t) V_{n2}^*(t)} \\ \overline{V_{n1}^*(t) V_{n2}(t)} & \overline{V_{n2}(t) V_{n2}^*(t)} \end{bmatrix} / (2B) \tag{2.31}$$

By using definitions (2.4), (2.16), and (2.17) it follows that

$$C_a = 2kT \, R_{nv} \begin{bmatrix} 1 & Y_{cor}^* \\ Y_{cor} & G_{ni} / R_{nv} \end{bmatrix} \tag{2.32}$$

The second term of (2.32) is derived as follows:

$$I_n(t) = I_{nu}(t) + Y_{cor} V_n(t)$$

multiplied by V_n^* and averaged over time leads to

$$\frac{1}{T} \int_0^T I_n(t) \, V_n^* \, dt = \frac{1}{T} \int_0^T I_u(t) \, V_n^*(t) + Y_{cor} V_n^*(t) V_n(t) \, dt$$

$$= 0 + \frac{1}{T} Y_{cor} \int_0^T V_n(t) \, V_n^*(t) \, dt$$

$$= 4kTBR_{nv} \, Y_{cor} \tag{2.33}$$

It is a simple matter to show that (2.32) is also equivalent to

$$C_a = 2 \, kT \, R_{nv} \begin{vmatrix} 1 & \dfrac{F_{min}-1}{2 \, R_{nv}} - Y_{s-opt}^{*} \\[3mm] \dfrac{F_{min}-1}{2 \, R_{nv}} - Y_{s-opt} & |Y_{s-opt}|^{2} \end{vmatrix} \tag{2.34}$$

It is possible to transform any of the correlation matrices defined above to any of the other types. The transformation matrices required for this purpose are summarized in Table 2.1 [15]. In this table Y, Z, and T are the Y-parameter, Z-parameter, and transmission parameter matrices, respectively, of the network under consideration. The transformation required is done by using the equation

$$C_{new} = X \, C_{ori} \, X'^{*} \tag{2.35}$$

where $'^{*}$ indicates the transposed conjugate of X.

The equations summarized in Table 2.1 can be derived easily by using the relationship between the noise voltages and currents in the different representations. Because of the principle of superposition, the equivalence can be derived by assuming the noise generators to be the only excitations present.

EXAMPLE 2.1   Derivation of expressions for the equivalent noise sources $I_{1n}(t)$ and $I_{2n}(t)$ in terms of $V_n(t)$ and $I_n(t)$.

Consider deriving expressions for the equivalent noise sources $I_{1n}(t)$ and $I_{2n}(t)$ (current representation) in terms of $V_n(t)$ and $I_n(t)$ in the cascade representation.

$I_n(t)$ is clearly part of $I_{1n}(t)$ and, therefore, only the equivalent current sources for $V_n(t)$ are required. Because of superposition and because there is no noise source on the output in the cascade representation, the load can be shorted and the currents resulting from $V_n(t)$ can be calculated by using the Y-parameters:

$$I_1 = -y_{11} \, V_n(t)$$

$$I_2 = -y_{21} \, V_n(t)$$

Adding $I_n(t)$ yields

$$I_{1n}(t) = -y_{11} \, V_n(t) + I_n(t)$$

$$I_{2n}(t) = -y_{21} \, V_n(t)$$

leading to

$$\begin{bmatrix} I_{1n}(t) \\ I_{2n}(t) \end{bmatrix} = \begin{bmatrix} -y_{11} & 1 \\ -y_{21} & 0 \end{bmatrix} \begin{bmatrix} V_n(t) \\ I_n(t) \end{bmatrix} \tag{2.36}$$

Equations (2.37)–(2.39) are generally used to calculate the equivalent correlation matrices of two networks connected in cascade, parallel, and series, respectively. The relevant equations [15] are

$$C_{aT} = C_{a1} + T \, C_{a2} T'^* \tag{2.37}$$

$$C_{yT} = C_{y1} + C_{y2} \tag{2.38}$$

$$C_T = C_{z1} + C_{z2} \tag{2.39}$$

T in (2.37) is the transmission matrix of the network closest to the generator (i.e., the network on the input side). The superscript used indicates the transposed conjugate of the transmission matrix.

When the noise parameters for a network are calculated, it is useful to know that

$$C_z = 2 \, kT \, \Re \, (Z) \tag{2.40}$$

Table 2.1 The Matrix (X) Required to Transform Any of the Noise Correlation Matrices to Another ($C_{\text{new}} = X \, C_{\text{ori}} \, X'^*$)

New \ Original	Y	Z	T
Y	$\begin{pmatrix} 1 & 0 \\ 0 & 1 \end{pmatrix}$	Y	$\begin{pmatrix} -y_{11} & 1 \\ -y_{21} & 0 \end{pmatrix}$
Z	Z	$\begin{pmatrix} 1 & 0 \\ 0 & 1 \end{pmatrix}$	$\begin{pmatrix} 1 & -z_{11} \\ 0 & -z_{21} \end{pmatrix}$
T	$\begin{pmatrix} 0 & B \\ 1 & D \end{pmatrix}$	$\begin{pmatrix} 1 & -A \\ 0 & -C \end{pmatrix}$	$\begin{pmatrix} 1 & 0 \\ 0 & 1 \end{pmatrix}$

and

$$C_y = 2\ kT\ \Re\ (Y) \tag{2.41}$$

for any passive network [15].

2.2.3   Calculating the Noise Figure of a Cascade Network

The noise figure of a cascade network (see Figure 2.5) is often of interest. Given the definition of the noise figure in terms of the available noise power at the input side of the network, it is a simple matter to prove that

$$F_T = F_1 + \frac{F_2 - 1}{G_{a1}} + \frac{F_3 - 1}{G_{a1}G_{a2}} + \dots \tag{2.42}$$

where F_1 is the noise figure of the first stage (input stage) and G_{a1} is its available power gain. Similarly, F_n is the noise figure of the nth stage when terminated on its input side with the output impedance of the previous stage, and G_{an} is its available power gain.

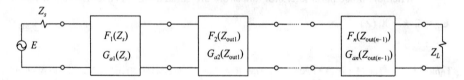

Figure 2.5   The circuit used to calculate the noise figure of a cascade network.

Equation (2.42) is known as Friss' formula [16]. It is clear from Friss' formula that the product of the gain of the stages preceding any given stage must be high in order for it to have a negligible contribution to the overall noise figure of the cascade.

It is also clear that any stage added will have a degrading effect on the noise figure. The contribution of any stage to the overall noise figure is a function of both its noise figure and its available power gain. The noise measure (M) of a network is a figure of merit for this effect and is defined as

$$M = F_\infty - 1 \tag{2.43}$$

where F_∞ is the noise figure of an infinite chain of identical stages each with noise figure F and available power gain G_a.

By using the identity

$$\frac{1}{1 - X} = 1 + X + X^2 + \dots \tag{2.44}$$

it can be shown that the noise measure, M, is given by

$$M = \frac{F - 1}{1 - 1/G_a} \tag{2.45}$$

where F is the noise figure of the stage of interest and G_a is its available power gain.

The associated noise figure is of greater interest and is given by substituting (2.43) into (2.45):

$$F_\infty = \frac{F - 1/G_a}{1 - 1/G_a} \tag{2.46}$$

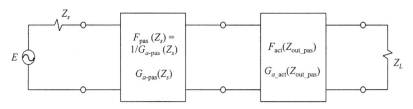

Figure 2.6   The effect of insertion loss on the noise figure of an amplifier stage.

EXAMPLE 2.2   Calculation of the effect of the losses of a passive cascade on the noise figure of a transistor.

The effect of the insertion loss of a lossy passive network on the noise figure of an active stage will be calculated by using Friss' formula.

The noise figure of a passive network is given by

$$F_{pas}(f) = P_{no}/P_{no\text{-}ide} = \frac{kTB}{kTB \ G_{a\text{-}pas}(f)} = 1/G_{a\text{-}pas}(f) \tag{2.47}$$

that is, if the passband is narrow enough for the available power gain and the mismatch from the output of the network to its load to be considered constant.

Entering this into Friss' formula for the cascade combination (see Figure 2.6) yields

$$F_T = F_{pas} + \frac{F_{act} - 1}{G_{a\text{-}pas}} = 1 \ / \ G_{a\text{-}pas} + \frac{F_{act} - 1}{G_{a\text{-}pas}}$$

$$= \frac{1 + F_{act} - 1}{G_{a\text{-}pas}}$$

$$= F_{act} / G_{a\text{-}pas} \tag{2.48}$$

Expressed in decibels, (2.48) becomes

$$10 \log F_T = 10 \log F_{\text{act}} - 10 \log G_{a\text{-pas}} \quad \text{(dB)} \qquad (2.49)$$

It follows from (2.49) that the noise figure of any stage is degraded proportionately with any losses directly preceding it [$G_{a\text{-pas}}$ in (2.49) will be negative for any passive network]. This is illustrated in Figure 2.6.

2.3   THE OUTPUT POWER OF LINEAR AMPLIFIERS

The maximum output power obtainable from a linear amplifier (1-dB compression point) will be considered in this section.

The transistors used in a linear amplifier are frequently biased in class A (360° conduction angle), class B (180° conduction angle), or class AB mode. Class AB is often used at microwave frequencies instead of class B, mostly because the gain obtainable in class B mode is usually too low at these frequencies. The voltage and current waveforms and the load lines associated with class-A and class-B stages will be considered in Section 2.3.1, and those corresponding to class AB, as well as class F, in Section 2.3.2. The class E case is considered in Section 2.3.3 and the Doherty amplifier in Section 2.3.4.

The 1-dB compression point and the third-order two-tone intercept point are usually used as measures of the linearity of an amplifier. In wireless applications the adjacent channel power ratio (ACPR) is also frequently used. The relevant definitions and the definition of the dynamic range of an amplifier will be considered in Section 2.3.5.

The 1-dB compression point and the third-order intercept point of an amplifier will be reduced by any driver stages added. This effect will also be considered in Section 2.3.5.

The maximum output power obtainable from a class-A amplifier can be estimated at RF, as well as at microwave frequencies, by using the approach introduced by Cripps [1]. The Cripps approach will be considered in Section 2.3.6.

The Cripps approach can be generalized and many of the inherent inaccuracies can be removed by using the power parameter approach introduced in [2]. This approach is outlined in Section 2.3.7.

The Cripps approach and the power parameter approach are based on the assumption that the maximum power obtainable from a linear stage is determined by the power level at which the intrinsic output current and/or voltage of the transistor(s) used starts to clip; that is, the power is limited mainly by the limited swing in the intrinsic output current and voltage.

The power parameter approach is sufficiently general to handle any loading effects, feedback, changes in the transistor configuration, cascade networks, and/or multistage amplifiers. All of these aspects will also be considered in Section 2.3.7.

The power parameter approach can also be used to initialize the fundamental tone quantities in a full harmonic balance nonlinear simulation of a linear amplifier.

2.3.1   Load-Line Considerations in Class-A and Class-B Amplifiers

When a transistor is biased for class-A operation, the average voltage across its output terminals (drain-source or collector-emitter) must be equal to the dc voltage (V_{DS} or V_{CE};

usually the supply voltage, V_s, if power is important), and the average current must be equal to the dc current (I_{DS} or I_{CE}) (the dc current may change as the drive level is increased). If the distortion in the waveforms is negligible (no harmonics or intermodulation products), the voltage and current will swing symmetrically around the average values.

The maximum possible voltage swing (V_{DS} or V_{CE}) is decreased in practice by the saturation voltage of the transistor (V_{sat}) and any saturation resistance (R_{sat}). The effect of the saturation resistance can be lumped with the load resistance presented at the transistor terminals.

The maximum output power obtainable from a class-A or a class-B stage at RF frequencies is given by [17]

$$P_{max} = \frac{(V_s - V_{sat})^2}{2 (R_L + \alpha R_{sat})} \frac{R_L}{R_L + \alpha R_{sat}} \tag{2.50}$$

where V_s is the supply voltage (assuming that no drain or collector resistor is used) and R_L is the parallel resistance presented to the output terminals of the transistor. α is equal to 2 for class-A amplifiers and equal to 1 for class-B amplifiers. It follows from this equation that the effective supply voltage is decreased by the saturation voltage and that the effective intrinsic load resistance is increased by the saturation resistance.

In deriving (2.50), it was assumed that any susceptance present at the output terminals of the transistor was removed.

2.3.1.1   Class-A Load Line

The output current and voltage and the associated load line in a class-A stage will be considered next.

In general, if

$$V_2(t) = |V_2| e^{j\omega t} \tag{2.51}$$

and

$$Y_L = -I_2 / V_2 = |Y_L| e^{j\theta} \tag{2.52}$$

the drain voltage and current (dc and ac components) are given by

$$I_d(t) = I_{DS} - Y_L V_2(t) \tag{2.53}$$

$$V_d(t) = V_{DS} + V_2(t) \tag{2.54}$$

With $V_2(t)$ replaced in terms of (2.51), it follows that

$$I_d(t) = I_{DS} - |V_2| e^{j\omega t} |Y_L| e^{j\theta} = I_{DS} - |Y_L V_2| e^{j(\omega t + \theta)} \tag{2.55}$$

$$V_d(t) = V_{DS} + |V_2| e^{j\omega t} \tag{2.56}$$

It follows from the last two equations that the dynamic load line is defined by

$$I_2(t) = I_d(t) - I_{DS} = -|Y_L V_2| e^{j(\omega t + \theta)} \tag{2.57}$$

$$V_2(t) = V_d(t) - V_{DS} = |V_2| e^{j\omega t} \tag{2.58}$$

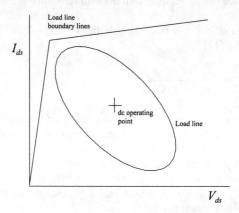

Figure 2.7   The dynamic load line of a transistor biased for class-A operation (complex load line).

When the load has a reactive component too (complex load), the load line will be similar to that shown in Figure 2.7.

The dc power dissipated in a class-A stage is constant and is given by

$$P_{dc} = V_{DS} I_{DS} \tag{2.59}$$

The power dissipated in the intrinsic load is given by

$$P_o = |V_2|^2 G_L / 2 \tag{2.60}$$

$$\leq P_{dc} / 2$$

or by

$$P_o = |I_2|^2 R_L / 2 \tag{2.61}$$

$$\leq P_{dc} / 2$$

If the voltage is clipped first, the maximum output power will be given by (2.60). If the current is clipped first, (2.61) will apply. In general, clipping can occur on any of the four line segments shown in Figure 2.8 (resistive load lines shown).

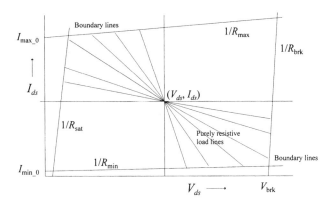

Figure 2.8   Clipping in a class-A amplifier can occur on any of the four boundary lines shown (resistive load lines shown; dc operating point: V_{ds}, I_{ds}).

2.3.1.2   Class-B Load Line

The conduction angle in a class-B amplifier is $180°$. A parallel-tuned circuit or a push-pull configuration is usually used to suppress the harmonics in the voltage waveform. When this is done, the output voltage can be assumed to be sinusoidal and can therefore be represented by using the same equations as in the class-A case.

The intrinsic transistor current $[I_T(t)]$ is a half-sinusoid. The peak amplitude of the fundamental tone (which determines the ac power) can be obtained from the Fourier series expansion for the half-sinusoid (refer to Figure 2.9).

The Fourier series expansion of a half-sinusoid is given by

$$I_T(t) = (I_{T\_peak} / \pi) [1 + (\pi/2) \cos \omega t + (2/3) \cos 2\omega t - (2/15) \cos 4\omega t + ...] \qquad (2.62)$$

Note that the half-sinusoid output current and its fundamental tone are in phase. This implies that if the amplitude of the fundamental tone output current is $I_{fund}(t)$ at any given moment in time, then the amplitude of the actual output current is $2\,I_{fund}(t)$. This can be used to translate the left-side boundary and the upper boundary for the transistor current on the I/V-plane to equivalent boundary lines for the fundamental tone component [2].

It follows from (2.62) that the peak amplitude of the fundamental tone is equal to half of the peak amplitude of the half-sinusoid (I_{T_peak}):

$$|I_2| = I_{T\_peak} / 2 \qquad (2.63)$$

The average value (dc component) of the transistor current $[I_T(t)]$ is given by

$$I_{dc} = I_{T\_peak} / \pi \qquad (2.64)$$

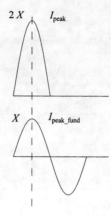

Figure 2.9   The relationship between the actual (intrinsic) output current and its fundamental tone component.

It follows that the dc dissipation in the transistor is given by

$$P_{dc} = V_{ds} I_{T\_peak} / \pi \tag{2.65}$$

while the output power is given by

$$P_o = |V_2|^2 G_L / 2 = |I_2 / Y_L|^2 G_L / 2$$

$$= |I_{T\_peak} / (2 Y_L)|^2 G_L / 2$$

$$= |I_{T\_peak} / Y_L|^2 G_L / 8 \tag{2.66}$$

$$= |I_{T\_peak}|^2 R_L / 8 \tag{2.67}$$

The efficiency is calculated as the ratio of the output power (P_o) or the effective output power ($P_o - P_{in}$) to the dc power (P_{dc}):

$$\eta = P_o / P_{dc} \tag{2.68}$$

$$\eta_a = (P_o - P_{in}) / P_{dc} \tag{2.69}$$

Note that η_a is referred to as the power-added efficiency.
     If (2.68) is used, the efficiency is given by

$$\eta = (V_{peak\_fund} / V_{ds}) (\pi / 4) \tag{2.70}$$

The efficiency (η) of a class-B amplifier increases linearly with increasing output voltage up to a maximum of 78.5% . If the intrinsic load termination is reactive, the efficiency will be lower.

When the output power is lower than the maximum possible, the efficiency of a class-B stage will be observed to vary with the angular position around the constant output power contours. The efficiency of a class-A amplifier is constant around a constant output power contour.

The dynamic load line for a class-B amplifier is shown in Figure 2.10. When the effective load line is purely resistive, the output current of the transistor and the voltage across it are constrained as shown in Figure 2.10(a). When the effective load line is reactive, the current and voltage are constrained as shown in Figure 2.10(b). Note that the current is zero during half of the cycle.

The I/V-constraints of a class-B stage apply to the total current through the transistor (half sinusoid) and the voltage across the transistor. The constraints on the fundamental tone quantities are, however, of greater interest. Because the voltage waveform was assumed to be a pure sinusoid and because of the fixed relationship between the total current and its fundamental tone (see Figure 2.9), the constraints on the fundamental tone quantities can be taken to be as illustrated in Figure 2.11. Note that the new origin (V_{ds}', I_{ds}') should be moved down enough to allow the fundamental tone current to swing symmetrically without clipping when the instantaneous voltage is higher than V_{dc}.

Under the transformation illustrated in Figure 2.11, a class-B stage can be treated as a class-A stage when its output power is calculated. This can also be done when a set of load-pull contours is generated for the transistor.

The dc I/V-constraints for a power transistor are often supplied by the manufacturer. These constraints can be taken to be the RF constraints of the intrinsic device too, if the current is interpreted as the sum of the current of the voltage-controlled current source and the intrinsic output resistance in the equivalent circuit.

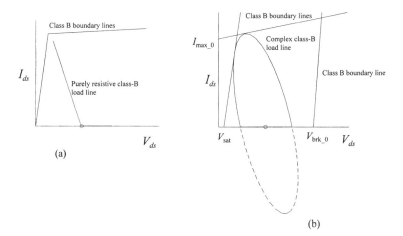

Figure 2.10   The dynamic load line of a transistor biased for class-B operation: (a) resistive load line and (b) reactive load line.

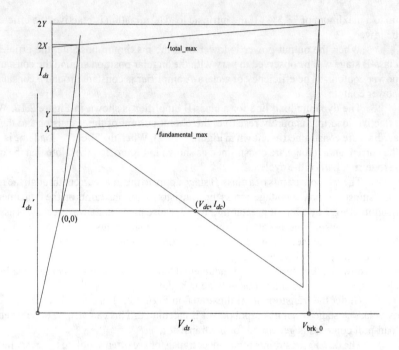

Figure 2.11   Illustration of the conversion of the I/V constraints on the total output current and the output voltage of a class-B amplifier to those applying to the fundamental tone quantities. Note the change in the slope of the boundary line on the left.

2.3.2   Class-AB and Class-F Load Line

As in the class-B case, the conduction angle of the transistor model current source in a class-F amplifier is 180°. The voltage across the current source, however, is or approximates a square wave and is out of phase with the current, which leads to improved efficiency. The conduction angle for a class-AB stage is more than 180° and for a class-C amplifier it is less than 180°. The dc and fundamental current in all of these cases are given by [5, 18]

$$I_{DC}(\alpha) = \frac{I_{max}}{2\pi} \frac{2\sin(\alpha/2) - \alpha\cos(\alpha/2)}{1 - \cos(\alpha/2)} \tag{2.71}$$

$$I_1(\alpha) = \frac{I_{max}}{2\pi} \frac{\alpha - \sin\alpha}{1 - \cos(\alpha/2)} = \frac{I_p}{2\pi}(\alpha - \sin\alpha) \tag{2.72}$$

where α is the conduction angle in radians, and the waveform oscillates between 0 and I_{max}, as show in Figure 2.12. The relationship between I_{max} and the amplitude, I_p, of the clipped

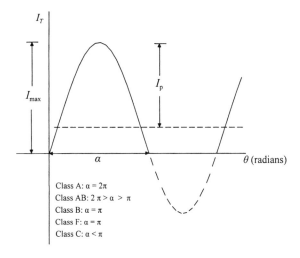

Figure 2.12   The current waveform associated with the current source in an ideal class-AB stage.

sine wave is given by

$$I_{max} = I_p \left(1 - \cos(\alpha / 2)\right) \tag{2.73}$$

The ideal voltage waveform for a class-B stage is a pure sinusoid, while that for a class-F stage is a square wave (see Figure 2.13), oscillating symmetrically around V_{dc}. An approximation to a square wave is obtained in a class-F stage by presenting high impedances to the odd harmonics and low impedances to the even harmonics. The Fourier series for a perfect square wave oscillating between V_{max} and $-V_{max}$ is [19]

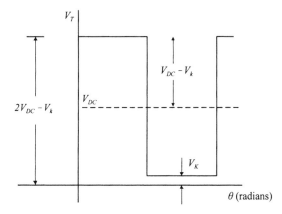

Figure 2.13   The ideal voltage waveform across the current source in a class-F stage.

$$v = V_{max} \left(\frac{4}{\pi}\sin(\omega t) + \frac{4}{3\pi}\sin(3\omega t) + \frac{4}{5\pi}\sin(5\omega t)+... \right) \tag{2.74}$$

It follows from this equation that the magnitude of the fundamental tone is larger than that of the square wave by a factor $4/\pi$ (1.273). This explains why the fundamental tone output power is increased when the output voltage is shaped towards a square waveform. It also follows that the amplitude of the fundamental tone can be larger than the supply voltage, that is, when the knee voltage is small.

It is difficult to obtain a perfect square wave in practice, first because of a limitation on the number of harmonics that can be controlled, and second because of the degree of control possible at each harmonic. Assuming ideal control, the peak value of the fundamental tone relative to the peak value of the square wave $[\delta_V(n)]$ will increase as shown below as the number of harmonics (n) is increased (assuming no contribution from the higher order harmonics) [20]

$$\delta_V(1) = 1.0$$
$$\delta_V(3) = 1.155$$
$$\delta_V(5) = 1.207$$
$$\delta_V(\infty) = 1.273$$

Note that when $n=3$, the optimum amplitude for the third harmonic component is given by $V_{o3} = V_{o1}/9$ [19].

Assuming that the total voltage (dc plus ac) across the transistor current source is a square waveform oscillating between the transistor knee voltage V_k and $2V_{DC} - V_k$; that is,

$$V_T = V_{DC} + (V_{DC} - V_k)\left(\frac{4}{\pi}\sin(\omega t) + \frac{4}{3\pi}\sin(3\omega t) + \frac{4}{5\pi}\sin(5\omega t)+... \right) \tag{2.75}$$

the amplitude of the fundamental tone is given by

$$V_{1max} = \frac{4}{\pi}(V_{DC} - V_k)$$

For the general case (limited harmonics), this becomes

$$V_{1max}(n) = \delta_V(n)\ (V_{DC} - V_k) \tag{2.76}$$

It follows from (2.72) and (2.76) that the optimum load line in the general case is given by

$$R_1(n,\alpha) = \frac{V_{1max}(n)}{I_1(\alpha)}$$

$$= \frac{\delta_V(n)\,(V_{DC} - V_k)}{I_{max}} \frac{2\pi\,(1 - \cos(\alpha/2))}{\alpha - \sin\alpha} \tag{2.77}$$

The maximum output power for the different cases are given by

$$P(n,\alpha) = \frac{1}{2} I_1(\alpha) \cdot \delta_V(n) \cdot (V_{DC} - V_k) \tag{2.78}$$

The dc power is given by

$$P_{DC}(\alpha) = I_{DC}(\alpha) \cdot V_{DC} \tag{2.79}$$

and the efficiency by

$$\eta(n,\alpha) = \frac{P(n,\alpha)}{P_{DC}(\alpha)} \tag{2.80}$$

Compared to the maximum class-B efficiency of 78.5%, the efficiency for an ideal class-F amplifier (square-wave voltage waveform and no knee voltage) is 100%.

2.3.3   Class-E Load Line

Similar to a class D amplifier [19], the transistor is operated as a switch in a class-E amplifier. The voltage and current waveforms in the class-E case is, however, shaped to eliminate the losses associated with discharging the shunt output capacitance of the transistor, as well as any extra loading capacitance added, when the transistor is switched on [4, 6, 21]. This is done by delaying the ensuring that the voltage returns to zero before the current begins to rise [6]. The rise of the transistor voltage is also delayed until after the current has decreased to zero. In doing so, the current and voltage waveforms are totally out of phase.

Detailed design equations for a typical class-E load network are provided in [6]. These equations were derived by solving time-domain equations set up for the network. Frequency domain equations and conditions for the design of class-E amplifiers are provided in [4, 22]:

$$Z_L = \begin{cases} R_L(1 + j1.152), & \text{at} \quad f_0 \\ \\ \infty, & \text{at} \quad nf_0, n > 1 \end{cases} \tag{2.81}$$

where R_L at the fundamental frequency, f_0, is given by

$$R_L = \frac{1}{34.225 f_0\, C_s} \tag{2.82}$$

where C_s is the shunt output capacitance of the transistor and any extra loading capacitance added to increase the output power. Note the impedance, Z_L, must be provided to the right of the transistor and the loading capacitance. Also note that the impedance as given by (2.81) ensures that the voltage across the transistor, as well as its derivative, will be zero when the transistor is switched on [4]. The high impedance at the harmonics ensures that the harmonics will be filtered and will not be reach the external load.

2.3.4   Doherty Amplifiers

The topology for a Doherty Amplifier is shown in Figure 2.14 [7]. The main amplifier is typically biased in class-AB mode, while the auxiliary amplifier is operated in class-C mode.

The auxiliary amplifier is biased to turn on when the output voltage of the main amplifier has saturated. The load resistance, R_L, and the quarter wavelength transformer is typically designed to set this transition point to around 6 dBm below the peak power level of the amplifier, but the high efficiency power range can be extended by changing the design parameters, as demonstrated in [7].

Instead of directly contributing to the load power, the current supplied by the auxiliary amplifier is used to force the output voltage, V_m, of the main amplifier to remain constant at its saturated level. This can only be done if the transformed output impedance of the main amplifier is low compared to the load resistance (R_L). While the output voltage of the main amplifier remains constant, its output current is increased as it is driven harder. Because of this, the power delivered to the load also increases until the final saturation point is reached.

The relevant design equations are derived in [7] by modeling the output of each amplifier as an ideal current source. The more realistic case where the output resistance is finite is also considered. The equations at the center frequency for the ideal case are

$$V_m = -\frac{Z_{0m}^2}{R_L} I_m + jZ_{0a} I_a$$

(2.83)

$$V_L = -jZ_{0m} I_m$$

(2.84)

Beyond the transition point where V_m has reached its saturated value, the main current can be written as the sum of a constant value, I_{mc}, and an extra component I_{mx}:

$$I_m = I_{mc} + I_{mx} = \frac{I_{m\_max}}{\gamma} + I_{mx}$$

(2.85)

where γ is the ratio chosen for the maximum main output current to the critical current.

With (2.85) substituted in (2.83), it follows that

$$V_m = -\frac{Z_{0m}^2}{R_L} I_{mc} - (\frac{Z_{0m}^2}{R_L} I_{mx} - jZ_{0a}\ I_a) \tag{2.86}$$

In order to force V_m to be constant beyond the transition point the last term in (2.86) must be equal to zero. It follows that

$$I_a = -j\frac{Z_{0m}}{R_L} I_{mx} \tag{2.87}$$

The maximum value of the auxiliary current is important. By using (2.85) and (2.87) it follows that

$$I_{a\_max} = -j\frac{Z_{0m}}{R_L} I_{mx\_max} = -j\frac{Z_{0m}}{R_L}\frac{\gamma-1}{\gamma} I_{m\_max} \tag{2.88}$$

It also follows from (2.86) that the output (drain) voltage of the main amplifier at, and beyond, the critical point is given by

$$V_{m\_max} = -\frac{Z_{0m}^2}{R_L} I_{mc} = -\frac{Z_{0m}^2}{R_L}\frac{I_{m\_max}}{\gamma} \tag{2.89}$$

The load line for the main amplifier at peak power ($R_{m\_max} = -V_{m\_max}/I_{m\_max}$) can also be obtained from this equation. It follows the load line has decreased by a factor γ from its small-signal value when the amplifier is operated at its peak power.

The output power in the idealized case is only a function of (2.84) and the load resistance, and is given by

$$P_o = \frac{|V_L|^2}{R_L} = \frac{Z_{0m}^2}{R_L}|I_m|^2 \tag{2.90}$$

The maximum output power of the Doherty amplifier is given by

$$P_{o\_max} = \frac{Z_{0m}^2}{R_L}|I_{m\_max}|^2 = \frac{|\gamma\ V_{m\_max}|^2}{\dfrac{Z_{0m}^2}{R_L}} \tag{2.91}$$

The critical point of the output current is typically chosen to be at half the maximum value of the output current of the main amplifier ($\gamma = 2$), but it can be set lower as is done in [7]. In the typical case Z_{0m} is also set to $2\ R_L$. With these choices, the maximum current (fundamental tone) of the auxiliary amplifier is equal to the maximum current of the main amplifier [see (2.88)].

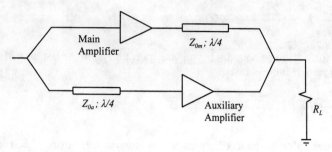

Figure 2.14   The Doherty topology.

Equations (2.83) and (2.84) were extended in [7] to allow for finite output resistance in the transistors. The results show that the load voltage is decreased and becomes a nonlinear function of I_m. Equation (2.88) is changed to

$$I_{a\_max} = -j \frac{Z_{0m}}{R_{Le}} I_{mx\_max} = -j \frac{Z_{0m}}{R_{Le}} \frac{\gamma - 1}{\gamma} I_{m\_max}$$

(2.92)

where R_{Le} is the parallel value of the load resistance R_L and the output resistance, R_{oa}, of the auxiliary amplifier. (2.89) becomes

$$V_{m\_max} = -\frac{R_{om} Z_{0m}}{R_{Le} R_{om} + Z_{0m}^2} Z_{0m} \frac{I_{m\_max}}{\gamma}$$

(2.93)

and (2.84) is changed to

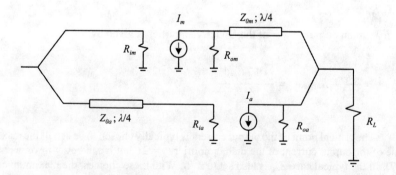

Figure 2.15   The equivalent circuit used for the simplified analysis.

$$V_L = \frac{R_{Le}}{R_{Le} + \frac{Z_{0m}^2}{R_{0m}}} (-jZ_{0m} I_m + I_a \frac{Z_{0m}^2}{R_{0m}}) \tag{2.94}$$

R_{om} is the output conductance of the main amplifier. The maximum value of V_L is given by

$$V_{L\_max} = -jZ_{0m} I_{m\_max} \frac{R_{Le}}{R_{Le} + \frac{Z_{0m}^2}{R_{0m}}} (1 + \frac{\gamma-1}{\gamma} \frac{Z_{0m}^3}{R_{Le} R_{0m}}) \tag{2.95}$$

The linearity of a Doherty amplifier can be improved by adjusting the input phase shift of the auxiliary amplifier from its nominal value of $90°$. This is demonstrated in [23].

2.3.5 Distortion in Linear Amplifiers

The 1-dB compression point (single tone) and the third-order intercept point for two-tone products are usually used as measures of the linearity of an amplifier. In wireless applications, where the signal is more complex, the adjacent channel power ratio (ACPR) is also used. In general, one can expect some correlation between the ACPR and the third-order intermodulation distortion (IM_3) [24].

The 1-dB compression point is defined as the level (usually expressed in terms of the output power) at which the operating power gain (G_ω) is 1 dB down from its small-signal level. The third-order two-tone intercept point (TOI) is defined as the power level at which each extrapolated third-order product ($2f_1 - f_2$ and $2f_2 - f_1$ components) is equal in magnitude to the extrapolated fundamental tone component. The definitions are illustrated in Figure 2.16.

At low signal levels the slope of the fundamental tone component (P_{out} in decibels versus P_{in} in decibels) is 1:1, and that for the third-order products is 3:1. The third-order intercept point of a linear amplifier is usually about 10 dB higher than the 1-dB compression point [25].

The dynamic range of an amplifier is usually defined as the difference between the 1-dB compression level and that of the minimum detectable signal, referenced to the output of the amplifier [5]:

$$DR = P_{1dB} - MDS_{out} \tag{2.96}$$

The minimum detectable signal could be defined as 3 dB above the noise floor of the amplifier, that is,

$$MDS_{out} = kTB + F + G_T + 3 \text{ (dB)} \tag{2.97}$$

where F is the noise figure of the amplifier and G_T is its transducer power gain.

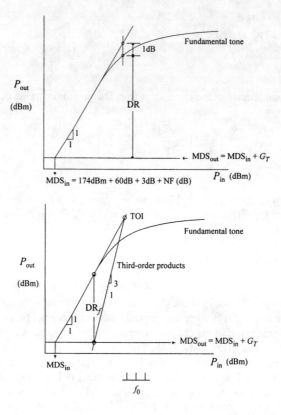

Figure 2.16   The dynamic range (DR) and the spurious free dynamic range (DR$_f$) of an amplifier.

The spurious free dynamic range (DR$_f$) is often also of interest. The definition is illustrated in the lower panel of Figure 2.16.

The ACPR is used in wireless applications and is a measure of the interference, or power, in the adjacent frequency channel. It is defined as the ratio of the total power in the adjacent frequency channel to the total power in the transmitted channel [26]:

$$\text{ACPR} = \frac{\int_{f_{s1}}^{f_{s2}} P(f)df}{\int_{f_{a1}}^{f_{a2}} P(f)df} \tag{2.98}$$

where f_{s1} and f_{s2} defines the signal bandwidth, and f_{a2} and f_{a1} the frequency limits of the adjacent band with the highest distortion. The channel bandwidth and spacing is set by the type of modulation considered. The ACPR is often estimated by taking the ratio of the average power level in the adjacent channel to the peak power level in the transmitted channel as shown in Figure 2.17 [27].

The adjacent channel power ratio is also referred to as the adjacent channel leakage power ratio, while the adjacent channel interference is also referred to as spectral regrowth.

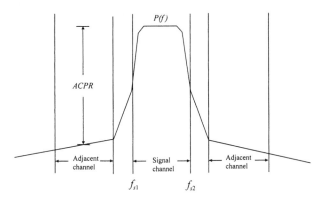

Figure 2.17   The ACPR is often estimated as illustrated.

2.3.5.1   The Third-Order Intercept Point of a Cascade

Gain compression and any additional frequency components generated are the result of the (weak) nonlinear transfer function of the amplifier [25]. At a given bias point (V_i, V_o), the output signal $(v_o = \delta V_o; v_i = \delta V_i)$ can be calculated by using Taylor's theorem:

$$v_o = \frac{\partial V_o}{\partial V_i} v_i + \frac{\partial^2 V_o}{\partial^2 V_i} \frac{v_i^2}{2} + \frac{\partial^3 V_o}{\partial^3 V_i} \frac{v_i^3}{6} + \dots \tag{2.99}$$

This can be simplified to

$$v_o = a_1 v_i + a_2 v_i^2 + a_3 v_i^3 + \dots \tag{2.100}$$

The coefficients in (2.100) are usually taken to be real, but they could be complex in general. If the coefficients are real, any distortion products generated will have a fixed phase relationship with the input signal.

If

$$v_i = a \cos \omega t \tag{2.101}$$

is substituted in (2.100), it can be shown that:

(a)   Odd-order harmonic components (3*f*, 5*f*, ...) are generated by the odd-order terms. In addition, each odd-order term will also generate a component at the fundamental frequency (f). These fundamental tone components are responsible for the gain compression observed in amplifiers.

(b)   Even-order components will generate even-order harmonics ($2f$, $4f$, ...). In addition, each even-order term will also generate a dc component. These components cause the shift in bias point observed when an amplifier is driven strongly.

The distortion created by the third-order term in (2.101) is usually of most interest:

$$a_3(a\cos\omega t)^3 = a_3 a^3 \cos^3\omega t = a_3 a^3 \cos\omega t\ 0.5(1 + \cos 2\omega t)$$

$$= a_3 a^3\ 0.5\cos\omega t + 0.5(\cos\omega t\ \cos 2\omega t)$$

$$= a_3 a^3 0.5\cos\omega t + 0.5(0.5(\cos\omega t + \cos 3\omega t))$$

$$= a_3 a^3 [\frac{3}{4}\cos\omega t + \frac{1}{4}\cos 3\omega t] \qquad (2.102)$$

Note that if a_3 in (2.102) is negative, the third-order contribution at the fundamental frequency will decrease the signal level; that is, the gain will be compressed. It is also clear that the third-order terms will be very small when the signal level is low [a^3-term in (2.102)].

If the contribution of the higher-order terms is ignored, the 1-dB gain compression point (v_{ic}) can be estimated by setting

$$a_1 v_{ic} - \frac{3}{4}a_3 v_{ic}^3 = a_1 v_{ic} 10^{-1/20}$$

from which it follows that

$$v_{ic}^2 = \frac{4a_1(1 - 10^{-1/20})}{3a_3} \qquad (2.103)$$

When a two-tone signal is used, the input signal (v_i) is given by

$$v_i = a\ [\cos\omega_1 t + \cos\omega_2 t] \qquad (2.104)$$

In this case, fundamental tone components ($f_1;f_2$) with amplitude ($2.25\,a_3 a^3$) are generated. The third harmonic components generated at each frequency are of the same amplitude as in the single tone case ($0.25 a_3 a^3$).

Apart from these components, additional components are generated at ($2f_2 - f_1$) and ($2f_1 - f_2$) in the two-tone case. The amplitude of these components is ($0.75 a_3 a^3$).

If displayed logarithmically, the two-tone products will increase at a 3:1 rate with increasing signal level, that is as long as the contribution of any higher order terms can be neglected. The fundamental tone will increase at a 1:1 rate as long as the compression can be neglected.

The two-tone intercept point can be estimated by using the results obtained:

$$a_1 v_{ip3} = \frac{3}{4} a_3 v_{ip3}^3$$

$$\Rightarrow$$

$$v_{ip3}^2 = \frac{4}{3} \frac{a_1}{a_3} \qquad (2.105)$$

Equations (2.105) and (2.103) can be used at this point to find the relationship between the third-order intercept point and the 1-dB compression point:

$$\frac{v_{ip3}^2}{v_{ic}^2} = \frac{1}{1 - 10^{-1/20}} = 9.195 = 9.6 \text{ dB} \qquad (2.106)$$

Note that the contribution of the higher-order terms (5*f*, 7*f*, ...) was ignored in this derivation. The estimation, however, is good enough to be of practical use.

Note that because of the fixed slopes (at least at lower signal levels), the level of third-order components associated with any signal level (*X*) can be calculated easily from the third-order intercept specification:

$$P_{o1}(X) = G + X \qquad (2.107)$$

$$\begin{aligned}
P_{o3}(X) &= P_{o3}(X_{IP3}) - 3\,(P_i(X_{IP3}) - X) \\
&= P_{o3}(X_{IP3}) - 3\,[(P_{o3}(X_{IP3}) - G) - X] \\
&= -2P_{o3}(X_{IP3}) + 3\,(X + G) \\
&= -2P_{o3}(X_{IP3}) + 3P_{o1}(X) \quad \text{(dBm)} \qquad (2.108)
\end{aligned}$$

where $P_{o1}(X)$ is the fundamental tone output power at signal level *X*, $P_i(X_{IP3})$ is the fundamental tone input power at the third-order intercept point, and $P_{o3}(X)$ is the power at $(2f_1 - f_2)$ or $(2f_2 - f_1)$ at the same signal level (*X*).

If the power is expressed as a number and not in dBm, (2.108) becomes

$$P_{o3}(X) = \frac{P_{o1}^3(X)}{P_{o3}^2(X_{IP3})} \qquad (2.109)$$

An interesting result follows directly from (2.109):

$$P_{o3}(X_{IP3}) = \sqrt{\frac{P_{o1}^3(X)}{P_{o3}(X)}} \qquad (2.110)$$

that is, the third-order intercept point can be calculated from the fundamental tone and the third-order intermodulation power level at any signal level. This result can be applied to calculate the third-order intercept point of a cascade too. For a cascade, P_{o1} would be the fundamental tone power level at the load, and P_{o3} the total third-order contribution associated with that signal level, also at the load.

Before deriving an equation for the third-order intercept point of a cascade, it is useful to consider the effect that an ideal amplifier stage will have on the intercept point. If the operating power gain of the ideal stage is G_{ω}, it follows from (2.110) that

$$P_{o3-A}(X_{IP3}) = \sqrt{\frac{(G_{\omega} P_{o1}(X))^3}{G_{\omega} P_{o3}(X)}} = G_{\omega} P_{o3}(X_{IP3}) \tag{2.111}$$

that is, the third-order intercept point is simply increased with the gain, as would be expected.

At this point the third-order intercept point of a cascade can be calculated easily. Consider the amplifier chain in Figure 2.18. It follows from (2.111) that the intercept point of each stage as calculated at the load is increased with the gain of the stages following it. The distortion component (power) contributed by stage j at the load is given by (2.110)

$$P_{o3-jA}(X) = \frac{P_{o1-jA}^3(X)}{P_{o3-jA}^2(X_{IP3})} \tag{2.112}$$

where all the power levels and the intercept point are referenced to the load. The normalized voltage contributed by stage j can be obtained by taking the square root of (2.112); that is,

$$v_{o3-jA}(X) = \frac{P_{o1-jA}^{3/2}(X)}{P_{o3-jA}(X_{IP3})} \tag{2.113}$$

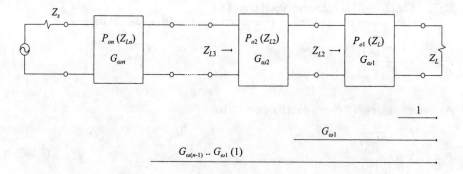

Figure 2.18   The circuit used to calculate the third-order two-tone intercept point of a cascade.

The fundamental tone power level at the load, $P_{o1\text{-}iA}(X)$, is the same for each stage and, therefore, it follows from (2.113) that the voltage contribution from each stage at the load (superposition) is inversely proportional to the modified (amplified) intercept point for that stage. By using (2.113) and assuming the worst case where all the third-order terms add in-phase, it follows that the intercept point for the cascade is given by (2.114)

$$P_T^2(X_{IP3}) = \frac{P_{o1\text{-}A}^3(X)}{\left[\dfrac{P_{o1\text{-}iA}^{3/2}(X)}{P_{o3\text{-}1A}(X_{IP3})} + \dots + \dfrac{P_{o1\text{-}iA}^{3/2}(X)}{P_{o3\text{-}nA}(X_{IP3})} \right]^2}$$

$$= \frac{1}{\left[\dfrac{1}{P_{o3\text{-}1A}(X_{IP3})} + \dots + \dfrac{1}{P_{o3\text{-}nA}(X_{IP3})} \right]^2} \tag{2.114}$$

that is,

$$\frac{1}{P_T(X_{IP3})} = \frac{1}{P_{o3\text{-}1A}(X_{IP3})} + \dots + \frac{1}{P_{o3\text{-}nA}(X_{IP3})} \tag{2.115}$$

If the assumption is made that the ratios of the 1-dB compression points and the corresponding third-order intercept points will remain invariant, (2.115) leads to

$$\frac{1}{P_{1dB\text{-}T}} = \frac{1}{P_{1dB\text{-}1A}} + \frac{1}{P_{1dB\text{-}2A}} + \dots + \frac{1}{P_{1dB\text{-}nA}} \tag{2.116}$$

where $P_{1dB\text{-}iA}$ is the 1-dB compression point of stage i referenced to the load (that is, increased with the operating power gain of the stages following it).

If an infinite chain of identical stages is considered, (2.115) becomes

$$P_T(X_{IP3}) = P_{o3}(X_{IP3}) \, (1 - 1/G_I) \tag{2.117}$$

where (2.117) is a figure of merit similar to the noise measure [17].

2.3.6 The Cripps Approach to Estimating the Maximum Output Power Obtainable from a Transistor

The constant output power contours for a transistor are usually closer to ellipses than circles. This leads to the assumption that even the linear power generated by a transistor (power up to the 1-dB compression point) is strongly influenced by the nonlinear components in the equivalent circuit. Cripps [1] demonstrated that the maximum linear output power obtainable from a transistor is mainly determined by the clipping of the intrinsic output voltage and current and that the elliptic form of the contours can be approximated as the intersection of two circles. This is illustrated in Figure 2.19.

In the Cripps approach the intrinsic output voltage is assumed to be bounded to a maximum V_{max}, while the current is bounded by I_{max}. Note that only the fundamental components of the voltage and current are considered. Under these assumptions, the optimum intrinsic load line is given by

$$Y_{Lopt\_i} = G_{Lopt\_i} = I_{max} / V_{max} \tag{2.118}$$

If R_{L_i} is smaller than R_{Lopt_i}, the current will be clipped and the voltage swing will be smaller than the maximum allowed. Relative to the maximum possible output power, the power (P_{oi}) is then given by

$$P_{oi} = (I_{max}{}^2 R_L) / (I_{max}{}^2 R_{Lopt}) = R_L / R_{Lopt} \tag{2.119}$$

In this case, reactance can be added in series with the resistance without changing the maximum output power obtainable until the magnitude of the load impedance is equal to R_{Lopt}, at which point the current and the voltage are both clipped.

When the intrinsic load resistance is higher than the optimum, the voltage will be clipped and the current will be lower than the maximum allowed. The output power relative to the maximum (P_{ov}) is given in this case by

$$P_{ov} = (V_{max}{}^2 G_L) / (V_{max}{}^2 G_{Lopt}) = G_L / G_{Lopt} \tag{2.120}$$

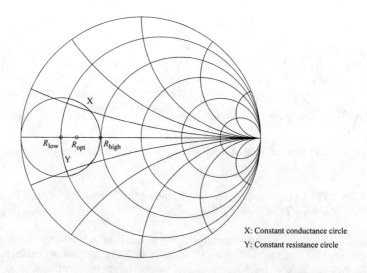

X: Constant conductance circle

Y: Constant resistance circle

Figure 2.19   The elliptical power contours obtained for linear amplifiers can be approximated as the intersection of circles when the intrinsic load line is considered [1].

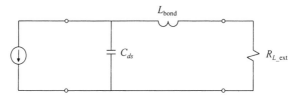

Figure 2.20   The equivalent circuit used in the Cripps approach.

Susceptance can be added in parallel with this resistance without changing the output power obtainable until the magnitude of the load admittance is equal to G_{Lopt}, at which point the voltage and the current are both clipped.

Equations (2.118)–(2.120) can be used to calculate the intrinsic power generated or to find the contours of constant output power as a function of the intrinsic load. If the $-X$ dB output power contour is of interest, the two resistance values are given by

$$R_{L\_low} / R_{Lopt} = 10.0^{-0.1X} \tag{2.121}$$

and

$$R_{L\_high} / R_{Lopt} = 10.0^{0.1X} \tag{2.122}$$

In order to find the constant output power contours in terms of the actual load impedance, the transforming effect of the transistor parasitics must be taken into account. If the simple equivalent circuit shown in Figure 2.20 is used, the load admittance corresponding to any given intrinsic load admittance can be calculated easily and the constant output power contours for the actual load can be obtained by adjusting the intrinsic load line contours appropriately.

Calculation of the external load associated with a given intrinsic termination can be a major task if feedback (parasitic or otherwise) is applied to the transistor. Any losses in the output circuit or a feedback loop will also be a problem.

2.3.7   Estimation of the Maximum Output Power of a Linear Network by Using the Power Parameters

Because of the simplifications in the equivalent circuit used for the transistor in the Cripps approach, the intrinsic load termination corresponding to any given external load could be calculated easily. With the intrinsic termination and the I/V constraints known, the output power could be estimated. An implicit assumption that all the intrinsic power generated ends up in the external load is made in the process.

The Cripps approach can be generalized by introducing a new set of parameters to map the intrinsic voltages to the external voltages and to the intrinsic output current [2]. Any reduction in the power caused by losses in the output circuit or in any feedback circuit is automatically tracked when this approach is followed.

The assumption that the intrinsic output current and voltage are constrained to a rectangular area on the I/V-plane can also be lifted. The allowable area on the I/V-plane can be restricted to the area defined by four boundary lines instead, as shown in Figure 2.21. If the goal is maximum power, the lines can be set to prevent voltage breakdown, operation in the resistive area, and forward conduction [field-effect transistors (FETs)]. The current limit can usually be set slightly above I_{dss}.

If the goal is linearity, the lines can be set to bound the area where the I/V curves are evenly spaced.

Because the purpose of these mapping parameters is to calculate the output power, they will be referred to as power parameters. The power parameters are defined by the following equations:

$$V_1 = MV_{1i} + NV_{2i} \tag{2.123}$$

$$V_2 = O\,V_{1i} + PV_{2i} \tag{2.124}$$

$$I_{2i} = RV_{1i} + SV_{2i} \tag{2.125}$$

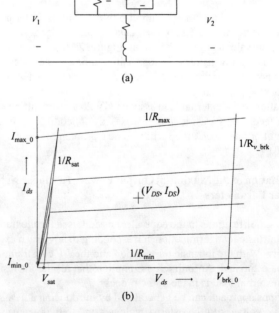

(a)

(b)

Figure 2.21   (a) The voltages and current of interest in the defined power parameters of a transistor; and (b) constraining of the allowable intrinsic load line area with four boundary lines [2]. Note that the small-signal model used must correspond to the dc operating point (V_{DS}, I_{DS}) around 1-dB compression, and not the bias point.

In these equations V_{1i} and V_{2i} are the intrinsic input and output voltages, respectively, while I_{2i} is the intrinsic output current, as shown in Figure 2.21 for an FET. V_1 and V_2 are the input and output voltages, respectively.

The power parameters can be used to calculate the intrinsic load associated with a given external load directly, as shown in Section 2.3.7.1. The external voltage and current associated with the maximum intrinsic voltage or current and the associated power at the load can then be calculated easily by using (2.123)–(2.125).

Similarly, the external load associated with a given intrinsic load can also be calculated easily by using the power parameters. This is useful when contours of constant output power are generated.

The main assumptions in this approach are linearity and hard clipping of the intrinsic output current and voltage at the boundary lines. The power parameter approach has proven to be useful up to at least the 1-dB compression point of class-A and class-B amplifiers.

The power parameters are quite general and can be manipulated to include the effect of any passive network in which the transistor may be imbedded. A set of power parameters should be calculated for each transistor used in the network. This is required because, in general, the output power may be limited by clipping in any of the transistors used in the circuit.

When the power parameters for any given two-port are calculated, it is assumed that no clipping will occur in any of the other active two-ports that may be present. The maximum (linear) output power is determined by the two-port in which clipping first occurs, that is, if the other stages are not close to clipping too.

When the output power is limited by clipping in more than one transistor, an approach similar to that used to calculate the third-order intercept point of a cascade can be followed. The assumption that the different intercept products all end up in-phase at the load would also be made in this case.

The effect of adding a stage on the input or the output side of a transistor, as well as the effect of adding series feedback, will be considered in the following sections. Adding a network in parallel (voltage-shunt feedback) has no effect on the power parameters. Note that a nodal approach to the calculation of the change in the power parameters can also be followed.

Note that (2.125) is only a function of the intrinsic voltages and not of the embedding network.

The following assumptions are implicit when the power parameters are used:

1. The maximum output power obtainable from a linear amplifier with a given load termination is determined by the power level at which the voltage and/or the current waveform starts to clip.
2. The behavior of the transistor is essentially linear up to the 1-dB compression point.
3. The output voltage of the transistor is sinusoidal, but an allowance can be made for a square-wave too (class-F case).

The last condition can be approximated in class-B stages by using a push-pull stage or by short-circuiting the harmonic currents of each transistor by using a resonant or lowpass circuit.

In order to calculate the linear behavior of the transistor with the load termination of interest, a linear model of the transistor is required. This is a simple matter in the class-A case (conventional small-signal models have been found to be adequate), but not necessarily so in the class-B case.

Acceptable results are usually obtained for the class-B case by using the class-A small-signal parameters at the rated class-B output current and reducing the g_m of the associated small-signal model by a factor of two.

Typical small-signal models are shown in Figure 2.22. Approximate values for the model components can often be obtained from the manufacturer. Optimization of the component values are usually required to ensure a tight fit of the model S-parameters to the measured S-parameters.

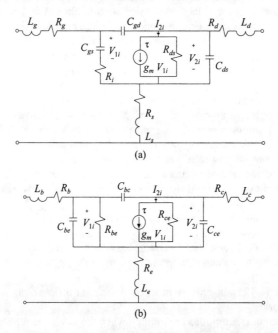

Figure 2.22   Typical small-signal models used for (a) FETs and (b) bipolar transistors.

2.3.7.1   Calculating the Intrinsic Load Associated with a Given External Load

The intrinsic load ($Y_{Li} = -I_{2i} / V_{2i}$) associated with any given external load can be calculated by using the Y-parameter expression for the voltage gain in terms of Y_L [2]:

$$A_v = \frac{V_2}{V_1} = -\frac{y_{21}}{y_{22} + Y_L} \tag{2.126}$$

Equations (2.123) and (2.124) can be used to replace V_2 and V_1 above, leading to

$$\frac{V_2}{V_1} = \frac{O\ V_{1i} + P\ V_{2i}}{M\ V_{1i} + N\ V_{2i}} = -\frac{y_{21}}{y_{22} + Y_L} = A_v \tag{2.127}$$

V_{1i} in (2.127) can now be replaced in terms of V_{2i} and I_{2i} by using (2.125):

$$\frac{O\ (\frac{1}{R}\ I_{2i} - \frac{S}{R}\ V_{2i}) + P\ V_{2i}}{M\ (\frac{1}{R}\ I_{2i} - \frac{S}{R}\ V_{2i}) + N\ V_{2i}} = A_v \tag{2.128}$$

The next step is to eliminate V_{2i} and I_{2i} in this expression in terms of the intrinsic load admittance ($Y_{Li} = -I_{2i}\ /\ V_{2i}$):

$$\frac{\frac{O}{R}\ I_{2i} + (P - \frac{S\ O}{R})\ V_{2i}}{\frac{M}{R}\ I_{2i} + (N - \frac{M\ S}{R})\ V_{2i}} = \frac{-\frac{O}{R}\ Y_{Li} + (P - \frac{S\ O}{R})}{-\frac{M}{R}\ Y_{Li} + (N - \frac{M\ S}{R})} = A_v \tag{2.129}$$

The required expression for Y_{Li} follows after rearranging this equation:

$$Y_{Li} = \frac{\frac{R}{O}\ (P - \frac{S\ O}{R}) - \frac{X_1\ R}{M}\ (N - \frac{M\ S}{R})}{1 - X_1} \tag{2.130}$$

where X_1 is given by

$$X_1 = A_v\ \frac{M}{R}\ \frac{R}{O} \tag{2.131}$$

Equation (2.130) can be used to find the intrinsic load associated with any external load, at which point the maximum output power can be calculated by finding the power level at which (hard) clipping will occur.

Equation (2.129) can also be used to find A_v in terms of Y_{Li}. With A_v known, the external load (Y_L) follows directly from (2.126).

Similar to the Cripps approach, these equations can be used to generate contours of constant output power or constant effective output power. The latter is of interest when an oscillator is designed.

An important difference between the Cripps approach and the power parameter approach is that the assumption that the output power will be a maximum when the intrinsic power generated is a maximum is inherent to the Cripps approach, while no such assumption is made when the power parameters are used. Note that if the circuit is loaded with its optimum load impedance, but the actual external load impedance is set to be a short circuit, the power generated will still be a maximum, but no power will be delivered to the actual load.

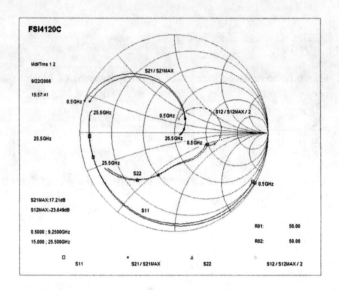

Figure 2.23   Comparison of the measured *S*-parameters of the transistor used in Figure 2.24 with the parameters associated with the small-signal model.

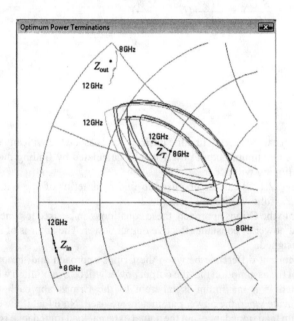

Figure 2.24   The load-pull contours (–1 dB; –2 dB) and the optimum load termination (Z_T trace) for the transistor in Figure 2.23 as predicted by using the power parameters [2]. The input reflection coefficients (Z_{in} trace) and the conjugate of the output reflection coefficients ($Z_{out}^{\bullet}$ trace) associated with the optimum power terminations (Z_T trace) are also displayed.

The power contours generated for a transistor (Texas Instruments Foundry FET) by using the power parameters [2] are shown in Figure 2.24. The optimum load line (maximum power) is also shown. The predicted contours at 10 GHz closely correspond with the measured load-pull contours provided by the manufacturer (the location and orientation of the contours are the same, but the measured contours are rounder). The S-parameters of the model used to calculate the power parameters are compared with the measured parameters in Figure 2.23.

Note that $s_{11\omega}$ in Figure 2.23 is the input reflection coefficient (referenced to $Z_{01} = 50\Omega$) associated with the optimum power load (S_L); $s_{22\omega}{}^*$ is the conjugate of the output reflection coefficient when the input side is conjugately matched (if possible).

If S_L and $s_{22\omega}{}^*$ were on top of each other, the optimum power and optimum output match ($\text{VSWR}_{\text{out}} = 1$) points would have been the same.

Voltage-shunt feedback can be added to this transistor to improve the output match associated with maximum power without losing too much power (around 1 dBm).

2.3.7.2   Modification of the Power Parameters of a Two-Port by Adding a Cascade Network on Its Output Side

When a passive network (two-port) is added in cascade on the output side of an active two-port, as shown in Figure 2.25, its power parameters are modified. The derivation for the new parameters is shown below.

The intrinsic voltages of the original network are mapped to the external voltages by

$$V_1 = M\,V_{1i} + N\,V_{2i} \tag{2.132}$$

$$V_2 = O\,V_{1i} + P\,V_{2i} \tag{2.133}$$

V_1 is also the input voltage of the combination, but the new output voltage is V_3 instead of V_2. It is therefore necessary to find V_2 as a function of V_3.

The input current and voltage of the cascade network are given in terms of the output quantities by

$$\begin{bmatrix} V_2 \\ I_2 \end{bmatrix} = \begin{bmatrix} A_2 & B_2 \\ C_2 & D_2 \end{bmatrix} \begin{bmatrix} V_3 \\ I_3 \end{bmatrix} \tag{2.134}$$

that is,

$$V_2 = A_2 V_3 + B_2 I_3 \tag{2.135}$$

$$I_2 = C_2 V_3 + D_2 I_3 \tag{2.136}$$

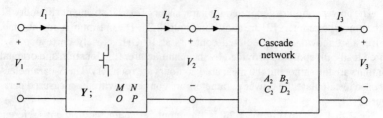

Figure 2.25   Adding a cascade network to the right of an active network.

Eliminating I_3 from the last two equations gives

$$V_2 = A_2 V_3 + B_2 \left(\frac{I_2}{D_2} - \frac{C_2}{D_2} \right) V_3$$

$$= A_2 V_3 + \frac{B_2}{D_i} I_2 - \frac{B_2 C_2}{D_2} V_3$$

leading to

$$V_2 = (A_2 - \frac{B_2 C_2}{D_2}) V_3 + \frac{B_2}{D_2} I_2 \qquad (2.137)$$

I_2 can be eliminated from this equation in terms of V_1 and V_2 by using the Y-parameters of the original network:

$$-I_2 = y_{21} V_1 + y_{22} V_2 \qquad (2.138)$$

leading to

$$V_2 = (A_2 - \frac{B_2 C_2}{D_2})V_3 + \frac{B_2}{D_2} [-y_{21} V_1 - y_{22} V_2]$$

Rearranging this equation yields

$$V_3 = (V_2 + \frac{B_2}{D_2} y_{21} V_1 + \frac{B_2}{D_2} y_{22} V_2) / (A_2 - \frac{B_2 C_2}{D_2})$$

$$= (\frac{B_2}{D_2} y_{21} V_1 + (1 + \frac{B_2}{D_2} y_{22}) V_2) / (A_2 - \frac{B_2 C_2}{D_2})$$

$$V_3 = \frac{\dfrac{B_2}{D_2} y_{21}}{A_2 - \dfrac{B_2 C_2}{D_2}} V_1 + \frac{1 + \dfrac{B_2}{D_2} y_{22}}{A_2 - \dfrac{B_2 C_2}{D_2}} V_2$$

$$= \frac{y_{21} B_2}{A_2 D_2 - B_2 C_2} V_1 + \frac{D_2 + y_{22} B_2}{A_2 D_2 - B_2 C_2} V_2 \tag{2.139}$$

After setting

$$\alpha_1 = \frac{y_{21} B_2}{A_2 D_2 - B_2 C_2} \tag{2.140}$$

and

$$\alpha_2 = \frac{D_2 + y_{22} B_2}{A_2 D_2 - B_2 C_2} \tag{2.141}$$

it follows that

$$V_3 = \alpha_1 V_1 + \alpha_2 V_2$$

$$= \alpha_1 (M V_{1i} + N V_{2i}) + \alpha_2 (O V_{1i} + P V_{2i})$$

$$= (\alpha_1 M + \alpha_2 O) V_{1i} + (\alpha_1 N + \alpha_2 P) V_{2i} \tag{2.142}$$

The new power parameters of the two-port are therefore given in terms of the original power parameters by (2.142):

$$V_1 = M V_{1i} + N V_{2i}$$

and

$$V_3 = (\alpha_1 M + \alpha_2 O) V_{1i} + (\alpha_1 N + \alpha_2 P) V_{2i} \tag{2.143}$$

2.3.7.3   Modification of the Power Parameters of a Two-Port by Adding a Cascade Network on Its Input Side

When a cascade network is added to the input side of a two-port, the input voltage for the combination is different from that of the original network, and the power parameters are therefore also changed. The effect of the cascade network is derived below.

The new input voltage and current (V_0 and I_0) are given in terms of the previous input voltage and current:

$$\begin{bmatrix} V_0 \\ I_0 \end{bmatrix} = \begin{bmatrix} A_1 & B_1 \\ C_1 & D_1 \end{bmatrix} \begin{bmatrix} V_1 \\ I_1 \end{bmatrix} \tag{2.144}$$

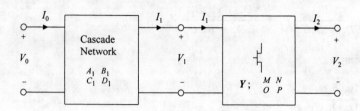

Figure 2.26   Adding a cascade network on the input side of a network.

Therefore,

$$V_0 = A_1 V_1 + B_1 I_1$$

I_1 in this equation can be replaced in terms of V_1 and V_2 in this equation by using the Y-parameters of the original two-port:

$$I_1 = y_{11} V_1 + y_{12} V_2 \qquad (2.145)$$

Therefore,

$$V_0 = A_1 V_1 + B (y_{11} V_1 + y_{12} V_2)$$

$$= (A + B y_{11}) V_1 + B y_{12} V_2 \qquad (2.146)$$

With

$$\alpha_{1L} = A + B y_{11} \qquad (2.147)$$

and

$$\alpha_{2L} = B y_{12} \qquad (2.148)$$

it follows that

$$V_0 = \alpha_{1L} V_1 + \alpha_{2L} V_2$$

$$= \alpha_{1L} (M V_{1i} + N V_{2i}) + \alpha_{2L} (O V_{1i} + P V_{2i})$$

$$= (\alpha_{1L} M + \alpha_{2L} O) V_{1i} + (\alpha_{1L} N + \alpha_{2L} P) V_{2i} \qquad (2.149)$$

The modified power parameters are, therefore, given by (2.149) and (2.133).

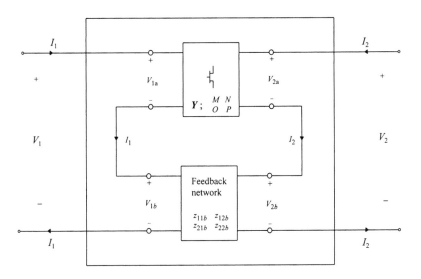

Figure 2.27   A network with series feedback.

2.3.7.4   The Influence of Series Feedback on the Power Parameters of a Two-Port

The influence of series feedback (see Figure 2.27) on the power parameters of a two-port will be derived here.

Before adding the series feedback network, assume that the power parameters are given by

$$V_{1a} = MV_{1i} + NV_{2i} \tag{2.150}$$

$$V_{2a} = OV_{1i} + PV_{2i} \tag{2.151}$$

The new input and output voltages are given by

$$V_1 = V_{1a} + V_{1b} \tag{2.152}$$

$$V_2 = V_{2a} + V_{2b} \tag{2.153}$$

From

$$V_{1b} = z_{11b}I_1 + z_{12b}I_2 \tag{2.154}$$

$$V_{2b} = z_{21b}I_1 + z_{22b}I_2 \tag{2.155}$$

and

$$I_1 = y_{11}V_{1a} + y_{12}V_{2a} \tag{2.156}$$

$$I_2 = y_{21}V_{1a} + y_{22}V_{2a} \tag{2.157}$$

it follows that

$$V_{1b} = z_{11b}\,(y_{11}V_{1a} + y_{12}V_{2a}) + z_{12b}\,(y_{21}V_{1a} + y_{22}V_{2a})$$

$$= (z_{11b}\,y_{11} + z_{12b}\,y_{21})\,V_{1a} + (z_{11b}\,y_{12} + z_{12b}\,y_{22})\,V_{2a} \tag{2.158}$$

and

$$V_{2b} = z_{21b}\,(y_{11}\,V_{1a} + y_{12}\,V_{2a}) + z_{22b}\,(y_{21}\,V_{1a} + y_{22}\,V_{2a})$$

$$= (z_{21b}\,y_{11} + z_{22b}\,y_{21})\,V_{1a} + (z_{21b}\,y_{12} + z_{22b}\,y_{22})\,V_{2a} \tag{2.159}$$

With

$$\alpha_{11s} = z_{11b}\,y_{11} + z_{12b}\,y_{21} \tag{2.160}$$

$$\alpha_{12s} = z_{11b}\,y_{12} + z_{12b}\,y_{22} \tag{2.161}$$

$$\alpha_{21s} = z_{21b}\,y_{11} + z_{22b}\,y_{21} \tag{2.162}$$

$$\alpha_{22s} = z_{21b}\,y_{12} + z_{22b}\,y_{22} \tag{2.163}$$

(2.158) and (2.159) reduce to

$$V_{1b} = \alpha_{11s}V_{1a} + \alpha_{12s}V_{2a} \tag{2.164}$$

$$V_{2b} = \alpha_{21s}V_{1a} + \alpha_{22s}V_{2a} \tag{2.165}$$

With V_{1b} and V_{2b} known in terms of the original power parameters, the modified power parameters can be calculated:

$$V_1 = V_{1a} + V_{1b}$$

$$= V_{1a} + \alpha_{11s}\,V_{1a} + \alpha_{12s}\,V_{2a}$$

$$= (1 + \alpha_{11s})\,V_{1a} + \alpha_{12s}\,V_{2a}$$

$$= (1 + \alpha_{11s}) (M V_{1i} + N V_{2i}) + \alpha_{12s} (O V_{1i} + P V_{2i})$$

$$= [(1 + \alpha_{11s}) M + \alpha_{12s} O] V_{1i} + [(1 + \alpha_{11s}) N + \alpha_{12s} P] V_{2i} \tag{2.166}$$

and

$$V_2 = V_{2a} + V_{2b}$$

$$= V_{2a} + \alpha_{21s} V_{1a} + \alpha_{22s} V_{2a}$$

$$= [\alpha_{21s} M + (1 + \alpha_{22s}) O] V_{1i} + [\alpha_{21s} N + (1 + \alpha_{22s}) P] V_{2i} \tag{2.167}$$

2.3.7.5 The Effect of Changing the Configuration on the Power Parameters

As was the case with the two-port parameters, the power parameters change when the configuration is changed. The change in the parameters is established below.

Common-Source to Common-Gate Case

If the power parameters for the common-source configuration (see Figure 2.28) are given by

$$V_{1s} = M_s V_{1i} + N_s V_{2i} \tag{2.168}$$

$$V_{2s} = O_s V_{1i} + P_s V_{2i} \tag{2.169}$$

the parameters for the common-gate configuration can be calculated from the voltage relationships:

$$V_{1g} = -V_{1s} \tag{2.170}$$

$$V_{2g} = V_{2s} - V_{1s} \tag{2.171}$$

$$\Rightarrow$$

$$V_{2g} = V_{2s} + V_{1g} \tag{2.172}$$

The first two parameters follow easily from (2.170) and (2.168):

$$V_{1s} = M_s V_{1i} + N_s V_{2i}$$

becomes

$$-V_{1g} = M_s V_{1i} + N_s V_{2i} \tag{2.173}$$

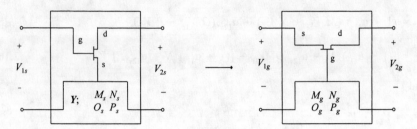

Figure 2.28   The effect of changing the configuration from common-source to common-gate on the power parameters.

which implies

$$V_{1g} = -M_s V_{1i} - N_s V_{2i} \tag{2.174}$$

By substituting this result in (2.172), it follows that

$$V_{2g} - V_{1g} = V_{2s} = O_s V_{1i} + P_s V_{2i}$$

and

$$V_{2g} = O_s V_{1i} + P_s V_{2i} + V_{1g}$$

$$= O_s V_{1i} + P_s V_{2i} + [-M_s V_{1i} - N_s V_{2i}]$$

and, therefore, that

$$V_{2g} = (O_s - M_s)V_{1i} + (P_s - N_s)V_{2i}$$

$$= (O_s + M_G)V_{1i} + (P_s + N_g)V_{2i} \tag{2.175}$$

The power parameters for the common-gate configuration are given by (2.174) and (2.175).

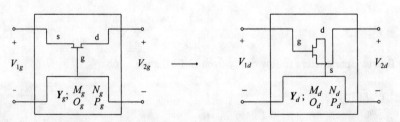

Figure 2.29   The effect of changing the configuration from common-gate to common-drain on the power parameters.

Common-Gate to Common-Drain Case

The common-drain power parameters (see Figure 2.29) can be calculated from the common-gate parameters as follows.

Starting with

$$V_{1g} = M_g V_{1i} + N_g V_{2i} \tag{2.176}$$

$$V_{2g} = O_g V_{1i} + P_g V_{2i} \tag{2.177}$$

and the voltage relationships

$$V_{1d} = -V_{2g} \tag{2.178}$$

$$V_{2d} = V_{1g} - V_{2g} \tag{2.179}$$

it follows that

$$-V_{1d} = V_{2g}$$

$$= O_g V_{1i} + P_g V_{2i}$$

and, therefore, that

$$V_{1d} = - O_g V_{1i} - P_g V_{2i} \tag{2.180}$$

and

$$V_{2d} = V_{1g} - V_{2g}$$

$$= M_g V_{1i} + N_g V_{2i} - O_g V_{1i} - P_g V_{2i}$$

$$= (M_g - O_g) V_{1i} + (N_g - P_g) V_{2i} \tag{2.181}$$

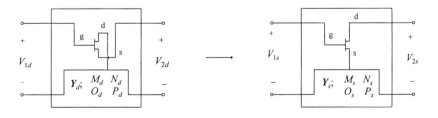

Figure 2.30   The effect of changing the configuration from common-drain to common-source on the power parameters.

Common-Drain to Common-Source Case

The common-source power parameters can be calculated from the common-drain parameters (see Figure 2.30) as follows.
    Starting with

$$V_{1d} = M_d V_{1i} + N_d V_{2i} \tag{2.182}$$

$$V_{2d} = O_d V_{1i} + P_d V_{2i} \tag{2.183}$$

and the voltage relationships

$$V_{2s} = -V_{2d} \tag{2.184}$$

$$V_{1s} = V_{1d} - V_{2d} \tag{2.185}$$

it follows that

$$V_{1s} = V_{1d} - V_{2d}$$

$$= M_d V_{1i} + N_d V_{2i} - O_d V_{1i} - P_d V_{2i}$$

$$= (M_d - O_d) V_{1i} + (N_d - P_d) V_{2i} \tag{2.186}$$

$$V_{2s} = -V_{2d} = -O_d V_{1i} - P_d V_{2i}$$

$$= -O_d V_{1i} - P_d V_{2i} \tag{2.187}$$

QUESTIONS AND PROBLEMS

2.1   The S-parameters and noise parameters of a NE32400 transistor at 4 GHz are

$s_{11} = -0.2645$ dB, $316°$ $F_{min} = 0.33$ dB
$s_{12} = -25.68$ dB, $65°$ $\Gamma_{s\text{-}n\text{-opt}} = -2.4987$ dB, $31°$
$s_{21} = 12.83$ dB, $144°$ $r_{nv} = 0.33$
$s_{22} = -4.2934$ dB, $331°$

where $\Gamma_{s\text{-}n\text{-opt}}$ is the reflection coefficient of the optimum noise termination, and r_{nv} is the normalized value of R_{nv} ($r_{nv} = R_{nv}/50$).

Calculate the available power gain associated with the optimum noise figure, as well as the noise figure and the transducer power gain with 50-Ω terminations. How much can the 50-Ω noise figure be improved by adding a shunt inductor across the input of the transistor?

2.2   Calculate the different noise correlation matrices for the transistor in Problem 2.1. Use the current matrix representation to calculate the noise parameters of two of these transistors connected in parallel. Also calculate the noise parameters for two of the transistors connected in cascade. Calculate the *S*-parameters for both cases too.

2.3   Use the current representation noise correlation matrix to establish the effect of a series inductor connected between the source of the transistor in Problem 2.1 and the ground. Set the inductor values to 0.5 nH, 1.0 nH, and 1.5 nH. Note the change in the optimum noise impedance, as well as the available power gain.

2.4   Calculate the effect of adding a series resistor of 8.2Ω adding in series on the input side of the transistor in Problem 2.1 on the 50-Ω noise figure, as well as the noise parameters.

2.5   The transistor in Problem 2.1 is biased at 2V 10 mA. Assuming that I_{dss} and the breakdown voltage are adequate to allow full swing in the voltage and the current, and assuming that the minimum voltage across the transistor should be 0.4V to avoid the resistive area on the *I*/*V*-curves, use the equations provided for the class A mode in Section 2.3.1.1 to calculate the maximum output power that can be expected, as well as the required load line.

2.6   Given the model for the transistor in Problem 2.1 as shown in Figure 2.31, estimate the optimum load impedance for maximum output power by using the Cripps approach. Compare the answer with the results in Problem 2.5.

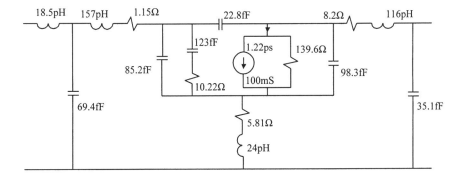

Figure 2.31   A small-signal model fitted to the *S*-parameters of the NE32400 at 2V 10mA.

2.7   Use the power parameter approach to show that the optimum class A load for the transistor in Problem 2.6 at 4GHz is $112 + j62$ Ω. Also show that for this load line the maximum output power before clipping is 8.6 dBm.

2.8   Add a voltage-shunt feedback resistor of 330 Ω to the transistor in Problem 2.7 and show that the maximum output power before clipping is reduced to 5.3 dBm. Also calculate the new optimum load line ($125 + j147$ Ω).

2.9   Load the output of the transistor in Problem 2.6 with a 50 Ω series resistance and show that the maximum output power before clipping is reduced to 6.6 dBm. Also calculate the new optimum load line ($107.5 + j30.9$ Ω).

2.10   Calculate the intrinsic load line presented to the transistor in Problems 2.7 through 2.9.

2.11   Assuming these same I/V-curve boundary lines as in Problem 2.7, and assuming that the same small-signal model can be used after reducing the g_m by a factor 2, calculate the optimum class-B load line and the associated output power for the transistor. Also calculate the expected efficiency.

2.12   Assuming the same model and constraints as in Problem 2.11, calculate the optimum class-F load line and the associated output power. Consider the two cases where the output voltage is shaped towards a square by using the third harmonic only, as well as the ideal case.

REFERENCES

[1]   Cripps, S. C., "GaAs Power Amplifier Design," Technical Notes 3.2, Palo Alto, CA: Matcom Inc.

[2]   *MultiMatch RF and Microwave Impedance-Matching, Amplifier and Oscillator Synthesis Software*, Stellenbosch: Ampsa (Pty) Ltd.; http://www.ampsa.com, 2009.

[3]   Nitronex Corporation, http://www.nitronex.com, NPT25015 Datasheet, 2008.

[4]   Negra, R., F.M. Ghannouchi, and W. Bächtold, "Study and Design Optimization of Multiharmonic Transmission-Line Load Networks for Class-E and Class-F K-Band MMIC Power Amplifier," *IEEE Trans. Microwave Theory and Techniques*, Vol. 55, No. 6, June 2007.

[5]   Kang, D., et al., "A Highly Efficient and Linear Class-AB/F Power Amplifier for Multimode Operation," *IEEE Trans. Microwave Theory and Techniques*, Vol. 56, No. 1, January 2008.

[6]   Sokal, N. O., "Class-E RF Power Amplifiers," *QEX*, January/February 2001.

[7]   Iwamoto, M., et al., "An Extended Doherty Amplifier with High Efficiency over a Wide Power Range," *IEEE Trans. Microwave Theory and Techniques*, Vol. 49, No. 12, December 2001.

[8]   Kim, J., et al., "Analysis of a Fully Matched Saturated Doherty Amplifier with Excellent Efficiency," *IEEE Trans. Microwave Theory and Techniques*, Vol. 56, No. 2, February 2008.

[9]   White, P. M., "Effect of Input Harmonic Terminations on High Efficiency Class-B and Class-F Operation of pHemt Devices," *IEEE MTT-S Digest*, 1998.

[10] Jeon, K., Y. Kwon, and S. Hong, "Input Harmonics Control Using Non-linear Capacitor in GaAs FET Power Amplifier," *IEEE MTT-S Digest*, 1997.

[11] Yum, T. Y., et al., "High-Efficiency Linear RF Amplifier—A Unified Circuit Approach to Achieving Compactness and Low Distortion," *IEEE Trans. Microwave Theory and Techniques*, Vol. 54, No. 8, August 2006.

[12] Ingruber, B., W. Pritzl, and G. Magerl, "High Efficiency Harmonic Control Amplifier," *IEEE MTT-S Digest*, 1996.

[13] Aparin, V., and C. Persico, "Effect of Out-of Band Terminations on Intermodulation Distortion in Common-Emitter Circuits," *IEEE MTT-S Digest*, 1999.

[14] Haus, H. A., and R. B. Adler, *Circuit Theory of Linear Noisy Networks*, New York: Wiley, 1959.

[15] Hillbrand, H., and P. H. Russer, "An Efficient Method for Computer Aided Noise Analysis of Linear Amplifier Networks," *IEEE Trans. Circuits and Systems*, Vol. CAS-23, No. 4, April 1976.

[16] Friss, H. T., "Noise Figure of Radio Receivers," *Proc. IRE*, Vol. 30, pp. 419–422, July 1944.

[17] Vendelin, G. D., A. M. Pavio, and U. L. Rohde, *Microwave Circuit Design Using Linear and NonLinear Techniques*, New York: John Wiley, 1990.

[18] Cripps, S. C., *Advanced Techniques in RF Power Amplifier Design*, Norwood, MA: Artech House, 2002.

[19] Kraus, H. L., C. W. Bostian, and F. H. Raab, *Solid State Radio Engineering*, New York: John Wiley, 1980.

[20] Colantonio, P., G. Leuzzi, and E. Limiti, "On the Class-F Power Amplifier Design," *Int. Journal of RF and Microwave Computer-Aided Engineering*, Vol. 9, No. 2, 1999, pp. 129–149.

[21]   Raab, F. H., "Class-E, Class-C, and Class-F Power Amplifiers Based upon a Finite Number of Harmonics," *IEEE Trans. Microwave Theory and Techniques*, Vol. 49, No. 8, August 2001.

[22]   Raab, F. H., "Idealized Operation of Class E Tuned Power Amplifier," *IEEE Trans. Circuits Syst.*, Vol. CAS-24, December 1977, pp. 725–735.

[23]   Zhao, Y., et al., "Linearity Improvement of HBT-Based Doherty Power Amplifiers Based on a Simple Analytical Model," *IEEE Trans. Microwave Theory and Techniques*, Vol. 54, No. 12, December 2006.

[24]   Hau, G., T. B. Nishimura, and N. Iwata, "Distortion Analysis of a Power Heterojunction FET Under Low Quiescent Drain Current for 3.5V Wide-Band CDMA Cellular Phones," *IEEE MTT-S Digest*, 1999.

[25]   Cripps, S. C., "Harmonic and Intermodulation Distortion in GaAsFET Amplifiers," Technical Notes 2.1, Palo Alto, CA: Matcom Inc.

[26]   Caverly, R. H., and J. C. Peyton Jones, "Contributions to Adjacent Channel Power in Microwave and Wireless Systems by PIN Diodes," *Microwave Symposium Digest*, 2006, IEEE MTT-S International.

[27]   Nagy, W., et al., "Linearity Characteristics of Microwave Power GaN HEMTs," *IEEE Trans. Microwave Theory and Techniques*, Vol. 51, No. 2, February 2003.

CHAPTER 3

RADIO-FREQUENCY COMPONENTS

3.1 INTRODUCTION

In order to design realizable radio-frequency and microwave circuits, some knowledge of the parasitics associated with practical components is essential. The characteristics of practical capacitors, inductors, magnetic materials, and microstrip transmission lines will be considered in this chapter.

The capacitors used in an RF circuit (impedance-matching networks, filters, coupling and decoupling networks) can usually be obtained from one of the many manufacturers of these components. Unfortunately, this does not always apply to inductors. The design of inductors will, therefore, also be considered in this chapter. Single-layer air-cored inductors and inductors with magnetic cores will be considered.

In order to get the circuit manufactured to perform as expected, care should be taken to ensure that the circuit realized is the same as the one designed. Apart from the parasitic effects of the components used, care should also be taken with any connections made between components. The effect of all the connections made should be included in the simulation.

Connections to the ground plane should also be made with care. Ground loops (unnecessary ground connections) should be avoided and connections cannot be made arbitrarily to the ground plane on the (false) assumption that all points on the ground plane are at the same potential (as would be the case on the circuit diagram). When any uncertainty arises as to exactly where a connection should be made to the ground plane, it is useful to realize that the electric signal is traveling as a wave through the circuit and ground at any point is where the wave is.

When an active circuit is manufactured, RF and microwave decoupling of the dc circuit is essential (introducing an RF ground). Parasitic resonances can easily be introduced inadvertently when this is done. It is often possible to eliminate such resonances by using small resistors in the decoupling circuit (the voltage across these resistors can also be used to check the dc current). A number of capacitors can also be used in parallel. The capacitance of the different capacitors is usually chosen to differ by a factor of 10 when this is done.

When different capacitors are used in parallel, the series resonating frequencies of the different capacitors should be taken into account when the values are chosen (the smaller the capacitance value, the higher the resonating frequency will be) and care should be taken to avoid parallel resonances between the components used.

The thin-film resistors and parallel-plate (single-layer) capacitors used at microwave frequencies cannot be accurately simulated as lumped components. The distributed nature of these components must be taken into account in the design. These components will be considered in Chapter 7.

Additional complications are introduced by the steps, T-junctions, and crosses associated with planar transmission lines. The ideal connection is a point junction, but these junctions are not point junctions. These effects will be considered in Chapter 9.

3.2   CAPACITORS

Capacitors differ in capacitance, resonant frequency, losses, temperature stability, tolerances, packaging, and size. Most of these characteristics are determined by the dielectric material used. The parasitic inductance is, however, also a function of the packaging and the lead lengths of the capacitor.

The equivalent circuit for a practical capacitor is shown in Figure 3.1.

The parasitic inductance causes the impedance of the capacitor to be lower than expected. The impedance at the series resonant frequency is equal to the series resistance of the capacitor. Above this frequency the impedance becomes inductive.

The effective capacitance below the resonant frequency is given by

$$C_{\text{eff}} = C_0 / [1 - (f / f_r)^2] \tag{3.1}$$

where C_0 is the capacitance at low frequencies and

$$f_r = \frac{1}{2\pi\sqrt{LC_0}}$$

where f_r is the resonant frequency of the capacitor.

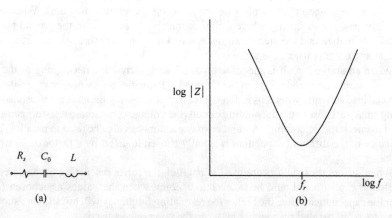

(a)   (b)

Figure 3.1   (a) An equivalent circuit for a capacitor; and (b) the effect of the parasitic inductance and resistance on the impedance (Z) of a capacitor.

Table 3.1 The Resonant Frequencies for Some Capacitors [1–4]

Capacitance	1 pF	10 pF	100 pF	1 nF	10 nF
Mica; disk ceramic	—	—	170 MHz	60 MHz	20 MHz
Porcelain chip capacitors	7–10 GHz	2–3 GHz	1 GHz	230 MHz	—
Parallel-plate capacitors	20 GHz	7 GHz	2 GHz	600 MHz	—

The resonant frequencies for some capacitors (with very short lead lengths or no leads) are shown in Table 3.1 [1–4].

As can be seen from Table 3.1, even chip capacitors have some parasitic inductance. There are two reasons for this: first, the finite dimensions (and therefore the inductance) of the capacitor plates, and second, the finite distance across the plates.

That there must be some inductance associated with the finite separation of the capacitor plates is obvious if Maxwell's law

$$\nabla \times H = i + \partial D / \partial t$$

is inspected. According to this equation, even a displacement current generates magnetic flux and, therefore, has inductance associated with it. The inductance can be minimized by choosing the smallest capacitor available (with voltage and power ratings taken into account).

The losses in a capacitor are usually specified by the quality factor (Q), where

$$Q = X_s / R_s \qquad (3.2)$$

R_s is the series resistance of the capacitor, and X_s is the effective reactance of the capacitor.

The quality factor (Q-factor) is frequency- and temperature-dependent. It is, therefore, important to specify the measuring frequency and the power level at which the measurement was made.

Table 3.2 The Dielectric Constants (ε_r) and Dissipation Factors for Some Commonly Used Materials

Material	ε_r	DF (low frequencies)	DF (@ 100 MHz)
BaT_1O_3	1200	0.01	0.03
NPO	30	0.0001	0.002
Porcelain	15	—	0.00007

While the losses of the component are specified in terms of the Q-factor, the losses of dielectric materials are specified in terms of the dissipation factor (DF) or the loss tangent (tan δ).

The dissipation factor specifies the ratio of the power dissipated to the power stored in the material:

$$DF = P_{\text{diss}} / P_{\text{stored}} \tag{3.3}$$

The relative power dissipation of dielectric materials is directly proportional to the dissipation factor. High losses are associated with high dielectric constants.

The dissipation factors for three commonly used materials are given in Table 3.2 [2]. Note the decrease in losses as the relative dielectric constant drops, as well as the increase in dissipation at higher frequencies.

It can be easily shown that if the parasitic inductance of a capacitor can be ignored, the dissipation factor and the Q-factor are related in the following way:

$$DF = 1 / Q \tag{3.4}$$

The losses of the dielectric materials and capacitors are sometimes specified in terms of the loss tangent (tan δ). The definition of the loss tangent is the same as that of the dissipation factor.

Dissipation factors are not only frequency dependent, but increase with temperature and, therefore, with power level. The power dissipation inside a typical chip capacitor only needs to be on the order of 40 mW to increase the temperature to that of commonly used soldering irons [2]. At high temperatures the dissipation factor can be an order of magnitude higher than at room temperature. As the temperature inside a capacitor increases, the dissipation factor increases, which causes a further increase in temperature with more losses. This thermal runaway phenomenon is particularly important at low impedance and high power level points in a circuit.

The series resistance and Q-factors of two high-quality capacitors at room temperature are given at two different frequencies in Table 3.3 [2]. Even for good capacitors, the Q-factor is surprisingly low at high frequencies.

Table 3.3   The Quality Factor and Resistance of Two Capacitors at High Frequencies

Frequency	100 MHz	500 MHz
10 pF	2200 (0.055Ω)	180 (0.169Ω)
100 pF	700 (0.018Ω)	60 (0.055Ω)

Not only the dissipation factor, but also the capacitance of a capacitor, are affected by a change in temperature. The change in capacitance can be very small [NPO (negative positive zero) capacitors] and linear (class 1 ceramics), or large and nonlinear (class 2 ceramics). Class 1 ceramics with positive (up to 150 ppm/°C) and negative (up to -5500 ppm/°C) temperature coefficients are available [5].

Note that capacitors with negative temperature coeffients may be marked with the letter N, while the letter P is used for positive temperature coefficients. NPO (instead of NP0) is used to indicate that the temperature coefficient is zero.

As a final remark on capacitors, it should be noted that the capacitance of capacitors with high dielectric constants is usually also voltage-sensitive. The capacitance of class 2 ceramics can change by more than 20% if the voltage is varied from 0% to 150% of the rated value [5].

Summary

The following points are important when choosing a capacitor for a particular purpose:

1. The parasitic inductance;

2. The tolerance of the capacitor;

3. The Q-factor at the desired frequency and power level;

4. The influence of voltage on the capacitor (capacitance changes, as well as the breakdown voltage);

5. The influence of temperature on the capacitor (ambient as well as increases due to the power dissipation in the capacitor);

6. The size and packaging of the capacitor.

3.3 INDUCTORS

The performance of practical inductors are degraded by parasitic capacitance and resistive losses.

The parasitic capacitance (see Figure 3.2) causes the resistance of the inductor to be higher than expected. This effect is very pronounced near the resonant frequency (f).

Inductor losses consist of copper losses (R_s) and, if magnetic material is used, hysteresis and eddy current losses (R_p). All of these losses are frequency-dependent. The copper losses increase above its dc value because of the skin and proximity effects.

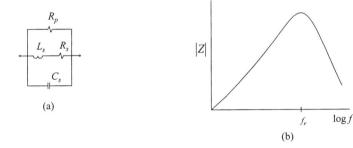

(a)

(b)

Figure 3.2   (a) The equivalent circuit of a practical inductor; and (b) the effect of parasitic capacitance and losses on its impedance (Z).

By using magnetic material, the size of the inductor can be reduced drastically and the parasitic capacitance will, therefore, also be considerably lower. Unfortunately, there will also be some losses in the material. These losses are mainly hysteresis losses in the case of ferrite materials.

The effect of parasitic capacitance on the Q-factor and the inductance of inductors, the skin and proximity effects, the design of air-cored solenoidal coils, the properties of magnetic materials, and the design of inductors with ferrite cores will be discussed in the following sections.

3.3.1   The Influence of Parasitic Capacitance on an Inductor

By using the equivalent circuit shown in Figure 3.2, it can be easily shown that the effective inductance (L_{eff}) of an inductor is given by

$$L_{eff} = L_s / [1 - (f / f_r)^2]$$ (3.5)

where f_r is the parallel resonant frequency of the inductor.
This equation applies only if the approximation

$$1 + 1/Q_s^2 \cong 1$$ (3.6)

where

$$Q_s = \omega L_s / R_s$$

can be made.

As can be seen from (3.5), the inductance increases rapidly as the resonant frequency (f_r) is approached.
Under the same conditions, the effective resistance (R_p ignored) is given by

$$R_{eff} = R_s / [1 - (f / f_r)^2]$$ (3.7)

Because the effective resistance has increased because of the parasitic capacitance present, the losses in the coil are higher if the input current to the inductor is considered to be the same. This happens because the current in the parasitic capacitor is out of phase with that in the inductive part of the inductor.

The effective Q-factor of the coil will therefore be lower than without parasitic capacitance. The effective Q-factor is given by

$$Q_{eff} = Q_s [1 - (f / f_r)^2]$$ (3.8)

When $f = 0.707 f_r$, the effective Q-factor will be half that of the inductive part of the inductor.

These effects can be minimized by keeping the parasitic capacitance as low as possible.

The capacitance of an air-cored solenoidal coil is given in Figure 3.3 as a function of the length-to-diameter ratio and the mean radius of the coil [6].

The capacitance of the coil is not a function of the number of turns as might be suspected; it is a strong function of the coil size (radius) and a weak function of the coil shape (length-to-diameter ratio, l/D). The capacitance can therefore be minimized by making the coil as small as possible. An initial value of 2 can be used for the length-to-diameter ratio.

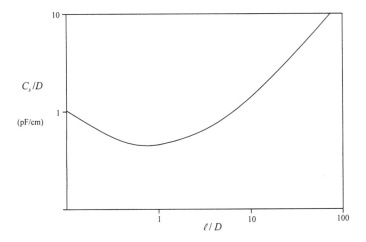

Figure 3.3   The self-capacitance of a single-layer solenoidal coil. (Source: [6].)

For high inductance, the turns of a coil should be spaced as closely as possible. It will be shown later that this distance is determined by the desired Q-factor of the coil.

When the coil capacitance is known, the resonant frequency can be found by using the equation

$$f_r = \frac{1}{2\pi \sqrt{L_s C_s}} \tag{3.9}$$

Typical resonant frequencies for some inductance values are given here as a guide to what can be achieved easily [1]:

100 nH:	400–800 MHz
1 µH:	100–200 MHz
10 µH:	25–60 MHz

Miniature chip coils (0805, 1008, ...) with self-resonant frequencies ranging from 250 MHz to above 6 GHz for values ranging from 1,500 nH to 2.2 nH are commercially available. The resonance frequency claimed for a 100 nH (22 nH) miniature chip inductor is 1.5 GHz (3.2 GHz) for a chip size of 0805 (8 mils × 5 mils) and 1 GHz (2.4 GHz) for a chip size of 1008 [7]. The minimum Q-values quoted at 150 MHz (250 MHz) and 100 MHz are 40 and 50, respectively [7].

Table 3.4   The Wire Diameter and Resistance for Wire Gauges 12–32 (20°C; Copper Material)

Gauge	Bare diameter (mm)		Double enamel-coated diameter (mm)		Resistance (Ω/km)	
	AWG	(SWG)	AWG	(SWG)	AWG	(SWG)
12	2.052	(2.64)	2.13	(2.73)	5.5	(3.1)
14	1.628	(2.03)	1.71	(2.12)	8.6	(5.2)
16	1.291	(1.63)	1.37	(1.71)	15.2	(8.2)
18	1.024	(1.22)	1.10	(1.29)	22.0	(14.5)
20	0.812	(0.914)	0.879	(0.984)	34.3	(25.8)
22	0.644	(0.711)	0.701	(0.774)	61.0	(42.6)
24	0.511	(0.559)	0.564	(0.617)	87.8	(69.1)
26	0.405	(0.457)	0.452	(0.512)	133.9	(103.2)
28	0.321	(0.376)	0.366	(0.424)	212.9	(152.6)
30	0.255	(0.315)	0.295	(0.361)	338.5	(217.4)
32	0.202	(0.274)	0.241	(0.316)	538.5	(286.6)

3.3.2   Low-Frequency Losses in Inductors

The resistive losses in a conductor are approximately constant at low frequencies. The resistance is a function of the material used and the wire diameter. The diameters and the resistance of copper wire with wire gauges ranging from 12 to 32 are given in Table 3.4. The American wire gauge (AWG) values are listed with the corresponding standard wire gauge (SWG) values. Note that the wire diameter doubles whenever the wire gauge decreases by a factor of 6.

It can be seen from the table that the diameter of AWG No.12 wire is approximately 2 mm and that of AWG No. 22 is 0.2 mm. The resistance of No. 12 wire is 5.5 Ω/km and that of No. 32 wire is 538 Ω/km. The increase of approximately 100 in resistance correlates well with the decrease in the diameter by a factor of 10 ($R \propto 1/A$, where A is the cross-section area of the wire).

3.3.3   The Skin Effect

A conductor can be viewed as a guide for the electrical and magnetic fields around it, as is shown in Figure 3.4. The current flowing in the conductor is caused by the changing magnetic flux that penetrates into the conductor. This current opposes the magnetic field that causes it. The result is that the magnetic field decreases in strength (exponentially) as it penetrates the conductor.

The induced electrical field within the conductor is given as a function of the penetration depth x by

$$E_z = E_{z0}\,e^{-\Gamma x} \tag{3.10}$$

where E_{z0} is the electric field strength at the surface of the conductor (in the direction of the conductor).

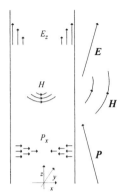

Figure 3.4   The electric, magnetic, and Poynting fields around and inside a circular conductor (after [8]).

The propagation constant of the electrical field in the wire is

$$\Gamma = \sqrt{j\omega\mu/\rho}$$

$$= (1+j)\sqrt{\pi f \mu/\rho}$$

$$= \alpha + j\beta \tag{3.11}$$

where ρ is the resistivity of the conductor.

The inverse of the attenuation constant α is defined as the skin depth δ:

$$\delta = 1/\alpha = 1/\sqrt{\pi f \mu/\rho} \tag{3.12}$$

Therefore, the amplitude of the electrical field at a distance x inside the conductor is

$$E(x) = E(0)e^{-x/\delta} \tag{3.13}$$

Because of the decrease in the field strength, the current density will be higher closer to the surface of the conductor. When the conductor is at least six skin depths (or depths of penetration) in diameter, all the current can be considered to flow uniformly in a layer one skin depth deep along the surface of the conductor.

Table 3.5   The Skin Depth of Some Materials as a Function of Frequency in Hertz

Material	Skin Depth (cm)
Brass	$12.7/f^{1/2}$
Aluminum	$8.3/f^{1/2}$
Gold	$7.7/f^{1/2}$
Copper	$6.6/f^{1/2}$
Silver	$6.2/f^{1/2}$
Mu-metal	$0.4/f^{1/2}$

The ac or RF resistance of the conductor can then be calculated within 10% by using the following equations [8]:

$$R_{ac} = \{\pi r^2 / [\pi r^2 - \pi(r - \delta)^2]\} R_{dc} \tag{3.14}$$

$$= \{\pi r^2 / [\pi r^2 - \pi(r^2 - 2\delta r + \delta^2)]\} R_{dc}$$

$$= \{\pi r^2 / [2\pi\delta r - \pi\delta^2]\} R_{dc} \tag{3.15}$$

where $2r$ is the outside diameter of the conductor. (The conductor could be hollow.)

At high frequencies, where $\delta \ll 2r$, this equation simplifies to

$$R_{ac} = [r / (2\delta)] R_{dc} \tag{3.16}$$

Because the skin depth is inversely proportional to the square root of the frequency, the resistance R_{ac} will increase proportionally to the root of the frequency, that is, if $\delta \ll d$ (where d is the diameter of the conductor).

The skin depths for some materials are given in Table 3.5 as a function of the frequency.

To illustrate the change in skin depth with frequency, consider the skin depth for copper at various frequencies:

$\delta = 0.66$ mm at 10 kHz
$\delta = 66$ μm at 1 MHz
$\delta = 6.6$ μm at 100 MHz

Because the skin depth is very small at high frequencies, it is important to ensure that conductor surfaces are smooth if the lowest possible resistance with a specific material is required. When materials with low conductivities are used (usually to ensure temperature stability), it becomes worthwhile to plate the conductors with silver above 100 MHz.

To get an idea of the increase in resistance with frequency caused by the skin effect, consider the resistance of 1m of AWG No. 22 wire as a function of frequency:

$R = 0.06\ \Omega$ at dc
$R = 0.60\ \Omega$ at 1 MHz
$R = 5.95\ \Omega$ at 100 MHz

Note that the resistance at 100 MHz is approximately $100^{1/2}$ times that at 1 MHz. It is obvious from these numbers that the increase in resistance caused by the skin effect cannot be ignored at high frequencies.

3.3.4   The Proximity Effect

A conductor carrying alternating current (main conductor) has a changing magnetic field around it. If another conductor (secondary conductor) is brought close to it (see Figure 3.5), the changing magnetic field through it will cause eddy currents in it (when $d>5\delta$, the penetration depth of the field is small compared to the diameter). These currents distort the magnetic field of the main conductor and cause current crowding it. The resistance of the main conductor is increased by the current crowding [9], as well as any eddy current losses in the secondary conductor.

Similar to the skin effect, the increase in resistance is proportional to the root of the frequency at high frequencies ($d>5\delta$).

When only two conductors are in close proximity, the influence of the proximity effect is relatively small compared to that of the skin effect, but when more conductors are used it should be taken into account. Because a solenoidal coil consists of many conductors close to one another, the proximity effect can significantly affect its resistance at high frequencies. As an example of this, the resistance of a single-layer solenoidal coil with turns touching and a length-to-diameter ratio of 0.7 is almost six times that of the same wire when straightened out (that is, if more than 10 turns are used).

When the turns of a coil are spaced well apart, the proximity effect can be ignored.

3.3.5   Magnetic Materials

The inductance of an air-cored coil can be increased significantly by using a magnetic material as the core. The reason for this is that the magnetic flux density increases substantially when the relative permeability of the material is high.

Typical values for the relative permeability (μ_r) of ferrite materials at radio frequencies are 10–150. The higher value is associated with cutoff frequencies on the order of 20 MHz, while lower value is associated with cutoff frequencies of around 1 GHz. Above the cutoff frequency, the relative permeability decreases sharply.

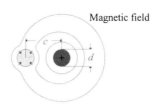

Figure 3.5   Illustration of the distortion in the magnetic field of a conductor carrying alternating current when another conductor (open-circuited) is placed close to it (proximity effect).

Apart from the relative permeability and its frequency dependence, losses in magnetic materials must also be taken into account, especially at high voltage points.

When ferrite materials are used, these losses are mainly hysteresis losses. When materials with higher conductivities are used, the eddy-current losses in the material also become significant.

Losses in a ferrite core are proportional to the energy stored in it. The energy stored is proportional to the energy density and the volume of the core. The volume is approximately equal to the product of the cross-sectional area and the mean path length. Therefore, losses in a ferrite core are given by an equation of the form

$$P_{loss} = k(\mu_r, f, B_{max}) B_{max}^2 Al \qquad (3.17)$$

where A is the average cross-sectional area of the core, l the mean path length of the core, B_{max} the maximum root mean square (rms) flux density in the core, and k a constant dependent on the frequency, relative permeability, flux density, and material used.

The power losses in a ferrite core are best specified in terms of the ratio $\mu_r R_p / L$ and not by (3.17). R_p is the loss resistance in parallel with the inductance (L) of the magnetic-cored inductor.

This ratio is independent of the core dimensions and is only a function of the material used and the maximum flux density. That the ratio $\mu_r R_p / L$ should be independent of the core size can be established as follows.

Because R_p represents the losses in the core, the power loss in the core is given by

$$P_{loss} = V_p^2 / R_p \qquad (3.18)$$

where V_p is the rms voltage across the inductor.

This voltage is related to the maximum flux density B_{max} by

$$V_p = j\omega(N\Phi) = j\omega NAB_{max} \qquad (3.19)$$

where N is the number of turns.

By using these two equations, the resistance R_p is found to be

$$R_p = V_p^2 / P_{loss}$$

$$= \frac{\omega^2 N^2 A^2 B_{max}^2}{P_{loss}} = \frac{\omega^2 N^2 A^2 B_{max}^2}{k Al B_{max}^2}$$

$$= [\omega^2 / k] N^2 A / l \qquad (3.20)$$

The resistance R_p is, therefore, proportional to the square of the number of turns and the cross-sectional area of the coil. It is inversely proportional to the mean path length.

This is also true for the inductance, which is given by

$$L = \frac{\Lambda}{I} = \frac{N\Phi}{I} = \mu_0 \mu_r N^2 \frac{A}{l} \qquad (3.21)$$

The ratio $\mu_r R_p / L$ is, therefore, independent of the core dimensions.

By using (3.20) and (3.21), it follows that

$$\mu_r R_p / L = \omega^2 / (k \mu_0) \qquad (3.22)$$

Because k is a function of the flux density and the frequency, the ratio $\mu_r R_p / L$ is also a function of the flux density and the frequency.

Curves for this ratio as a function of frequency are shown in Figure 3.6 [10]. These curves apply at small-signal conditions (that is, when B_{max} is small).

By using these curves and a value of 120 for the relative permeability, it can be shown easily that the highest unloaded Q [$Q_u = R_p / (\omega L)$] that can be expected at 6 MHz by using 4C6 material is approximately 125.

When the flux density increases, the losses in the core increase as well. Curves for the ratio $\mu_r R_p / L$ as a function of the product $B_{max} f$ are shown for 4C4 material at different frequencies in Figure 3.7.

The product $B_{max} f$ is used because it is independent of the frequency if the maximum voltage across the inductor (V_p) is assumed to be constant.

By using the curve for 1.6 MHz, it follows that the losses double from their small-signal value when the flux density is approximately 14 mT (140 Gauss).

As a final remark on magnetic materials, it should be noted that the relative permeability of magnetic materials is temperature-dependent. Materials with higher permeabilities are influenced more by temperature changes.

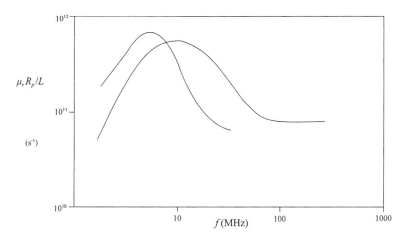

Figure 3.6   Curves of the ratio $\mu_r R_p / L$ ($\omega \mu_r / \tan \delta$) plotted against frequency for two ferrite materials ($B_{max} \to 0$). (Source: [10].)

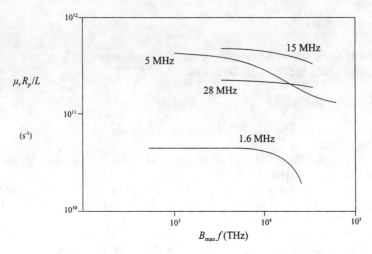

Figure 3.7   Curves of $\mu_r R_p /L$ ($\omega\mu_r/\tan\delta$) plotted against the product ($B_{max}f$) for 4C4 material at various frequencies. (Source: [10].)

Because the temperature of the material changes when heat is dissipated in it, the relative permeability will also change when more power is dissipated in it.

Summary

The following points should be taken into account when a magnetic material is selected for a particular purpose:

1. The highest frequency of operation;

2. The maximum allowable amount of losses;

3. The size of the inductor and, therefore, the relative permeability;

4. The temperature dependence of the magnetic material.

3.3.6   The Design of Single-Layer Solenoidal Coils

Single-layer solenoidal coils are often used at radio frequencies. Their use is limited by the inductance values and unloaded Q-factors obtainable, as well as by the associated parasitic capacitance.

       The inductance of a single-layer solenoidal coil is given approximately by

$$L = N^2 r / [22.9 l / r + 25.4] \quad (\mu H) \tag{3.23}$$

where r is the mean radius of the coil (in centimeter), l the length of the coil (in centimeter), and N the number of turns.

The parasitic capacitance of these coils is given in Figure 3.3 as a function of the length-to-diameter ratio (l/D) and the radius of the coil. The capacitance is small when the coil radius is small.

The unloaded Q of air-cored coils is a function of the frequency, inductance, dc resistance, skin effect, proximity effect, and self-capacitance of the coil.

At frequencies where the self-capacitance can be neglected, the unloaded Q is given by [6]

$$Q_u = kr\sqrt{f} \tag{3.24}$$

where the radius must be specified in centimeters and the frequency in hertz.

The factor k depends on the length-to-diameter ratio of the coil and the relative spacing of the turns. Its value is plotted in Figure 3.8 for various coil shapes and wire spacing ratios (d/c), where c is the distance between the centers of two adjacent turns and d is the diameter of the wire used.

The following facts can be deduced from the curves in Figure 3.8 and (3.24):

1. Higher unloaded Q-factors can be obtained by using coils with larger diameters and length-to-diameter ratios (l/D).

2. The turns of an air-cored solenoidal coil should be spaced close enough to ensure that the d/c ratio is larger than $0.4\,d$, and in shorter coils ($l/D \cong 1$) they should be spaced far enough apart to ensure that the d/c ratio is smaller than $0.8\,d$.

    When larger coils are used the turns can touch without any significant reduction in the unloaded Q (less than 25%).

By using the curves in Figure 3.8 and the equations given, solenoidal coils can be designed to have a specified inductance and unloaded Q. The parasitic capacitance can be determined by using the curve in Figure 3.3. The design can be done as summarized below.

A Design Procedure for Controlling the Inductance and Quality Factor of an Air-Cored Solenoidal Coil

1. Choose the length-to-diameter ratio (l/D) equal to 1.

2. Calculate the radius (r) of the coil (in centimeter) by using the equation

    $$r = Q_u / (k\sqrt{f}) \tag{3.25}$$

    where Q_u is the unloaded Q required, and $k = 0.1$ for $l/D = 1.0$ (see

Figure 3.8).

3.   Find the parasitic capacitance of the coil by using Figure 3.3. Calculate the
      resonant frequency by using the equation

$$f_r = 1 / \sqrt{LC} / (2\pi) \tag{3.26}$$

where $C/D = 0.45$ pF/cm for $l/D = 1$.

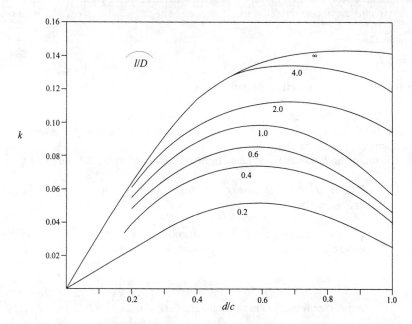

Figure 3.8   Curves for calculating the unloaded Q of single-layer solenoidal coils at high frequencies.
                  (Source: [6].)

4.   If the resonant frequency is too low, the specifications cannot be reached
      and it will have to be changed.

5.   Calculate the required number of turns by using the equation

$$N = [L(22.9(l/r) + 25.4) / r]^{1/2} \tag{3.27}$$

6.   Calculate the required wire thickness by using the d/c ratio used in step 2:

$$d = (d/c)[l/(N-1)] = (l/D)(d/c)[2r/(N-1)] \tag{3.28}$$

where d is the wire diameter to be used, and $d/c = 0.55$ for $l/D = 1$ (see Figure 3.8).

7. If the required wire thickness is small, a coil former will be needed for mechanical support. If the coil is to be self-supporting, it can be redesigned.

In order to increase the wire diameter, it will be necessary to increase the size of the coil. When the resonant frequency is a potential problem, the l/D ratio can be increased. The resonant frequency will decrease if the radius is increased.

Where the resonant frequency is not a problem, the radius of the coil can be increased in order to increase the wire diameter. The maximum value of the radius is

$$r_{max} = C_m / (2C) \tag{3.29}$$

where C_m is the maximum self-capacitance allowable, and C is the capacitance per centimeter as given by Figure 3.3.

With $l/D = 1$, $C = 0.45$ pF/cm.

EXAMPLE 3.1   Designing a single-layer air-cored solenoidal coil to have a specified Q and resonant frequency.

As an example of the application of the procedure outlined, a 1-µH coil was designed to have a minimum unloaded Q of 300 at 50 MHz and resonant frequency above 250 MHz. The results of the different steps are as follows:

1. $l/D = 1$

2. $r = 0.42$ cm

3. $f_r = 256$ MHz

4. —

5. $N = 13$

6. $d = 0.36$ mm

7. Because the wire diameter is small, it will be necessary to use a coil former.

It is not possible to increase the wire diameter by increasing the coil radius in this case ($f_r = 250$ MHz). It is possible, however, to increase it by increasing the l/D ratio of the coil.

Unfortunately, it is not possible to increase sufficiently the wire thickness to make the coil self-supporting.

The results for different l/D ratios are compared in Table 3.6. Note that the wire diameter can be doubled if the length-to-diameter ratio is chosen to be equal to 4.

Although the wire thickness is a strong function of the length-to-diameter ratio, the resonant frequency of coils with length-to-diameter ratios from 0.6 to 4 does not vary significantly if they are designed to have the same unloaded Q-factor.

The volumes of the coils in Table 3.6 increase with increasing l/D ratio. When a small coil is required, the length-to-diameter ratio can therefore be chosen to be equal to 0.6.

Table 3.6   The Dimensions, Unloaded Q, and Resonant Frequency for a 1-µH Coil as a Function of the l/d Ratio

l/D	r	N	d	d/c	f_r	Q_u
	(cm)		(mm)		(MHz)	
0.6	0.48	10	0.31	0.55	252	300
1.0	0.42	13	0.36	0.55	256	300
2.0	0.37	18	0.52	0.63	255	300
4.0	0.32	26	0.63	0.63	242	300

The capacitance, k-factor, and optimum d/c ratio for coils with the l/D ratios used in Table 3.6 are tabulated in Table 3.7 for convenience.

When resonant circuits with high Q-factors are designed, the unloaded Q-factors of the coils and capacitors used must be as high as possible. In order to determine the maximum realizable unloaded Q possible for a single-layer air-cored coil, it is necessary to determine the optimum l/D ratio. Because, in (3.24),

$$Q = k r \sqrt{f}$$

[see (3.24)] the length-to-diameter ratio influences the unloaded Q directly through the associated value of the constant, k, and indirectly (through r) because of the limit that exists on the self-capacitance of the coil.

The maximum radius corresponding to a particular l/D ratio can be determined by using (3.29).

By substituting the value for r as given by (3.29) into (3.24), the maximum Q corresponding to a particular l/D ratio is found to be

$$Q_{max} = (k_m / C) \sqrt{f} \, C_{max} \tag{3.30}$$

where k_m is the maximum value of k corresponding to the particular l/D ratio, C is the capacitance per centimeter as given by the curve in Figure 3.3, and C_{max} is the maximum value of the self-capacitance as determined from the specified resonant frequency.

Table 3.7 The Self-Capacitance, d/c Ratio, Optimum Value of k (k_{opt}), and the Ratio of the k and the Self-Capacitance per Centimeter for Coils with Different l/D Ratios

l/D	C	d/c	k_{opt}	k/C
	(pF/cm)		(Hz$^{-0.5}$/cm)	(Hz$^{-0.5}$/pF)
0.6	0.44	0.55	0.088	0.200
1.0	0.45	0.55	0.100	0.222
2.0	0.53	0.63	0.115	0.216
4.0	0.68	0.63	0.133	0.196

The influence of the l/D ratio on the unloaded Q is clearly limited to the first term in (3.30). The k/C ratios for different l/D ratios are compared in the last column of Table 3.7. It follows from this comparison that the highest Q will be obtained when the length-to-diameter ratio of the coil is equal to 1.

At this stage, the highest Q realizable with a single-layer solenoidal air-cored coil can be determined for any particular inductance value if the operating frequency and the self-resonant frequency are specified. The following procedure can be followed in order to determine the Q.

Design Procedure for Maximum Q and Specified Inductance

1. Choose $l/D = 1$.
2. Determine the maximum value of the self-capacitance (C_{max}). If the coil is to be used in a parallel resonant circuit, the self-resonant frequency can be chosen close to the resonant frequency of the circuit.

    Calculate the maximum allowable radius of the coil by using the equation

    $$r_{max} = C_{max} / (2C) = C_{max} / 0.9 \quad \text{(cm)} \tag{3.31}$$

    with C_{max} specified in picofarads (pF).

    If the value of the radius is unrealistically high, reduce it to an acceptable value.

3. Determine the maximum realizable unloaded Q by using (3.24):

    $$Q_{max} = 0.10 r_{max} \sqrt{f} \tag{3.32}$$

4. Calculate the required thickness of the wire:

    $$N^2 = 71.2 L / r_{max} \tag{3.33}$$

where the inductance (L) is specified in microhenries and r_{max} in centimeters.

$$c = l / (N - 1) = 2r_{max} / (N - 1) \qquad (3.34)$$

$$d = 0.55c \quad \text{(cm)} \qquad (3.35)$$

If the wire thickness turns out to be unrealistic, change the radius.

EXAMPLE 3.2 Designing a single-layer solenoidal coil for maximum Q.

The highest possible Q will be determined for a coil of 10 µH at 5 MHz with self-resonant frequency at 10 MHz by following the procedure outlined above.

1. $l/D = 1$

2. $C_{max} = 1 / [(2\pi \times 10 \times 10^6)^2 \, 10 \times 10^{-6}] = 25.3$ pF

   $r_{max} = 25.3 / 0.9 = 28.1$ cm

3. $Q_{max} = 0.10 \, r_{max} (f)^{\frac{1}{2}} = 6286!$

4. $N^2 = 71.2 \times 10 / 28.1 = 25.3$

   $N = 5.0$

   $c = 2 \times 28.1 / (5 - 1) = 14.1$ cm

   $d = 0.55 \, c = 7.73$ cm!

If the coil size is limited to 3 cm × 3 cm × 3 cm, the maximum realizable Q will be 335.

3.3.7 The Design of Inductors with Magnetic Cores

Smaller inductors with less parasitic capacitance can be designed by using magnetic materials.

The core can be a rod, a toroid, a balun, or stacked toroids (see Figure 3.9).

Rods are often used if the inductor is to be tuned, while toroids and baluns are used for fixed-value inductors. Stacked cores can be used as an alternative to a balun.

The type of material used is a function of the frequency range over which the inductor is to be used, the desired unloaded Q, the available space, and the temperature range.

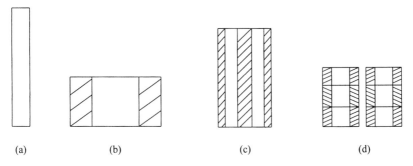

Figure 3.9 Different types of magnetic cores: (a) rod core, (b) toroid, (c) balun core, and (d) stacked toroids.

The unloaded Q is determined by the flux density in the core and, therefore, by the maximum voltage across the inductor, the number of turns, and the frequency.

Materials with a high relative permeability are usually very sensitive to changes in temperature.

With the material and type of core selected, the size of the core must be determined. The core must be large enough for the flux density to be sufficiently low to ensure that the desired unloaded Q is realized and that the required number of turns can be accommodated.

The selection of the minimum core size for toroidal (single and stacked) inductors will be discussed in the next two sections. If a balun is to be used, the results for stacked cores can be applied to get an idea of the size required.

3.3.7.1　The Design of an Inductor with a Single Toroidal Core

The inductance of a toroidal core inductor is given by

$$L = \partial\phi / \partial i = \mu_0 \mu_r N^2 A / l \tag{3.36}$$

The value of the product $\mu_r R_p / L$ can be determined from the unloaded Q (Q_u) by using the following equation:

$$\mu_r R_p / L = \mu_r \omega R_p / (\omega L) = 2\pi \mu_r f Q_u \tag{3.37}$$

The flux density corresponding to this ratio (B_{maxA}) can be determined, where given, from the manufacturer's specifications (see Figure 3.7 for an example).

The flux density in the core is given by

$$B_{max} = V_{max} / (\omega A N) \tag{3.38}$$

The flux density in the core must be less than or equal to the maximum allowable value B_{maxA}.

If the number of turns in (3.36) is replaced by using (3.38), the product of the cross-section area of the core (A) and the mean path length (l) is found to be

$$Al = [\mu_r \mu_0 / (\omega B_{maxA}^2)] \, V_{max}^2 / (\omega L) \tag{3.39}$$

The required core size can now be found by comparing this product with that of available cores.

If a core with the required Al-product is not available, a core with a larger Al-product can be chosen. The number of turns required must then be calculated by using (3.36). The alternative is to use more than one core (smaller) to obtain the required Al-product.

With the core dimensions known, the number of the turns required can be found by using (3.38).

The following procedure can be followed to design an inductor with a toroidal core.

A Design Procedure for an Inductor with a Toroidal Core

1. Select a suitable material. Take the frequency range, temperature range, required unloaded Q, and inductor size into account.

2. Calculate the $\mu_r R_p / L$ ratio at the lowest frequency by using (3.37):

   $$\mu_r R_p / L = 2\pi \mu_r f \, Q_u$$

   where Q_u is the desired value of the unloaded Q.

3. Find the flux density corresponding to the calculated $\mu_r R_p / L$ ratio from the manufacturer's specifications.

4. Calculate the required Al-product by using (3.39):

   $$Al = \mu_r \mu_0 / (\omega B_{max}^2 A) V_{max}^2 / (\omega L)$$

5. Compare this product to that of available cores. Select a core with an Al-product equal or close to it. If the difference in Al-product is significant, choose the core with an Al-product greater than that required.

   Alternatively, smaller cores can be combined to obtain the required Al-product (see Section 3.3.7.2).

6. Calculate the required number of turns by using (3.36):

   $$L = \mu_r \mu_0 N^2 A / l$$

7. Check if there is enough space to accommodate the required number of turns of the conductor with the required thickness. If the core is too small, a larger toroid must be used.

Table 3.8 A List of Typical Magnetic Core Sizes

Core	A (μm^2)	l (mm)	Al (μm^3)	Size (mm^3)
1	12.5	36	0.44	14×9×5
2	31.5	57	1.80	23×14×7
3	37.5	75	2.81	29×19×7.5
4	65.0	92	5.98	36×23×10
5	97.5	92	8.97	36×23×15

EXAMPLE 3.3 Finding the core size required for an inductor.

As an example of the application of the procedure outlined here, the core size for a magnetic-cored inductor with 31.4-Ω reactance at 2 MHz, and loss resistance equal to 392 Ω, will be determined. The maximum rms voltage across the inductor will be 20 V and 4C4 material is available. Note that $\mu_r = 120$.

The $\mu_r R_p / L$ ratio for the inductor is

$$\frac{\mu_r R_p}{L} = \frac{120 \times 392}{31.4/(2\pi \times 2 \times 10^6)} = 1.88 \times 10^{10} \text{ s}^{-1}$$

By using the 1.6-MHz curve given for 4C4 material in Figure 3.7, the ($B_{max} f$) product corresponding to a $\mu_r R_p /L$ ratio of 1.8×10^{10} s^{-1} is found to be 2×10^4 THz. The maximum allowable flux density in the core is, therefore, 0.01 T.

The Al-product of the required core can be found by using (3.39). The required Al-product is 1.53×10^{-6} m^3.

By comparing this value to those in the list of some Al-products given in Table 3.8, it can be seen that the core with Al-product equal to 1.8 μm^3 ($A = 31.5 \mu m^3$; $l = 57$mm; 23×14×7 mm^3) can be used.

The number of turns required is

$$N = \sqrt{Ll/(\mu_0 \mu_r A)} = 5.5$$

The selected core can accommodate the required number of turns with ease.

3.3.7.2 The Design of an Inductor with a Stacked Toroidal Core

The design of an inductor with a stacked toroidal core is similar to that of an inductor with a single core, except for the fact that the cross-sectional area (A) used in the previous section must now be taken as $N_c A$, where N_c is the number of toroids used (an even number) and A is the cross-sectional area of a single toroid. The mean path length is that of a single toroid.

The inductance of a stacked core inductor is given by the equation

$$L = \mu_r \mu_0 N^2 (N_c A / l) \tag{3.40}$$

The maximum flux density is

$$B_{max} = V_{max} / [\omega (N_c A) N] \tag{3.41}$$

and the required Al-product is obtained from

$$N_c Al = [\mu_r \mu_0 / (\omega B^2_{maxA})] V^2_{max} / (\omega L) \tag{3.42}$$

3.4   TRANSMISSION LINES

The transmission lines used at radio frequencies are usually coaxial cables (flexible (F) or semirigid (SR)), microstrip lines, or twisted pairs. The important characteristics of these lines are the characteristic impedance, the insertion loss, and the power-handling capability.

The characteristics of coaxial cables, microstrip lines, and twisted pairs will be discussed briefly.

3.4.1   Coaxial Cables

The characteristic impedance of a coaxial cable is given by

$$Z_0 = (138 / \sqrt{\varepsilon_r}) \log_{10} (b / a) \tag{3.43}$$

where a is the outer diameter of the inner conductor (centimeter) and b is the inner diameter of the outer conductor (in centimeter).

The attenuation of the cable is given by [11]

$$\alpha = (3.615 / Z_0) (K_1 / a + K_2 / b) \ T \sqrt{f} + 9.121 f \sqrt{\varepsilon_r} \ \tan \delta \tag{3.44}$$

where
       α is the attenuation in decibels/ 100m;
       K_1 is the square root of the ratio of the resistivity of the inner conductor to that of copper;
       K_2 is the square root of the ratio of the resistivity of the outer conductor to that of copper;
       f is the operating frequency in megahertz;

$T = [1 + 0.0039 \, (t - 20)]^{1/2}$, where t is the operating temperature in degrees Celsius;
a is the inner conductor outer diameter in centimeter;
b is the outer conductor inner diameter in centimeter;
$\tan \delta$ is the loss tangent of the inner conductor insulation.

The attenuation is increasing with frequency because of the skin effect and the losses in the dielectric material.

The power-handling capability of a coaxial cable is limited by the maximum allowable temperature. This is a function of the insulation used (200°C for polytetrafluoroethylene), the diameter of the cable, and the environmental temperature.

The power-handling capability and attenuation along some coaxial cables are given in Table 3.9 [12]. Note the decrease in power-handling capability with increasing frequency.

Semirigid coaxial cable is often used for transmission-line transformers in the VHF and UHF ranges.

Coaxial lines with characteristic impedances of 50Ω and 25Ω are freely available. Lower impedances can be obtained by connecting cables in parallel, while higher impedances can be obtained by connecting lines with lower impedances in series (using semirigid cable). By doing the latter, the effective capacitance is decreased. The series connection is shown in Figure 3.10.

Table 3.9   The Attenuation and Average Power-Handling Capabilities of Some Coaxial Cables at Different Frequencies

Cable Type	α (dB/m) (P (W))			
	@ 1 MHz	@ 10 MHz	@ 100 MHz	@ 500 MHz
50Ω; 1.7 mm (F)	0.04 (1k)	0.14 (300)	0.44 (90)	—
50Ω; 2.8 mm (F)	0.03 (1k)	0.08 (800)	0.27 (250)	—
50Ω; 1.1 mm (SR)	—	—	0.35 (68)	0.75 (32)
50Ω; 2.2 mm (SR)	—	—	0.18 (330)	0.43 (140)
50Ω; 6.4 mm (SR)	—	—	0.11 (1.17k)	0.25 (515)

Note: F is flexible cable; SR is semi-rigid cable.

3.4.2   Microstrip Transmission Lines

The characteristic impedance $[Z_0(f)]$ and the effective relative dielectric constant $[\varepsilon_{r\text{-eff}}(f)]$ of a microstrip line is a function of the width-to-height ratio (W/h), the conductor thickness (t), cover height (H_2), and frequency (f). The characteristic impedance is also a function of the effective dielectric constant.

The characteristic impedance $[Z_0(f)]$ and effective relative dielectric constant $[\varepsilon_{r\text{-eff}}(f)]$ can be computed by using the following set of equations [13, 14]:

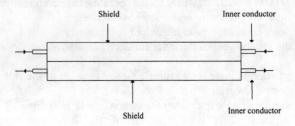

Figure 3.10   Increasing the characteristic impedance of a coaxial cable by connecting two cables in series.

$$W_{eff} = W + \frac{t}{\pi}\left\{1 + \ln 4 - 0.5\ln\left[\left(\frac{t}{h}\right)^2 + \left(\frac{t}{\pi W}\right)^2\right]\right\} \tag{3.45}$$

$$f(W/h) = 6 + (2\pi - 6)\,\mathrm{EXP}\left[-\left(\frac{30.666}{W/h}\right)^{0.7528}\right] \tag{3.46}$$

$$Z_{0a\infty} = 60\ln\left[\frac{f(W/h)}{W/h} + \left[1 + \left(\frac{2h}{W}\right)^2\right]^{1/2}\right] \tag{3.47}$$

$$P = 270\left\{1 - \tanh\left[1.192 + 0.706(1 + H_2/h)^{1/2} - \frac{1.389}{1 + H_2/h}\right]\right\} \tag{3.48}$$

$$Q = 1.0109 - \tanh^{-1}\{[0.012\,W/h + 0.177(W/h)^2 - 0.027(W/h)^3]$$
$$/[1 + H_2/h]^2\} \tag{3.49}$$

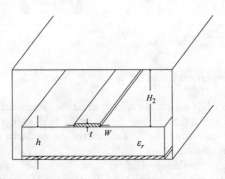

Figure 3.11   The geometry of a microstrip line.

$$Z_{0a} = Z_{0a\infty} - PQ \tag{3.50}$$

$$b = -0.564 \left(\frac{\varepsilon_r - 0.9}{\varepsilon_r + 3.0} \right)^{0.053} \tag{3.51}$$

$$a = 1 + (1/49) \ln \{(W/h)^2 [(W/h)^2 + 1/52^2] / [(W/h)^4 + 0.432]\}$$
$$+ (1/18.7) \ln \{1 + [W/(18.1h)]^3\} \tag{3.52}$$

$$j = ab \tag{3.53}$$

$$q = \{[1 + 10h/W]^j - 2[(\ln 2)/\pi] (t/h)/(W/h)^{1/2}\}$$
$$\tanh[1.043 + 0.121(H_2/h) - 1.164/(H_2/h)] \tag{3.54}$$

$$\varepsilon_{\text{eff}} = \frac{\varepsilon_r + 1}{2} + q \frac{\varepsilon_r - 1}{2} \tag{3.55}$$

$$Z_0 = Z_{0a} / \sqrt{\varepsilon_{\text{eff}}} \tag{3.56}$$

$$v_p = c / \sqrt{\varepsilon_{\text{eff}}} \tag{3.57}$$

where v_p is the phase velocity in the microstrip,

$$f_p = Z_0 / [2\mu_0 h] \tag{3.58}$$

$$G = (\pi^2 / 12)[(\varepsilon_r - 1) / \varepsilon_{\text{eff}}](Z_0 / 60)^{1/2} \tag{3.59}$$

$$\varepsilon_{r-\text{eff}}(f) = \varepsilon_r - \frac{\varepsilon_r - \varepsilon_{\text{eff}}}{1 + G(f/f_p)^2} \tag{3.60}$$

$$s = \frac{c^2}{4f^2[\varepsilon_{r-\text{eff}}(f) - 1]} \tag{3.61}$$

$$y = s/3 - (W/3)^2 \tag{3.62}$$

$$W_{\text{eff}}(0) = 120\pi h / [Z_0 \sqrt{\varepsilon_{\text{eff}}}] \tag{3.63}$$

$$P = (W/3)^3 + (s/2)[W_{\text{eff}}(0) - W/3] \tag{3.64}$$

$$r = (p^2 + y^3)^{1/2} \tag{3.65}$$

$$W_{\text{eff}}(f) = W/3 + [r+p]^{1/3} - [r-p]^{1/3} \tag{3.66}$$

$$Z_0(f) = \frac{120\pi h}{W_{\text{eff}}(f)\sqrt{\varepsilon_{r-\text{eff}}(f)}} \tag{3.67}$$

The frequency dependence (dispersion) of the characteristic impedance and the effective dielectric constant of a microstrip line result from the non-TEM nature (inhomogeneity) of the mode of propagation along the microstrip.

As an example of the application of (3.45) to (3.67), the width-to-height ratios and the effective dielectric constants of a 50Ω line on an alumina ($\varepsilon_r = 10.2$) and a Teflon ($\varepsilon_r = 2.5$) substrate at 2 GHz with $H_2/h = 20.0$ were calculated. The results, respectively, are as follows:

$W/h = 0.85$ with $\varepsilon_{r,\text{eff}} = 6.6945$

and

$W/h = 2.75$ with $\varepsilon_{r,\text{eff}} = 2.0775$

At microwave frequencies it also becomes necessary to take into account the losses (conductor and dielectric) in microstrip lines. The main source of these losses is usually conductor loss. The conductor loss attenuation constant α_c is given by the following set of equations [15, 16]:

$$\alpha_c = \frac{8.68 R_s M}{2\pi Z_0 h}[1 + \frac{h}{W_{\text{eff}}} + \frac{h}{\pi W_{\text{eff}}}(\ln\frac{4\pi W}{t} + \frac{t}{W})], \quad (W/h) < 1/(2\pi)$$

$$\alpha_c = \frac{8.68 R_s MN}{2\pi Z_0 h}, \quad 1/(2\pi) < (W/h) < 2$$

$$\alpha_c = \frac{8.68 R_s N}{Z_0 h}\left\{\frac{W_{\text{eff}}}{h} + \frac{2}{\pi}\ln\left[2\pi\,\text{EXP}\left(\frac{W_{\text{eff}}}{2h} + 0.94\right)\right]\right\}^{-2} \cdot$$

$$\left[\frac{W_{\text{eff}}}{h} + \frac{W_{\text{eff}}/(\pi h)}{W_{\text{eff}}/(2h) + 0.94}\right], \quad (W/h) > 2 \tag{3.68}$$

with α_c in decibels per centimeter.

$$M = 1 - \left[\frac{W_{\text{eff}}}{4h}\right]^2 \tag{3.69}$$

$$N = 1 + h / W_{\text{eff}} + \frac{h}{\pi W_{\text{eff}}} \left[\ln\frac{2h}{t} - t / h\right] \tag{3.70}$$

$$R_s = \sqrt{\pi f \mu_0 / \sigma} \tag{3.71}$$

where σ is the conductivity of the strip conductor.

At high frequencies, the copper losses are higher than those predicted by the equations above. This is due to the coarse interface between the dielectric material and the conductor. These losses are included in the dielectric losses by some manufacturers.

With the loss tangent ($\tan \delta$) known, the attenuation constant corresponding to the dielectric losses in a microstrip line can be calculated by using the following equation [15, 17]:

$$\alpha_d = 27.3\frac{\varepsilon_r[\varepsilon_{r\text{-eff}} - 1]\tan\delta}{\lambda_0\sqrt{\varepsilon_{r\text{-eff}}}[\varepsilon_r - 1]} \quad (\text{dB / cm}) \tag{3.72}$$

where λ_0 is the operating wavelength.

Materials with dissipation factors of 0.00085 at 1 MHz and 0.0018 at 10 GHz are available. With such low values for the dissipative factor, the dielectric losses are usually small compared to the conductor losses. Silicon is an example of a material where the dielectric losses cannot be neglected.

As an example of the dissipative losses in a microstrip line, the insertion loss of an 8-in 50Ω line on an Epsilam-10 substrate is specified by the manufacturer to be approximately 0.1 dB at 100 MHz and 0.21 dB at 500 MHz (0.19 dB/wavelength).

The power-handling capability of a microstrip line is a function of the insertion loss, the breakdown voltage of the dielectric material, and the maximum allowable temperature of the line. If the thermal resistance of the substrate is known as a function of the line width, the maximum power-handling capability can be computed easily.

3.4.3 Twisted Pairs

Transmission lines with a wide range of characteristic impedances can be realized by twisting lengths of wire together.

The characteristic impedances of these twisted-pair lines decrease when thicker wire is used. For example, the characteristic impedance is 35Ω when No. 20 (AWG) enamel-insulated wire with three twists per centimeter is used and is 120Ω when No. 30 vinyl-coated wire (0.05 cm outside diameter) with 3.6 twists per centimeter is used [18].

Increasing the number of twists per centimeter also decreases the characteristic impedance of these transmission lines. For example, the characteristic impedance obtained by twisting two No. 20 enamel-insulated wires together decreases from approximately 42Ω to 30Ω when the number of twists is doubled from two to four [18].

A line with 50-Ω characteristic impedance can be obtained by twisting two No. 22 enamel-insulated wires together to have 2.5 twists per centimeter.

Characteristic impedances lower than 10Ω are often required in the HF range. These impedances can be realized by twisting together many wires (using two-wire lines) with smaller diameters.

It is difficult to calculate the losses in these transmission lines because the dielectric losses, skin effect, proximity effect, and the fact that the current is flowing in both directions along the line must be taken into account. It is, therefore, easier to determine the attenuation of these lines practically.

The losses in twisted-wire transmission lines are usually not a problem below 100 MHz.

QUESTIONS AND PROBLEMS

3.1   At high frequencies the reactance of a capacitor is lower than expected. What is the reason for this?

3.2   At what frequency would you expect a 10-nF silver-mica capacitor (with very short lead lengths) to resonate?

3.3   The resonator frequency of a 1-nF disk ceramic capacitor is 60 MHz. What is its effective capacitance at 40 MHz?

3.4   Is there any parasitic inductance associated with a chip capacitor?

3.5   (a) What is the definition of the Q-factor of a component?
        (b) What is the definition of the dissipation factor?
        (c) How are these two factors related?

3.6   Is the phenomenon of thermal runaway possible in a capacitor?

3.7   What would you expect the Q-factor of a good capacitor to be at

        (a) 1 MHz;
        (b) 100 MHz;
        (c) 500 MHz?

3.8   Is it true that the parasitic capacitance of an inductor actually increases its losses? Find the ratio of the operating and resonant frequencies for which the Q-factor of an inductor will be 10% lower than expected.

3.9   Two single-layer solenoidal coils have exactly the same dimensions, but one has more turns than the other. Which coil has more capacitance? Which coil has the lowest resonant frequency?

3.10 At what frequency would you expect a 1-μH inductor to resonate?

3.11 What is the skin depth for copper at 100 MHz, at 10 kHz?

3.12 Why is the resistance of a solenoidal coil higher at high frequencies than at low frequencies (three reasons)?

3.13 The losses in magnetic materials can be specified by plotting the ratio $\mu_r R_p / L$ as a function of the product $B_{max} f$ as a function of frequency. Show that the ratio $\mu_r R_p / L$ is independent of the core dimensions and that the product $B_{max} f$ is independent of the frequency, if the voltage across the inductor remains constant.

3.14 Suppose you need a 1-μH inductor with low losses. You are given a choice between two toroidal-cored inductors, one with a small core and the other with a larger one. Exactly the same material is used for both cores. Which inductor would you choose?

3.15 Is it true that larger air-cored inductors have higher Q?

3.16 Design a single-layer air-cored inductor to have 10-μH inductance with an unloaded Q of 200 at 10 MHz. The resonant frequency must be higher than 30 MHz.

3.17 Determine the highest Q attainable by using a single-layer solenoidal coil to obtain 500-nH inductance at 50 MHz, if the resonant frequency is to be above 200 MHz and the coil size is to be smaller than 3cm ×3cm ×3cm.

3.18 Design an inductor to have 1-μH inductance. The unloaded Q must be higher than 40 at 1.6 MHz. A toroidal core of 4C4 material must be used. The maximum voltage across the inductor will be 20V. Assume the relative permeability to be equal to 100.

The available cores have the following dimensions:

Core 1: A=12.5 μm³; *l*=36 mm; *Al*=0.44 μm³ (14 ×9 ×5 mm³)
Core 2: A=31.5 μm³; *l*=57 mm; *Al*=1.80 μm³ (23 ×14 ×7 mm³)
Core 3: A=37.5 μm³; *l*=75 mm; *Al*=2.81 μm³ (29 ×19 ×7.5 mm³)
Core 4: A=65.0 μm³; *l*=92 mm; *Al*=5.98 μm³ (36 ×23 ×10 mm³)
Core 5: A=97.5 μm³; *l*=92 mm; *Al*=8.97 μm³ (36 ×23 ×15 mm³)

3.19 If a core consisting of four stacked toroids is to be used, design an inductor to meet the same specifications as the inductor in the previous problem.

3.20 How much power can a 2.8-mm flexible coaxial cable handle at 100 MHz?

3.21   If the input power to the cable in the previous question is 100W, what will the output power be if the length of the cable is 10m?

3.22   Show the series connection for two transmission lines.

REFERENCES

[1]   Krauss, H. L., W. B. Bostian, and F. H. Raab, *Solid State Radio Engineering*, New York: John Wiley and Sons, 1980.

[2]   *The RF Capacitor Handbook*, American Technical Ceramics, 1979.

[3]   "RF & Microwave Porcelain Capacitors," Cazenovia, NY: Dielectric Laboratories, Inc., 1998.

[4]   "Di-Cap Microwave Ceramic Capacitors," Cazenovia, NY: Dielectric Laboratories, Inc., 1998.

[5]   Hardy, K. H., *High Frequency Circuit Design*, Reston, VA: Preston Publishing Company, 1979.

[6]   Medhurst, R. G., "High Frequency Resistance and Capacity of Single-Layer Solenoids," *Wireless Engineer,* March 1947, p. 35.

[7]   Hilbers, A. H., "On the Design of H. F. Wideband Power Transformers (ECO 6907)," Eindhoven: Philips C.A.B. Group, 1969.

[8]   "High-Performance Chip Coils for the Wireless Communication Industry," Franklin Park, IL: Stetco, Inc., 1998.

[9]   Kuhn, W. B., and N. M. Ibrahim, "Analysis of Current Crowding Effects in Multiturn Spiral Inductors," *IEEE Trans. Microwave Theory Tech.*, Vol. MTT-49, January 2001.

[10] Skilling, H. H., *Fundamentals of Electric Waves*, New York: John Wiley and Sons, 1948.

[11] Coaxitube Semi-Rigid Coaxial Cable, North Wales, UK: Precision Tube Company, Inc., (n.d.).

[12] Welsby, V. G., *The Theory and Design of Inductance Coils*, London: MacDonald and Co. Ltd., 1960.

[13] March, S., "Microstrip Packaging: Watch The Last Step," *Microwaves*, December 1981.

[14] Pues, H. F., and A. R. van de Capelle, "Approximate Formulas for Frequency Dependence of Microstrip Parameters," *Electron. Letters*, Vol. 16, November 6, 1980, pp. 870–872.

[15] Bahl, I. J., and D. K. Trevedi, "A Designer's Guide to Microstrip Line," *Microwaves*, May 1977.

[16] Pucel, R. A., D. J. Masse, and C. P. Hartwig, "Losses in Microstrip," *IEEE Trans. Microwave Theory Tech.*, Vol. MTT-16, June 1968, pp. 342–350; "Correction to Losses in Microstrip," *Ibid.*, (Corresp.), Vol. MTT-16, December 1968, p. 1064.

[17] Welsh, J. D., and H. J. Pratt, "Losses in Microstrip Transmission Systems for Integrated Microwave Circuits," *NEREM Rec.*, Vol. 8, 1966, pp. 100–101.

[18] Krauss, H. L., and C. W. Allen, "Designing Toroidal Transformers to Optimize Wideband Performance," *Electronics*, August 16, 1973.

SELECTED BIBLIOGRAPHY

Snelling, E. C., *Soft Ferrites: Properties and Applications*, London: Iliffe Books Ltd., 1969.

Howe, H., *Stripline Circuit Design*, Dedham, MA: Artech House, 1974.

Gupta, K. C., et al., *Microstrip Lines and Slotlines*, Second Edition, Norwood, MA: Artech House, 1996.

CHAPTER 4

NARROWBAND IMPEDANCE-MATCHING WITH LC NETWORKS

4.1   INTRODUCTION

Impedance-matching networks are used for transforming impedances to certain required values, which may or may not be the conjugate of the source or the load impedance. When a source is conjugately matched to a load (i.e., $Z_L = Z_s^*$), maximum power is transferred between them. This is important when the power gain of a transistor is low, as is the case with most transistors at higher frequencies.

Apart from matching, impedance-matching networks are also often used to control the gain, the noise figure, the linearity and/or the output power of the different stages in an amplifier. When a matching network is designed for maximum power transfer, the terminations are usually known. The considerations establishing the terminations to be used when the active performance of a transistor is controlled will be covered in Chapter 9.

Independent of how the terminations are established, the design procedure for the matching network remains the same. The design of narrowband impedance-matching networks, mostly for maximum power transfer, will be considered in this chapter.

Narrowband impedance matching is done with two or more components. Where two components are used to bring about an impedance transformation, the matching network is called an L-section. Three-element matching networks are usually T- or PI-sections. The names are descriptive of the configuration formed by the reactive elements.

The design of L-, T-, and PI-sections will be discussed in this chapter. Transformation of real and reactive loads will be considered.

When T- and PI-sections are used, it is possible to bring about the required transformation and to control the bandwidth of the network. Although the 3-dB bandwidth of an L-section can be determined easily, it is not a design parameter.

It is sometimes necessary to know the bandwidth resulting from a transforming section more accurately than is possible with the approximation method that is usually used. In these cases, as well as in instances where a bandwidth other than the 3-dB bandwidth is of interest, the procedure outlined in Section 4.9 can be used.

It was shown in Chapter 3 that lossless reactive components do not exist. For this reason, all impedance-matching networks will have some insertion loss. These losses can be quite pronounced when the bandwidth of a circuit is very narrow. A simple procedure for calculating the insertion loss caused by a cascaded LC network will be outlined in Section 4.8.

Apart from matching and transforming impedances, impedance-matching networks are sometimes also used to reject unwanted signals outside the passband. (This practice is not recommended when wideband impedance-matching networks are designed.) The rejection required can often be obtained by using impedance-matching networks with high Q-factors, that is, if the rejection required is not too great.

The rejection obtainable by using parallel and series resonant circuits will be considered in Section 4.2.

When the required rejection becomes very high, the Q of the components, their temperature stability, and any tuning required can become a problem. If the associated insertion loss can be tolerated and the filtering occurs at low power levels, the required rejection can often be obtained by using surface acoustic wave (SAW) devices, ceramic filters, or crystal filters. These components are very stable and can provide extremely sharp rejection. Because of the impedances presented by these devices, some (low Q) impedance matching is usually also required.

4.2   PARALLEL RESONANCE

A parallel resonant circuit is shown in Figure 4.1. Although it is not an impedance-matching network, it is of interest here because of its frequency response.

The frequency response of this circuit is determined by the zero at the origin, the zero at infinity, and the two poles. That is,

$$V_o(s) = Z(s) I$$

$$= I / [1 / R_L + sC + 1 / (sL)]$$

$$= \frac{sL \, I}{s^2 LC + sL / R_L + 1} \tag{4.1}$$

From Figure 4.1 it is obvious that the highest possible output voltage will occur where

$$\omega L = 1 / (\omega C)$$

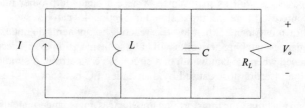

Figure 4.1   A parallel resonant circuit.

that is, when

$$\omega_0 = 1/\sqrt{LC} \tag{4.2}$$

The 3-dB frequencies of the circuit can be determined by using (4.1) and (4.2). These frequencies occur where

$$|1/R_L + j\omega C + 1/(j\omega L)| = \sqrt{2}\,|1/R_L + j\omega_0 C + 1/(j\omega_0 L)|$$

$$= \sqrt{2}/R_L \tag{4.3}$$

After some manipulation the solutions of this equation are found to be

$$\omega_{3dB} = \omega_0\sqrt{1 + 1/(4Q^2)} \pm 1/(2RC) \tag{4.4}$$

Therefore, the bandwidth of the circuit is

$$B = \omega_{3dB\_2} - \omega_{3dB\_1} = 1/(RC) \quad (\text{rad}/s) \tag{4.5}$$

It can be seen from (4.4) that the circuit response is not symmetrical around the resonant frequency ω_0. It can be proved easily, however, that the resonant frequency is the geometric mean of the two cut-off frequencies by multiplying the two solutions given by (4.4), that is,

$$\omega_0 = \sqrt{\omega_{3dB\_1} \cdot \omega_{3dB\_2}} \tag{4.6}$$

The Q-factor of the circuit is defined as the ratio of the center frequency to the bandwidth; that is,

$$Q = \omega_0/B \tag{4.7}$$

$$= \omega_0 CR \tag{4.8}$$

$$= R/(\omega_0 L) \tag{4.9}$$

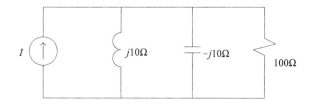

Figure 4.2   A parallel resonant circuit with $Q = 10$.

A high Q, therefore, implies a very small relative bandwidth and, in the case of parallel resonance, reactances with low impedance compared to that of the load resistance. The reactances are shown in Figure 4.2 for a Q of 10.

When the Q of the circuit is high, the arithmetic and the geometric mean of the cut-off frequencies are approximately the same [see (4.4)].

Extremely sharp rejection can be obtained by using a parallel or series resonant circuit. When the components can be considered to be ideal, it can be shown that the ratio of the power transmitted to the load at resonance ($P_{o\text{-max}}$) and that at any other frequency [$P_o(f)$] can be calculated by using the following equation:

$$\frac{P_{o\text{-max}}}{P_o(f)} = (1 - 2Q^2) + Q^2[(\omega / \omega_0)^2 + (\omega_0 / \omega)^2]$$

(4.10)

As an illustration of the rejection characteristics of the circuit, the attenuation is given in Table 4.1 as a function of the normalized frequency (f/f_0) for different values of the circuit Q. Only the frequencies above resonance are considered, because the response is close to symmetrical when the Q is high. Where the response curve levels off to a single-pole response, no more entries were made into the table.

In order to appreciate the rate of rejection a -30-dB quality factory (Q_{-30}) is defined here as

$$Q_{-30} = f_0 / B_{-30}$$

(4.11)

where B_{-30} is the -30-dB "bandwidth" of the circuit (in hertz).

The -30-dB Q-factors for the three Q-factors used in Table 4.1 are 0.315 ($Q = 10$), 3.15 ($Q = 100$), and 7.90 ($Q = 250$), respectively.

It follows by observation of the results obtained that the -30-dB Q-factor of a resonant circuit is related to the 3-dB Q-factor in a simple way when the 3-dB Q-factor is greater than 10:

$$Q_{-30} \cong 0.0315Q$$

(4.12)

The normalized -30-dB bandwidth of the circuit is therefore given to good approximation by the following equation:

$$B_{-30} = 31.75 / Q$$

(4.13)

It can be shown that the two normalized -30-dB rejection frequencies are given to good approximation by the following equation:

$$f_{-30} = \pm 15.875 / Q + \sqrt{\left(\frac{15.875}{Q}\right)^2 + 1}$$

(4.14)

By using (4.13), the Q-factor required for a specified -30-dB bandwidth can be calculated easily.

Table 4.1　　The Frequencies at Which the Output Signal of an Ideal Parallel or Series Resonant Circuit Is Attenuated as Listed for Some Values of the Circuit Quality Factor

Attenuation (dB)	Normalized frequencies (f/f_0)		
	$Q = 10$	$Q = 100$	$Q = 250$
0	1.0000	1.000	1.000
−3	1.0512	1.005	1.002
−10	1.1615	1.015	1.006
−20	1.615	1.051	1.020
−30	3.46	1.171	1.065
−40	—	1.620	1.22
−50	—	3.46	1.82
−60	—	4.24	—
−70	—	—	—

EXAMPLE 4.1　　Establishing the Q-factor required for −30-dB rejection at two specified frequencies.

As an example of the application of (4.13), the Q-factor necessary to provide −30-dB rejection at 40 and 60 MHz with a parallel resonant circuit will be determined.

The resonant frequency of the circuit is

$$f_0 = \sqrt{40 \times 60} = 48.99 \, \text{MHz}$$

The normalized −30-dB bandwidth is

$$B_{-30} = (60 - 40)/48.99 = 0.4082$$

The required Q is obtained by using (4.13):

$$Q = 31.75/B_{-30} = 77.76 \tag{4.15}$$

Up to this point the losses in the components of the parallel resonant circuit have been ignored. When the required Q of the circuit becomes of the same order as the unloaded Qs of the components used, this cannot be done.

When the losses are taken into account, the effective load resistance (R_T) at the resonant frequency is then given by

$$1/R_T = 1/R_L + 1/(Q_L X_L) + 1/(Q_C X_C) \tag{4.16}$$

where X_L and X_C are the reactance of the inductor and capacitor, respectively, at the resonant frequency (see Figure 4.3). Q_L and Q_C are the unloaded Q-factors of the inductor and capacitor, respectively.

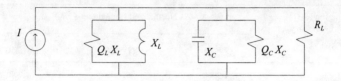

Figure 4.3   A parallel resonant circuit with lossy components.

At resonance the capacitive and inductive reactances are equal, and (4.16) can be simplified to

$$X_L / R_T = X_L / R_L + [1 / Q_L + 1 / Q_C] \tag{4.17}$$

The last term in this equation is defined as the unloaded Q (Q_u) of the circuit:

$$1 / Q_u = 1 / Q_L + 1 / Q_C \tag{4.18}$$

The effective Q of the circuit (Q_{eff}) is therefore given by

$$1 / Q_{\text{eff}} = 1 / Q_I + 1 / Q_u \tag{4.19}$$

where Q_I is the Q when the components are assumed to be lossless.

The highest Q obtainable with a parallel resonant circuit is limited by component losses and the temperature stability of the components.

4.3   SERIES RESONANCE

The results obtained for a parallel resonant circuit can be applied directly to a series resonant circuit by using the principle of dualism.

According to this principle, for every circuit there is another circuit for which whatever applies to the current of one circuit, also applies to the voltage of the other circuit, and vice versa.

This equivalent can be obtained by following the procedure illustrated in Figure 4.4. A node is placed in every loop of the first circuit, as well as in the space outside it. These nodes are then connected by passing from one loop to another through the components of the different loops. Inductors are replaced with capacitors, capacitors with inductors, resistors with conductors, and conductors with resistors. The values assigned to the new components (H, F, Ω, S) are numerically equal to those of the original components.

The output voltage of the parallel resonant circuit in Figure 4.4 is given by the following equation:

$$V_o = I / [1/R + sC + 1/(sL)] = 0.2 / [0.5 + 3s + 1/(5s)] \tag{4.20}$$

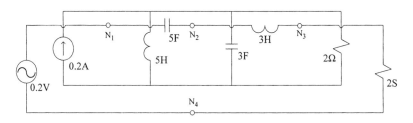

Figure 4.4 The principle of dualism applied to a parallel resonant circuit.

The output of the series resonant circuit is obtained by replacing V_o with I_o, I with E, R with G, C with L, and L with C.

Thus,

$$I_o = E / [1/G + sL + 1/(sC)] = 0.2 / [0.5 + 3s + 1/(5s)] \tag{4.21}$$

It follows from Figure 4.4 that the output current of the series resonant circuit is indeed given by this equation.

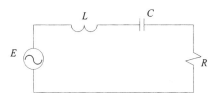

Figure 4.5 The series resonant circuit.

By applying the principle of dualism to the results deduced in the previous section, the following equations are found to apply to the series resonant circuit of Figure 4.5:

$$Q = \omega_0 L / R = 1 / (\omega_0 CR) \tag{4.22}$$

$$\omega_{3dB} = \omega_0 \sqrt{1 + 1/(4Q^2)} \pm R/(2L) \tag{4.23}$$

$$B = R / L \text{ (rad / s)} \tag{4.24}$$

$$R_T = R_L + X_L / Q_L + X_C / Q_C \tag{4.25}$$

$$1 / Q_{eff} = R_L / X_L + (1 / Q_L + 1 / Q_C) \tag{4.26}$$

It follows from (4.26) that similar to the parallel resonant circuit, the unloaded Q for the series resonant circuit is given by

$$1 / Q_u = 1 / Q_L + 1 / Q_C \tag{4.27}$$

The reactance for a series resonant circuit with $Q = 10$ are shown in Figure 4.6 at the resonant frequency. Note that the reactance values are high compared to the load resistance.

With the same loaded Q, the frequency response of the series resonant circuit is identical to that of the parallel resonant circuit.

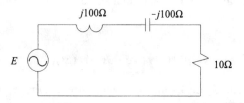

Figure 4.6   A series resonant circuit with $Q = 10$.

4.4   L-SECTIONS

An L-section is a two-element matching network. The four possible configurations are shown in Figure 4.7.

Depending on the position of the first component (as viewed from the load), the load resistance can be transformed upwards or downwards with an L-section.

When the first reactive component is a series component, the transformation is upward; and when it is a parallel element, the transformation is downward.

The second element in the L-section is used to remove the residual reactance caused by the transformation element (i.e., the first element). This second element is therefore the compensating element.

The basic principle used in narrowband impedance matching is that the resistance of a complex load is not the same when viewed in impedance or admittance form. This is illustrated in Figure 4.8.

When a reactive element (X_1) is added in series with a resistor (R) and the equivalent parallel combination is considered (series to shunt transformation), the resistance increases with a factor

$$D_1 = 1 + Q_1^2 \tag{4.28}$$

where

$$Q_1 = X_1 / R \tag{4.29}$$

When a reactive element (X_1) is added in parallel with a resistor (R) and the equivalent series combination is considered (parallel to series transformation), the resistance decreases with the same factor (D_1). In this case, however, the Q-factor in (4.28) is defined by

$$Q_1 = -R / X_1 = B_1 / G \tag{4.30}$$

The ratios defined in (4.29) and (4.30) are similar in form to the Q-factors of the series or parallel resonant circuits, respectively. These ratios are referred to as transformation Qs.

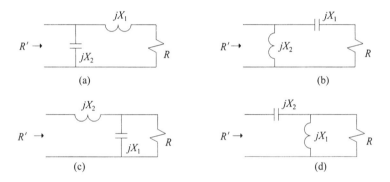

Figure 4.7   The four possible configurations for an L-section.

The sign of the reactance or susceptance is carried over to the transformation Q. It follows that the transformation Q is positive when the effective series reactance is inductive (impedance format) or when the effective shunt susceptance is capacitive (admittance format).

The reactance changes by a factor

$$E_1 = 1 + 1 / Q_1^2 \tag{4.31}$$

in the transformation step.

As is the case with the resistance, the reactance increases after a series to shunt transformation and decreases when a shunt to series transformation is considered.

Figure 4.8   A complex impedance displayed in impedance and admittance form.

The reactance of the first element used in an L-section is determined by the transformation Q required to transform the load resistance (R) to the value required ($R\,'$). The Q value can be calculated by using the relationship

$$R' = D_1 R = (1 + Q_1^2) R \tag{4.32}$$

A positive or a negative sign can be assigned to the transformation Q.

The second element in the L-section is used to achieve the desired reactance level. If a purely resistive input impedance is required, the reactance of this element is given by

$$X_2 = -X_1(1 + 1/Q_1^2) = R'/Q_1 \tag{4.33}$$

if the first element is a series element, and by

$$X_2 = -X_1 / (1 + 1/Q_1^2) = -R'Q_1 \tag{4.34}$$

if the first element is a shunt element.

Equations (4.33) and (4.34) can be verified easily by using the relationships $Z = 1/Y$ and $Y = 1/Z$, respectively.

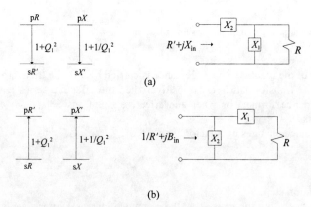

Figure 4.9   Illustration of the transformation properties of an L-section when the first element on the load side is (a) a shunt element and (b) a series element.

The formulas relevant to the design of an L-section are given below. Equations (4.35)–(4.37) apply when the first element on the load side is a shunt element [see Figure 4.9(a)], while (4.38)–(4.40) are used when the first element is a series element [see Figure 4.9(b)].

$$R' = R / (1 + Q_1^2) \tag{4.35}$$

$$X' = X_1 / (1 + 1 / Q_1^2) \tag{4.36}$$

$$X_2 = X_{in} + Q_1 R' \tag{4.37}$$

$$R' = R (1 + Q_1^2) \tag{4.38}$$

$$X' = X (1 + 1 / Q_1^2) \tag{4.39}$$

$$B_2 = B_{in} + Q_1 / R' \tag{4.40}$$

When the transformation Q is high, the frequency response of an L-section near the resonant frequency will be similar to that of a simple series or parallel resonant circuit.

The circuit Q of the L-section is approximately equal to half the transformation Q.

If a more accurate value for the Q of the circuit is required, the procedure outlined in Section 4.9 can be followed.

EXAMPLE 4.2   Illustration of the transformation properties of an L-section.

The transformation and compensation properties of an L-section will be illustrated here by using the L-section shown in Figure 4.7(a) as an example. In this example, the load resistance and L-section reactance values at the transformation frequency are taken to be

R = 1Ω, ωL = 1Ω and 1/ωC = 2Ω.

The first element in the matching network (see Figure 4.10) is a series inductor of $j1\Omega$. Because the load resistance is also equal to 1Ω, the transformation Q is equal to +1, and the 1Ω load resistance is transformed upwards with a factor $1 + 1^2 = 2$ to a value of 2Ω. The transformation Q has the same magnitude before and after the transformation (the sign of the Q changes in the process), and the reactance in parallel with the transformed resistance is therefore $+j2\Omega$ (still inductive). This reactance is removed by resonating it off with a capacitor ($-j2\Omega$ reactance), after which the original 1Ω resistor has been transformed to 2Ω at the frequency of interest.

Note that the series inductor is a transformation element, while the capacitor is a compensation element.

EXAMPLE 4.3   Designing an L-section.

An L-section will be designed to transform a load of 50Ω to 250Ω at 50 MHz as an example of the application of the theory discussed above.

Because the transformation is upwards, the first element of the L-section must be a series element. The diagram in Figure 4.11(a) applies.

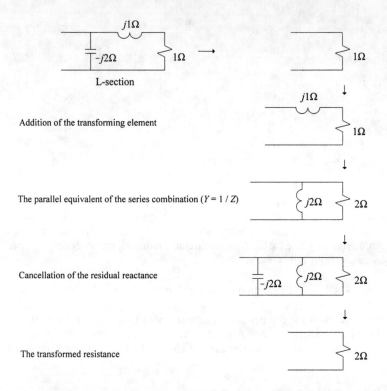

Addition of the transforming element

The parallel equivalent of the series combination ($Y = 1 / Z$)

Cancellation of the residual reactance

The transformed resistance

Figure 4.10 Illustration of the transformation and compensation properties of an L-section.

It follows from the diagram that the transformation Q must be equal to 2; it follows that a series inductor or capacitor with reactance equal to

$$X = Q\,R = 2 \times 50 = 100\,\Omega$$

is required.

If the first element is chosen to be an inductor, the required inductance is

$$L = 100 / (2\pi \times 50 \times 10^{6}) = 0.318\,\mu H$$

The parallel equivalent of the transforming section is the required 250Ω in parallel with a reactance

$$X' = R' / Q_1 = 250 / 2 = 125\Omega$$

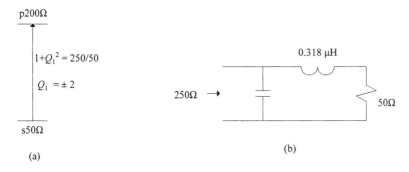

Figure 4.11   (a) The transformation diagram and (b) the L-section relevant to Example 4.3.

Note that the Q-factors of the series combination and its parallel equivalent must be equal in magnitude (the sign of the Q changes when the transformation is done).

The capacitance required to remove the reactive part of the input admittance of the resistor and inductor combination is

$$C = 1 / [125(2\pi \times 50 \times 10^6)] = 125\,\text{pF}$$

The designed network is shown in Figure 4.11(b).

4.5   PI-SECTIONS AND T-SECTIONS

PI-sections and T-sections are three-element matching networks. A PI-section has two parallel elements, and the T-section has two series elements, as shown in Figure 4.12.

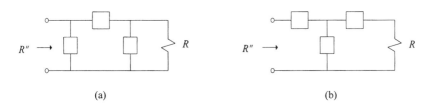

(a)   (b)

Figure 4.12   Topology for (a) a PI-section and (b) a T-section.

The first two elements in these sections are transforming elements. One of these elements causes the resistance to increase, while the other causes it to decrease.

The reactance level is set by the last element in the section (the compensating element).

Because the resistance is transformed twice, there are two transformation Qs in these sections. The highest transformation Q can be chosen to have any value higher than that required in an equivalent L-section.

As in the case of L-sections, the bandwidth of PI-sections and T-sections are also determined by the transformation Qs. Where the two Q-factors are different, the Q of the network will be approximately equal to half of the highest transformation Q.

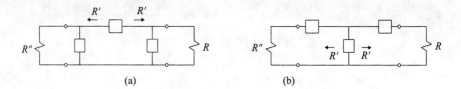

(a)   (b)

Figure 4.13   An alternative view of the transformation process in (a) a PI- or (b) a T-section.

Because the highest transformation Q is adjustable, the bandwidth of a PI-section or a T-section can be controlled.

The transformation properties of a PI-section or a T-section can also be considered, as illustrated in Figure 4.13. The fact that the source termination (R'') and the load termination (R) must be transformed to the same intermediate value (R') is considered in this case. Both terminations are transformed downwards in a PI-section and upwards in a T-section. The second element (as counted from the load side) is the compensation element in this case. The bandwidth is determined by the side with the highest transformation Q.

4.5.1   The PI-Section

The resistance transformations caused by a PI-section (cascade approach) are illustrated in Figure 4.14. The resistance is first transformed downwards by a factor $(1+Q_1^2)$ and then upwards with a factor $(1+Q_2^2)$.

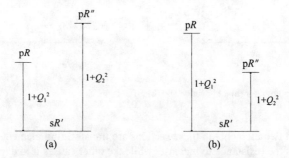

(a)   (b)

Figure 4.14   (a) Upward transformation of the load resistance with a PI-section and (b) downward transformation of the load resistance with a PI-section.

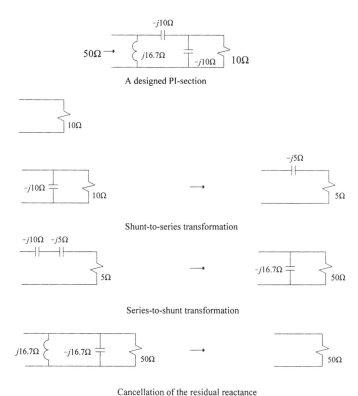

Figure 4.15 The transformation of a 10-Ω load to 50Ω with a PI-section.

Q_1 is the first transformation Q and is associated with the load resistance and the first element of the network. Q_2 is the second transformation Q.

The second transformation Q is equal to the ratio of the effective reactance in series with the transformed resistance (R') and the transformed resistance itself.

The transformed resistance will be lower than the load resistance when the first transformation Q is higher than the second. An upward transformation requires the second transformation Q to be higher than the first.

The value of the highest transformation Q is determined by the required bandwidth of the network. The Q of the network is approximately equal to one-half of the highest transformation Q when the transformation Q factors are sufficiently different.

The transformation of a 10-Ω load to 50Ω by using a PI-section is illustrated in detail in Figure 4.15.

Equations (4.41)–(4.46) can be used to design a PI-section when the load resistance must be transformed downward [see Figure 4.16(a)], while (4.47)–(4.49) should be used with (4.44)–(4.46) when the load resistance must be transformed upwards [see Figure 4.16(b)].

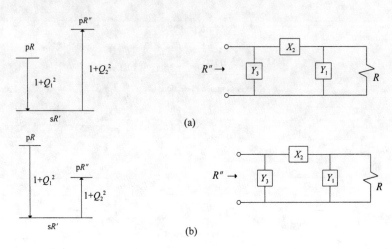

Figure 4.16   The transformation properties of a PI-section are illustrated for (a) a downward and (b) an upward transformation of the load resistance.

$$Q_1 = Q_{max} = 2Q \tag{4.41}$$

$$R' = R / (1 + Q_1^2) \tag{4.42}$$

$$1 + Q_2^2 = R'' / R' \tag{4.43}$$

$$Y_1 = Q_1 / R \tag{4.44}$$

$$X_2 = R' (Q_1 + Q_2) \tag{4.45}$$

$$Y_3 = Q_2 / R'' \tag{4.46}$$

$$Q_2 = Q_{max} = 2Q \tag{4.47}$$

$$R' = R'' / (1 + Q_2^2) \tag{4.48}$$

$$1 + Q_1^2 = R / R' \tag{4.49}$$

EXAMPLE 4.4   A PI-section example.

A matching network for transforming 50Ω to 12.5Ω will be designed. The maximum transformation Q will be taken to be 5.

     Because the transformation is downwards, the first transformation Q will be the highest.

     The next step is to choose the network topology to be used. The network is arbitrarily assumed to have an inductor as the first element, while the other components are chosen to be capacitors. (It is not possible to choose both of the first two components to be inductors or capacitors. If this is done, the second transformation Q will be the highest.)

Because the first transformation Q is

$$Q_1 = -5$$

the reactance of the inductor must be

$$X_1 = 50/5 = 10\Omega$$

The second component must change the transformation Q to 2.35. The Q must be positive (inductive) if the last component is to be a capacitor:

$$X_2 = R'(Q_1 + Q_2) = 1.92[2.35 + (-5)] = -5.1\Omega$$

The reactance of the last component is

$$X_3 = -R'' / Q_2 = -12.5 / 2.35 = -5.3\Omega$$

The designed network is shown in Figure 4.17(b).

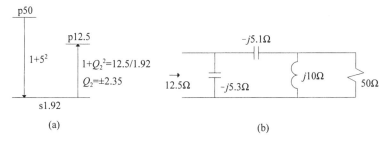

(a)   (b)

Figure 4.17   (a) The transformation diagram corresponding to Example 4.4; and (b) a PI-section for matching 50Ω to 12.5Ω.

4.5.2   The T-Section

The dual of a PI-section is a T-section. Therefore, the formulas for designing a PI-section can also be used to design a T-section. In order to do this, it is necessary to replace the resistance and reactance in these formulas with conductance and susceptance, respectively. The terminations used must also be inverted (i.e., if the actual terminations for the T-section are 50Ω terminations, the terminations for the equivalent PI-section should be 1/50Ω).

The reactance results of the PI-section apply directly to the T-section if these are interpreted to be susceptances. To illustrate this, if the components required for a PI-section are $j10\Omega$, $-j5.1\Omega$, and $j5.3\Omega$, the components required in the T-section are $j10$S, $-j5.1$S, and $j5.3$S (see Figure 4.18).

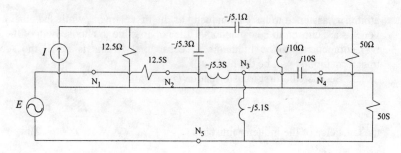

Figure 4.18   An example of finding the dual of a PI-section.

This approach is useful when a program to design PI- and T-sections is developed. The program can be written to design PI-sections only, and by entering the specifications correctly it can also be used to design T-sections. When the design is not done by computer, it is better to follow the equations provided here. Equations (4.50)–(4.55) can be used when the load resistance must be decreased, while (4.56)–(4.58) and (4.53)–(4.55) can be used when an upward transformation is required (see Figure 4.19).

$$Q_2 = Q_{max} = 2Q \tag{4.50}$$

$$R = R'' (1 + Q_1^2) \tag{4.51}$$

$$1 + Q_1^2 = R' / R \tag{4.52}$$

$$X_1 = Q_1 R \tag{4.53}$$

$$Y_2 = (Q_1 + Q_2) / R' \tag{4.54}$$

$$X_3 = Q_2 R'' \tag{4.55}$$

$$Q_1 = Q_{max} = 2Q \tag{4.56}$$

$$R' = R (1 + Q_1^2) \tag{4.57}$$

$$1 + Q_2^2 = R' / R'' \tag{4.58}$$

4.6   THE DESIGN OF PI-SECTIONS AND T-SECTIONS WITH COMPLEX TERMINATIONS

The procedures outlined in the previous sections can be extended easily to the general case where the load and source impedances are complex. The approach is illustrated in Figure 4.20.

The reactive parts of the load and source impedance (T-section) or admittance (PI-section) are ignored initially, and the network is designed to match the load and source resistance to each other. The first and last components are then changed to take the imaginary parts of the load and source impedance or admittance into account.

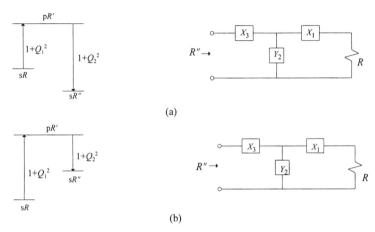

(a)

(b)

Figure 4.19   The transformation properties of a T-section are illustrated for (a) a downward and (b) an upward transformation of the load resistance.

Because a T-section transforms a series load resistance to a new series value, the load and the required input impedance must be specified in series form, that is, as impedances. The specifications for a Π-section must be of parallel form. The first step in designing a matching network when the terminations are complex, therefore, is to get the terminations in the right form.

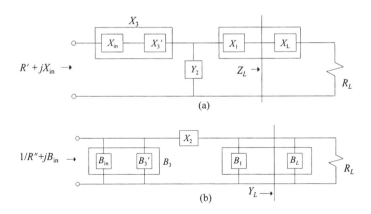

Figure 4.20   The design of (a) a T-section and (b) a PI-section when the terminations are complex.

The following equations apply to Figure 4.20:

$$X_3' = Q_2 R'' \tag{4.59}$$

$$X_1 = Q_1 R_L - X_L \tag{4.60}$$

$$B_3' = Q_2 / R'' \tag{4.61}$$

$$B_1 = Q_1 / R_L - B_L \tag{4.62}$$

EXAMPLE 4.5   Designing a T-section with complex terminations.

A T-section for matching a $10 + j10\Omega$ load to $50 + j40\Omega$ (see Figure 4.22) with a maximum transformation Q equal to 5 will be designed. These specifications are in impedance form, as required for a T-section.

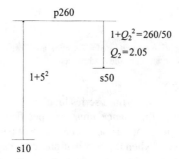

Figure 4.21   The transformation diagram for the T-section of Example 4.5.

The transformation diagram for this problem is shown in Figure 4.21. Because the transformation is upwards, the first transformation Q must be the highest in this case. The second transformation Q must be equal to 2.05.

With the Q-factors known, the next step is to choose a topology. If the network shown in Figure 4.22 is chosen, the bandwidth of the circuit can be calculated as was done before. Because the Q-factors of the load and source impedances are low compared to the maximum transformation Q, predictable results can also be obtained with other topologies. The only major difference will be in the rate at which the slope outside the pass band levels off because of the higher number of poles.

The component values of the chosen network can now be determined by using the values calculated for the transformation Qs.

In order to have a transformation Q of 5 at the load, a reactance of $+j40\Omega$ must be added to the existing $+j10\Omega$; that is,

$$X_1 = (5 \times 10) - 10 = 40\Omega$$

After the first transformation, the transformation Q is still equal to 5. In order to change it to 2.05, a capacitor with susceptance

$$Y_2 = (Q_1 + Q_2) / R' = (5 + 2.05) / 260 = 27.1\,\text{mS}$$

must be used; that is, both Q_1 and Q_2 must be positive.

It is not possible in this case to use a capacitor with susceptance

$$Y = (Q_1 - |Q_2|) / R'$$

because the last component of the network was chosen to be an inductor.

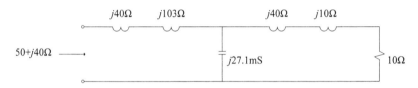

Figure 4.22   A T-section for transforming a $10 + j10\Omega$ load to $50 + j40\Omega$.

The last component of the T-section must remove the residual reactance and change it to the required level of $+j40\Omega$. In order to do this, a 14.3-Ω inductor is required.

The designed network is shown in Figure 4.22.

4.7   FOUR-ELEMENT MATCHING NETWORKS

When four elements are used, the control over the frequency response of the impedance-matching network increases. The bandwidth can be lower or higher than that of an L-section.

One possible approach to designing a network to have a very high Q is shown in Figure 4.23.

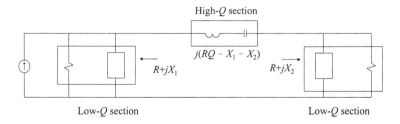

Figure 4.23   A high-Q, easily tunable, four-element impedance-matching network.

The two low-Q sections transform the load impedance and the source impedance to have the same resistance ($R < R_1$, $R < R_2$ with the network, as shown in Figure 4.23), and the high-Q section sets the reactance level and provides the required rejection.

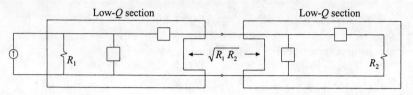

Figure 4.24   A wideband four-element impedance-matching network.

Strictly speaking, only one downward-transforming section is required in this network. When two downward-transforming sections are used, however, it is often possible to decrease the insertion loss of the circuit. This follows because it is often possible to use higher Q components in the circuit when this is done.

When the approach illustrated in Figure 4.24 is followed, the bandwidth can be wider than that obtainable with an L-section (R_2 is assumed to be smaller than R_1). In this case the source and the load resistance are transformed to their geometric mean by using two L-sections.

4.8   CALCULATION OF THE INSERTION LOSS OF AN LC IMPEDANCE-MATCHING NETWORK

It was shown in Chapter 3 that the ideal component does not exist. For this reason, all practical circuits will have some insertion loss. If the insertion loss is to be kept low, the unloaded Q-factors of the components must be significantly higher than the Q of the circuit.

The insertion loss of any cascaded LC network can be computed by following the procedure outlined here:

1.   Model each reactive component in the network as an ideal component with a resistor in series or in parallel with it, depending on whether it is a series or shunt element, respectively.

   The value of this resistance can be determined from the unloaded Q-factor estimated for the component.

   Unloaded Q-factors for capacitors and magnetic-core inductors can usually be found by using the data given by the manufacturer, while those of air-cored solenoidal coils can be determined by following the procedure outlined in Section 3.3.6.

   The unloaded Q of a component may be a strong function of the frequency.

2.   Assume that the power dissipated in the load is equal to 1W.

3. If the first component of the network (as viewed from the load) is a series element, calculate the power dissipated in it by using the equation

$$P_D = (R_Q / R_L)P_L \qquad (4.63)$$

where R_Q is the series resistance associated with the element and R_L is the (effective) load resistance. P_L is the power dissipated in the load (1W in this case).

If the first component is a parallel element, calculate the power dissipated in it by using the equation

$$P_D = (G_Q / G_L)P_L \qquad (4.64)$$

where G_Q is the parallel conductance associated with the component, and G_L is the (effective) conductance of the load. P_L is the power dissipated in the load.

4. Add the power dissipated in the first component to that dissipated in the load:

$$P_T = P_L + P_D \qquad (4.65)$$

5. Consider the first component to be part of the load and calculate the new (effective) load admittance or impedance.

6. Repeat steps 3 to 5 until the power entering the matching network (P_T) and the effective input impedance of the network (Z_{in}) are known.

7. Calculate the transducer power gain of the network (G_T) by using the equation

$$G_T = (1 - |S_s|^2) \frac{P_L}{P_T} \qquad (4.66)$$

$$= \left[1 - \left| \frac{Z_{in} - Z_s^*}{Z_{in} + Z_s} \right|^2 \right] P_L / P_T \qquad (4.67)$$

$$= \frac{4R_{in}R_s}{(R_{in} + R_s)^2 + (X_{in} + X_s)^2} P_L / P_T \qquad (4.68)$$

where S_s is the input reflection parameter with Z_s as normalizing impedance,

$$Z_{in} = R_{in} + jX_{in} \qquad (4.69)$$

and

$$Z_s = R_s + jX_s \tag{4.70}$$

where Z_s is the internal impedance of the source driving the network.

EXAMPLE 4.6   Calculating the insertion loss of a PI-section.

As an example of the application of this procedure, the insertion loss of the PI-section designed in Example 4.4 [see Figure 4.17(b)] will be calculated at the center frequency. The unloaded Q-factors of the capacitors are assumed to be 500, while that of the inductor is taken as 100.

It follows from the specified unloaded Q-factors that the conductance associated with the shunt inductor is 1 mS, the resistance associated with the series capacitor is 0.01Ω, and the conductance associated with the shunt capacitor is 0.38 mS.

If the power delivered to the load is taken to be 1 W, the power dissipated in the inductor is 50 mW. [This result is obtained by using (4.64).] The power entering the last section of the network is therefore 1.05 W. The input impedance of this section is

$$Z = 2.0 + j9.6\Omega$$

By using (4.63), the power dissipated in the series capacitor is found to be 5 mW. The power entering the network at this point is therefore 1.055 W. The input admittance at this point is

$$Y = (84 - j186)\,\text{mS}$$

The power dissipated in the shunt capacitor is obtained by using (4.64) again, and is also 5 mW. The total power entering the network is therefore 1.06 W. The input impedance is

$$Z_{in} = 11.9 - j0.45\Omega$$

By using (4.68), the transducer power gain of the network was calculated to be 0.94, that is, an insertion loss of 0.3 dB.

4.9   CALCULATION OF THE BANDWIDTH OF CASCADED LC NETWORKS

The bandwidth of a network can be found iteratively if its transducer power gain is determined as a function of frequency. The transducer power gain of any cascaded LC network can be found by following the procedure outlined in the previous section.

Because the cutoff frequencies (3-dB) of L-, PI-, and T-sections are known to good approximation ($f_{-3dB} = f_0 \pm f_0 / Q_{max}$), the exact bandwidth of these circuits can be determined quickly by following this procedure.

EXAMPLE 4.7   The 3-dB bandwidth of a matching network.

By following the procedure described, the 3-dB cutoff frequencies of the network in Example 4.6 are found to be 83 and 130 MHz, that is, if the center frequency is selected as 100 MHz. The exact Q of the circuit is therefore 2.3 instead of the estimated 2.5.

If a bandwidth other than the 3-dB bandwidth is required, it can be found easily by following the same procedure.

QUESTIONS AND PROBLEMS

4.1   What is the definition of the Q-factor of a circuit?

4.2   (a) Find the -30-dB frequencies of a parallel resonant circuit with $Q = 300$ at 10.7 MHz. (b) A parallel resonant circuit is to be used to obtain 30-dB rejection at 5 MHz and 7 MHz (center frequency 5.92 MHz). What is the Q required to meet these specifications?

4.3   The Q of a parallel resonant circuit is 50 when ideal components are used. What will it be if an inductor and capacitor with unloaded Q-factors of 150 and 250, respectively, are used?

4.4   Differentiate between the Q of a circuit and a transformation Q.

4.5   Is it true that the load will always be transformed downward by an L-section if the first element of the section is in parallel with the load?

4.6   What is the relationship between the transformation Q of an L-section and the Q of the network?

4.7   Design an L-section to transform a 50-Ω load to 12.5Ω at 50 MHz. Use the information in Chapter 3 to determine the unloaded Q of the inductor in the design, that is, if it is to be a single-layer air-cored coil (dimensions 2 cm×2 cm×2 cm, realistic wire thickness). Do this for both of the topologies that can be used.

4.8   What is the relationship between the transformation Q of a T- or PI-section and the Q of the network?

4.9   Design a PI-section to transform a load of 500Ω to 50Ω. The bandwidth of the network must be 2 MHz with the resonant frequency at 10 MHz. A capacitor must be used as the first element and the other two elements must be inductors.

4.10   Design a T-section to achieve the transformation in the previous problem.

4.11   Design a lowpass L-section to transform the input impedance of a 200-W push-pull amplifier to 12.5Ω. The input impedance of the amplifier is $2-j2\Omega$ and the frequency of interest 30 MHz.

4.12   (a) Design a T-section to transform a load of $3-j6\Omega$ to $50+j25\Omega$. The Q of the circuit must be approximately 8; $f_o = 100$ MHz. (b) Design a PI-section to meet the specifications in (a).

4.13   (a) Use the material in Section 3.3.6 to determine the range of inductance values for which an unloaded Q of 250 can be obtained at 50 MHz with standard wire thicknesses and coil sizes below 2cm × 2cm × 2cm. (b) Design a four-element matching network to match a load of 200Ω to a source with 50-Ω internal resistance. The Q of the circuit must be 25 and the resonant frequency 50 MHz. (c) Determine the – 30-dB frequencies of the circuit. (d) Calculate the insertion losses of the circuit at the resonant frequency.

4.14   Transform a load of 10Ω to 50Ω. The bandwidth is to be as large as possible. Use a four-element matching network; $f_0 = 20$ MHz.

4.15   Determine realistic values for the unloaded Q-factors of the components in Problems 4.9 and 4.10. Compare the insertion losses of the two networks.

4.16   Calculate the insertion loss of the network designed in Problem 4.11 if the unloaded Q-factors of the capacitor and inductor used in the network are 150 and 50, respectively.

4.17   Calculate the exact 3-dB bandwidths of the circuits designed in Problems 4.9, 4.10, and Problems 4.12–4.14. Compare the expected and actual bandwidths.

SELECTED BIBLIOGRAPHY

RF Power Transistor Manual, Somerville, NJ: RCA Corporation (Solid State Division), 1971.

CHAPTER 5

COUPLED COILS AND TRANSFORMERS

5.1 INTRODUCTION

When the parasitics can be ignored, transformers are ideally suited for impedance scaling. Ideal and practical transformers will be considered in this chapter.

A practical transformer differs from the ideal in that it has leakage flux, finite magnetizing inductance, losses, and parasitic capacitance, all of which degrade its performance. Several equivalent circuits for practical transformers will be presented here.

Transformers are often used when wideband transformation of resistance is required (when possible, this is usually a better option than using LC networks). It will be shown here that the wideband performance of a transformer is mainly determined by the coupling factor. This is also the reason why stacked toroids or balun cores are usually used to realize such a transformer.

Although the finite magnetizing inductance and the leakage inductance are undesirable in a wideband transformer, they can be put to good use in narrowband matching networks. Several narrowband matching networks using transformers will be considered in detail.

Because it is important to adjust the coupling factor of a transformer to the required value when narrowband matching networks are used, methods to measure the coupling factor will be also be considered in this chapter.

5.2 THE IDEAL TRANSFORMER

The equivalent circuit of the ideal transformer is shown in Figure 5.1. The number of turns on the primary and secondary sides of the transformer are, respectively, n_1 and n_2.

The ideal transformer has the following characteristics:

1. The magnetic flux in the two windings is the same. Therefore, there is no leakage flux.

   The voltage induced by the changing flux in each winding is given by Faraday's law:

   $$V = n_x \, \partial \Phi / \partial t \tag{5.1}$$

   where n_x is the number of turns in the winding under consideration. Because the flux coupling each winding is the same, the ratio of the primary to secondary voltage of the transformer is

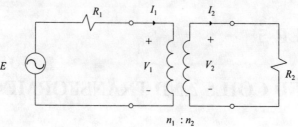

Figure 5.1   The ideal transformer.

$$V_1 / V_2 = n_1 / n_2 \tag{5.2}$$

This relationship is more significant if it is written in the form

$$V_1 / n_1 = V_2 / n_2 \tag{5.3}$$

The voltage per turn, therefore, is the same for both sides of the transformer.

2.   The primary current necessary to establish the flux in the ideal transformer is negligible. The input impedance of the transformer with the load open-circuited, therefore, is infinite.

3.   There are no losses in the ideal transformer. The average power dissipated in the load, therefore, is exactly the same as the average power entering the transformer.

Because the ideal transformer has no reactive components, the instantaneous power dissipated in the load is also equal to the instantaneous power entering the transformer; that is,

$$v_1 i_1 = v_2 i_2 \tag{5.4}$$

By using (5.2) to replace the voltages in this equation, the relationship between the primary and secondary currents is found to be

$$n_1 I_1 = n_2 I_2 \tag{5.5}$$

The demagnetizing force (magnetomotive force) of the current in the secondary winding is, therefore, balanced by that of the current in the primary winding.

By using this equation and (5.3), the relationship between the primary and secondary impedances is found to be

$$Z_1 = V_1 / I_1 = [n_1 / n_2]^2 Z_2 \tag{5.6}$$

The impedance ratio is, therefore, only a function of the turns ratio of the transformer.

4.   The permeability of the ideal transformer is independent of the flux density in the core. This implies that the ideal transformer is a perfectly linear device.

From an impedance-matching viewpoint, the ideal transformer is very useful in that it can be used to scale impedances by any factor.

5.3   EQUIVALENT CIRCUITS FOR PRACTICAL TRANSFORMERS

A practical transformer deviates from the ideal in the following ways:

1.   There is some leakage flux and, therefore, leakage inductance.

2.   The magnetizing inductance is finite.

3.   There are losses in the windings of the transformer (copper losses), as well as in the core (hysteresis and eddy current losses).

4.   The relative permeability of the magnetic material changes with signal level and dc current (saturation), as well as with frequency and temperature.

5.   Apart from the effect of the leakage inductance, the high-frequency response is degraded by the presence of parasitic capacitance between the windings and turns of each winding.

A circuit model for the practical transformer, ignoring the capacitance and nonlinearities, is shown in Figure 5.2 [1]. The two dots indicate the sides of the two windings that have the same voltage polarity.

The two series inductances represent the leakage flux, the series inductance to the left together with the shunt inductance are the magnetizing inductances, the resistances r_1 and r_2 represent the copper losses, and R_p represents the losses in the magnetic material.

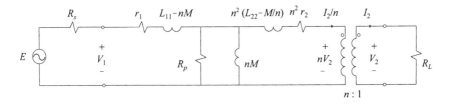

Figure 5.2   An equivalent circuit for a practical transformer.

The mutual inductance M can be determined as a function of the magnetizing inductances L_{11} and L_{22} by using the relationship

$$M = k(L_{11}L_{22})^{1/2} \tag{5.7}$$

where k is the coupling factor of the transformer.

The symbol n in Figure 5.2 can have any arbitrary value, but it is usually chosen to be equal to the turns ratio of the two windings of the transformer. However, a better choice for it is

$$n = \frac{1}{k}\sqrt{\frac{L_{11}}{L_{22}}} \tag{5.8}$$

If the losses in the magnetic material can be ignored, the equivalent circuit for two coupled coils [see Figure 5.3(a)] can be used for the transformer. The circuit shown in Figure 5.3(b) is equivalent to the coupled coil circuit [2]. This can be proven by setting up the Z-parameter matrices for the two circuits. The transformation ratio shown in Figure 5.3(b) is that for the impedances.

The following characteristics of the transformer are immediately evident from the equivalent circuit in Figure 5.3(b):

1.   The load impedance is transformed to the primary side of the transformer as

$$Z'_L = [L_{11} / (k^2 L_{22})][Z_L + r_2] \tag{5.9}$$

that is, as long as

$$[\omega L_{22}(1 - k^2)]^2 << |Z_L + r_2|^2 \tag{5.10}$$

This equation can be changed to a more useful form by substituting L_{22} by using (5.9):

$$[\omega L_{11}(1 - k^2) / k^2]^2 << |Z'_L|^2$$

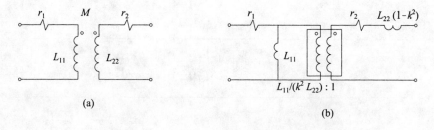

(a)

(b)

Figure 5.3   (a, b) Two equivalent circuits for a practical transformer (R_p neglected).

leading to

$$[1/k^2 - 1] << |Z'_L / (\omega L_{11})|^2 \qquad (5.11)$$

This inequality will always apply at low frequencies, and, if the coupling between the windings is good, it will also apply at higher frequencies.

It follows from (5.9) that the impedance transformation factor at low frequencies is always

$$n^2 = L_{11} / (k^2 L_{22}) \qquad (5.12)$$

2. The high-frequency response of the transformer is limited by the leakage reactance ωL_{22} $(1 - k^2)$.

3. The low-frequency response of the transformer is limited by the magnetizing inductance L_{11}.

4. The input inductance at low frequencies is equal to the magnetizing inductance L_{11}.

If the frequency is low enough for (5.11) to apply, the equivalent circuit of Figure 5.3(b) can be simplified to that shown in Figure 5.4.

By using this equivalent circuit, the low cutoff frequency is found to be

$$\omega_{L\_3dB} = R_s / (2 L_{11}) \qquad (5.13)$$

This equation applies when r_1 and r_2 can be ignored and the load resistance R_L is transformed to be equal to the source resistance R_s.

It follows that, under these circumstances, the required primary inductance can be determined by using the specifications for the cutoff frequency and the source resistance. The required secondary inductance is given by the equation

$$L_{22} = \frac{R_L}{R_s} \frac{L_{11}}{k^2} \qquad (5.14)$$

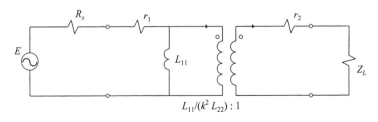

$L_{11}/(k^2 L_{22}) : 1$

Figure 5.4 An equivalent circuit for the practical transformer at low frequencies.

This inductance is clearly a function of the coupling factor, which is usually not known at the design stage.

The easiest way to overcome this problem is to ensure that the coupling factor is close to unity, if possible.

Very good coupling can usually be obtained by using materials with high relative permeabilities. The coupling between the windings of the transformer is also better if balun or stacked toroidal cores, instead of a single toroidal core, are used.

5.4   WIDEBAND IMPEDANCE MATCHING WITH TRANSFORMERS

Impedance matching over very wide bandwidths (decades) is possible with transformers when the coupling between the windings is good and when the parasitic capacitance between the windings and turns of each winding is small.

If the parasitic capacitance can be ignored, the 3-dB bandwidth of a transformer can be determined by finding the poles of the equivalent circuit in Figure 5.5. This equivalent circuit can be derived from the one in Figure 5.2 by setting n equal to 1.

The equivalent circuit has two poles, a zero at the origin and 1 at infinity. The poles can be found by setting the impedances in either of the two loops of the equivalent circuit equal to zero. It this is done for the first loop, the following equation is obtained:

$$(R_s + r_1) + s(L_{11} - M) + sM \parallel [sL_{22} - sM + (R_L + r_2)] = 0 \qquad (5.15)$$

This equation simplifies to

$$s^2 L_{11} L_{22} (1 - k^2) + s[L_{22} R_s' + L_{11} R_L'] + R_s' R_L' = 0 \qquad (5.16)$$

The poles, and therefore the cutoff frequencies, can be obtained by solving this equation.

When the load resistance and the source resistance are matched, as is generally the case, the following equation applies:

$$R_L' = R_L + r_2 = (k^2 L_{22} / L_{11}) \cdot R_s' \qquad (5.17)$$

Because the coupling factor is usually unknown at the outset of the problem and is usually close to unity at low frequencies, the transformer is normally designed for

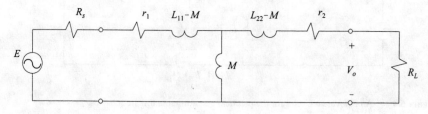

Figure 5.5   A T-section equivalent for the transformer.

$$R'_L = (L_{22} / L_{11}) \cdot R'_s \qquad (5.18)$$

If this value for R_L' is substituted in (5.16), it simplifies to

$$s^2 L_{11} L_{22} (1 - k^2) + 2s L_{22} R'_s + (L_{22} / L_{11}) R'_s R'_s = 0 \qquad (5.19)$$

This equation can be written as

$$[sL_{11}(1 + k) + R'_s][sL_{11}(1 - k) + R'_s] = 0 \qquad (5.20)$$

The cutoff frequencies are, therefore, given by the following equations:

$$\omega_L = R'_s / [L_{11}(1 + k)] \qquad (5.21)$$

$$\omega_H = R'_s / [L_{11}(1 - k)] \qquad (5.22)$$

that is, when (5.18) applies.

It is clear from these equations that the low cutoff frequency of the transformer is determined by the effective resistance in parallel with the magnetizing inductance and that the high cutoff frequency is a strong function of the coupling factor.

The relative bandwidth of the transformer with the load and source impedances matched is given by

$$\omega_H / \omega_L = [1 + k] / [1 - k] \qquad (5.23)$$

The relative bandwidth of the transformer, therefore, is only a function of the coupling factor, that is, when the load and source impedances are matched and the parasitic capacitance can be ignored.

Because the coupling factor is frequency-dependent and close to unity at the lower frequencies, (5.23) can be simplified to

$$\omega_H / \omega_L = 2 / [1 - k] \qquad (5.24)$$

EXAMPLE 5.1 Designing a transformer to transform a 50-Ω load to 300Ω.

The primary inductance, secondary inductance, and coupling factor required to transform a load of 50Ω to 300Ω with 3-dB cutoff frequencies at 1 and 20 MHz will be determined as an example.

Assuming that the copper losses in the windings and the hysteresis losses in the magnetic core can be ignored, the required inductance ratio of the transformer can be obtained by using (5.18):

$$L_{11} / L_{22} = R_s / R_L = 50 / 300 = 0.1667$$

The required magnetizing inductance L_{11} can be obtained by using (5.21):

$L_{11} = 300.0/[2.0\,\pi \times 1.0 \times 10^6(1+1)] = 23.9\ \mu\text{H}$

The secondary inductance, therefore, is

$L_{22} = L_{11}/6 = 4.0\ \mu\text{H}$

The required coupling factor at 20 MHz can be obtained by using (5.24):

$1 - k = 2/[20/1] = 0.1$

$k = 0.9$

The coupling factor required to obtain a relative bandwidth of 20, therefore, is 0.9.

5.5   SINGLE-TUNED TRANSFORMERS

The single-tuned transformer shown in Figure 5.6 can be used to step the load impedance up or down and to obtain a frequency response identical to that of a parallel resonant circuit.

If the coupling between the windings of the transformer is good, the leakage inductance of the transformer can be ignored.

The required magnetizing inductance of the transformer can be found in terms of the Q-specification, and the source and load resistance, by using the equation

$$\omega_0\,L_{11} = (R_s \,\|\, R'_L)/Q = R_s/(2Q) \tag{5.25}$$

where R_L' is the load resistance transformed to the input side. Q in (5.25) is the circuit Q.

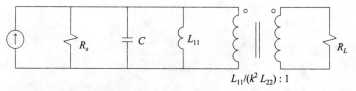

$$L_{11}/(k^2\,L_{22}) : 1$$

Figure 5.6   The single-tuned transformer.

The capacitance necessary to provide resonance is

$$C = 1/(\omega_0^2 L_{11}) \tag{5.26}$$

Because the coupling is assumed to be good, the required secondary inductance is given by the equation

$$L_{22} = (R_L/R_s)\,L_{11}/k^2 \cong (R_L/R_s)\,L_{11} \tag{5.27}$$

The next step in the design of the single-tuned transformer is to select a suitable magnetic material and to determine the type and size of the core required (see Chapter 3).

If necessary, the required number of turns around the core can be found by measuring the inductance of a few turns of wire around the core to be used. Because the inductance is proportional to the square of the number of turns, the number of turns required can be found easily.

5.6 TAPPED COILS

The tapped coil (Figure 5.7) is a very useful narrowband matching network. Independent adjustment of the center frequency and the transformation ratio of the transformer is possible. It is also very easy to manufacture once it has been designed.

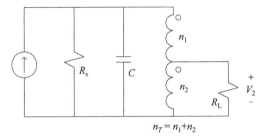

$$n_T = n_1 + n_2$$

Figure 5.7   The tapped-coil resonant circuit.

The frequency response of this transformer is usually similar to that of the single-tuned transformer.

Analysis of the tapped coil is possible by considering it to be two coupled coils.

It follows from (3.23) that the self-inductance of the upper section of the coil is given by

$$L_1 = r n_1^2 10^{-6} / \left[22.9 \frac{l_T}{r} \frac{n_1}{n_T} + 25.4 \right] \tag{5.28}$$

and that of the lower section by

$$L_2 = r n_2^2 10^{-6} / \left[22.9 \frac{l_T}{r} \frac{n_2}{n_T} + 25.4 \right] \tag{5.29}$$

where l_T is the total length of the coil, n_1 the number of windings in the upper section, n_2 the number in the lower section, and n_T the total number of turns ($n_T = n_1 + n_2$).

The inductance of the coil as a whole is given by the equation

$$L_T = r n_T^2 10^{-6} / \left[22.9 \frac{l_T}{r} + 25.4 \right] \tag{5.30}$$

If viewed in terms of its component parts, the total inductance can also be written as

$$L_T = L_1 + L_2 + 2M \tag{5.31}$$

By using this equation together with (5.7),

$$M = k(L_1 L_2)^{1/2}$$

the coupling factor is found to be

$$k = [L_T - L_1 - L_2] / [2(L_1 L_2)^{1/2}] \tag{5.32}$$

By substituting (5.28) through (5.31) into (5.32), the following can be deduced:

1. The coupling factor is dependent on the length-to-radius ratio of the coil, as well as the relative position of the tap-point.

2. The coupling factor is independent of the total number of turns (n_T).

The coupling factor of the tapped coil is given in Table 5.1 as a function of the relative position of the tap-point for a number of l_T/r ratios. The coupling factors are not shown for relative positions greater than 0.5, since they are the mirror image (arithmetic) of the lower values (0.8 corresponds to 0.2).

Table 5.1   The Coupling Factor of the Tapped Coil as a Function of the l_T/r Ratio of the Coil and the Relative Position of the Tap-Point

n_1/n_T	$l_T/r = 1$	$l_T/r = 1.5$	$l_T/r = 2$	$l_T/r = 3$	$l_T/r = 4$	$l_T/r = 5$
0.1	0.543	0.449	0.386	0.304	0.253	0.218
0.2	0.535	0.438	0.372	0.288	0.235	0.200
0.3	0.530	0.431	0.363	0.277	0.225	0.189
0.4	0.529	0.426	0.358	0.272	0.219	0.183
0.5	0.526	0.425	0.357	0.270	0.217	0.182

It can be seen from Table 5.1 that the coupling factor is not a strong function of the relative position of the tap-point.

The input admittance of the coupled coil can be found by using the equation

$$\begin{bmatrix} j\omega L_{1+2+2M} & -j\omega L_{2+M} \\ -j\omega L_{2+M} & j\omega L_2 + R_L \end{bmatrix} \begin{bmatrix} I_1 \\ I_2 \end{bmatrix} = \begin{bmatrix} E \\ 0 \end{bmatrix} \tag{5.33}$$

By applying Cramer's rule [3] to this equation, it follows that

$$Y_{in} = I_1 / E$$

$$= \begin{vmatrix} 1 & -j\omega L_{2+M} \\ 0 & j\omega L_2 + R_L \end{vmatrix} / D$$

$$= (R_L + j\omega L_2) / [j\omega L_{1+2+2M}(R_L + j\omega L_2) - j\omega L_{2+M} j\omega L_{2+M}]$$

After some manipulation, it follows that

$$Y_{in} = \frac{L_2 L_T R_L + L_1 L_2 (k^2 - 1)R_L}{\omega^2 L_1^2 L_2^2 (k^2 - 1)^2 + L_T^2 R_L^2} - j \frac{L_T R_L^2 - \omega^2 L_1 L_2 (k^2 - 1)}{\omega^3 L_1^2 L_2^2 (k^2 - 1)^2 + \omega L_T^2 R_L^2} \qquad (5.34)$$

If the coupling is perfect, this equation reduces to

$$Y_{in} = \frac{L_2}{L_T} / R_L - j / (\omega L_T) \qquad (5.35)$$

as expected.

When the coupling is not perfect, the resistive part of the input admittance will be frequency-independent if

$$\omega^2 L_1^2 L_2^2 (k^2 - 1)^2 \ll L_T^2 R_L^2$$

that is, if

$$[\omega L_1 L_2 (1 - k^2)]^2 \ll (L_T R_L)^2 \qquad (5.36)$$

When this is true, the input conductance will be

$$G_L' = \frac{L_2 L_T R_L + L_1 L_2 (k^2 - 1)R_L}{L_T^2 R_L^2}$$

$$= \frac{L_2}{L_T} G_L - \frac{L_1 L_2 (1 - k^2)}{L_T^2} G_L \qquad (5.37)$$

It can be seen from (5.34) that the actual conductance of the coil will be 10% higher than the value predicted by this equation if

$$[\omega L_1 L_2 (k^2 - 1)]^2 = 0.1 (L_T R_L)^2$$

For the parallel input inductance to be approximately L_T, it is necessary that both

(5.36) and the following equation apply:

$$\omega^2 L_1 L_2 (1 - k^2) \ll L_T R_L^2 \tag{5.38}$$

It follows that the equivalent circuit of the tapped coil can be simplified to that shown in Figure 5.8 when these two equations apply.

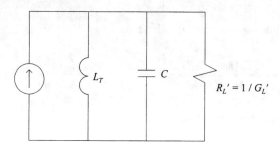

Figure 5.8   A simplified equivalent circuit for a tapped coil resonant circuit.

It should be noted that, from a practical viewpoint, it is more important that (5.36) applies rather than (5.38). This is because the resonant frequency of the circuit can be adjusted easily to the desired frequency by stretching or compressing the coil slightly (the coupling factor is only a weak function of the l/r ratio of the coil).

The transformation factor of the load [as given by (5.37)] is a function of the inductance L_T, the ratios L_1/L_T, L_2/L_T and the coupling factor k. The last three variables are all determined by the relative position of the tap-point and the l_T/r ratio of the coil. The transformation factor itself is therefore only a function of the l/r ratio of the coil and the relative position of the tap-point.

The transformation factor for different l/r ratios and positions of the tap-point was determined and is tabulated in Table 5.2. This table can only be used when (5.36) applies. When it does not, the transformation factor can be calculated by using (5.28) to (5.30), (5.32), and (5.34).

It can be seen from Table 5.2 that the transformation factor of the tapped coil is only a weak function of the l/r ratio of the coil. It is also evident that the transformation factor becomes very sensitive when the coil is tapped close to its ground point.

A tapped-coil resonant circuit can be designed to match a source to a load with a specified Q by following the procedure outlined next.

Design Procedure for a Tapped-Coil Resonant Circuit

1. Assume that the input inductance of the tapped coil will be equal to the inductance (L_T) of the coil itself (refer to Figures 5.7 and 5.8). Find the required inductance by using (5.25):

$$\omega L_T = [R_L' \| R_s] / Q = R_s / [2Q]$$

This equation will only apply if the unloaded Q-factor of the inductor and capacitor used are high compared to the specified circuit Q.

2. Design the inductor (L_T) to have the required unloaded Q as described in Chapter 3.

      Check if the self-capacitance of the designed coil is lower than the capacitance required for resonance.

      At this stage L_T, l_T, r, and n_T are known.

3. The next step is to determine the relative position of the tap-point. If the inequality (5.36)

$$[\omega L_1 L_2 (1 - k^2)]^2 \ll (L_T R_L)^2$$

applies, Table 5.2 can be used for this purpose. If it does not apply, then it will be easier to determine the tap-point practically.

Table 5.2   The Transformation Factor of the Tapped Coil as a Function of the l/r Ratio and the Relative Position of the Tap-Point from the Upper End of the Coil

n_1/n_T	R_L'/R_L		
	$l_T/r = 2$	$l_T/r = 4$	$l_T/r = 6$
0.10	1.18	1.18	1.18
0.20	1.46	1.46	1.48
0.30	1.92	1.92	1.94
0.40	2.66	2.68	2.70
0.50	4.00	4.00	4.00
0.55	5.08	5.07	5.05
0.60	6.64	6.62	6.56
0.65	9.04	8.99	8.86
0.70	12.90	12.80	12.60
0.75	19.70	19.60	19.10
0.80	33.00	33.00	32.10
0.85	64.00	64.70	62.80
0.90	160.00	166.00	162.00
0.95	732.00	505.00	808.00

EXAMPLE 5.2   Designing a tapped coil resonant circuit to transform a 50-Ω load to 1,000Ω.

A tapped coil resonant circuit will be designed to meet the following specifications, if possible: $f_0 = 10$ MHz, $R_L = 50\Omega$, $R_s = 1,000\Omega$, $Q = 20$, and $IL < 10\%$.

In order to have less than 10% losses, the loss resistance must be higher than (approximately) 10 times the transformed load resistance $(P=V^2/R)$; that is,

$$R_{loss} = 10.0 \times 1.0 \times 10^3 = 10k\Omega$$

This implies that the unloaded Q of the inductor must be at least

$$Q_u = R_{loss} / (\omega L) = 10k / 25 = 400$$

It is assumed here that the losses in the capacitor can be ignored. The required Q, however, is high in this design and it will be necessary to use a very good capacitor and even a parallel combination of smaller values to obtain a Q that is high compared to that of the inductor.

Assuming the l_T/r ratio of the coil to be equal to 2.0 and a d/c ratio of 0.55, the required radius of the coil is found to be [using (3.25)]

$$r = Q_u / [k(f)^{1/2}] = 1.26 \text{ cm}$$

The required length of the coil is 2.53 cm ($l_T/r = 2$). The number of turns can be found by using (3.27):

$$n = [L (22.9 \cdot l / r + 25.4) / r]^{1/2} = 4.75$$

The required wire thickness is [using (3.28)]

$$d = (l / D) \times (d / c) \times 2r / (N - 1) = 0.37 \text{ cm}$$

Because the thickness of No.12 SWG wire is 0.264 cm, the required wire thickness is unrealistic and a Q of 400 cannot be obtained with a standard value of the wire diameter.

If the radius of the coil is decreased to $r_2 = 1.01$ cm, the required wire thickness will be 0.26 cm.

The ratio of the former radius (r_1) to that of the required radius was obtained iteratively by using the equation

$$\frac{r_1}{r_2} \left[\sqrt{\frac{r_1}{r_2}} - 1 / N_1 \right] = \frac{d_1}{d_2} [1 - 1 / N_1] \qquad (5.39)$$

This equation can be derived easily from (3.27) and (3.28).

The number of turns required is

$$N_2 = [r_1 / r_2]^{1/2} N_1 = 5.3 \qquad (5.40)$$

Because of the reduction in radius, the unloaded Q of the inductor will decrease to 317. The insertion loss will therefore increase to approximately 13%.

The parasitic capacitance of the coil (0.51 pF) is much smaller than the capacitance required to provide resonance (637 pF).

With the coil designed and realizable, the tap-point can be determined. This can be done by using Table 5.2, that is, if inequality (5.36) applies. It follows from the $l_T/r = 2$ column that the coil must be tapped where

$$N_1 / N_T = 0.75$$

that is, where

$$N_1 = 4$$

In order to establish whether the inequality does apply, it is necessary to calculate the values of the inductances L_1 and L_2 and to determine the coupling factor of the tapped coil by using (5.28):

$$L_1 = r n_1^2 \ 10^{-6} / \left[22.9 \frac{l_T}{r} \frac{n_1}{n_T} + 25.4 \right]$$

$$= 0.269 \, \mu H$$

By using (5.29), the value of L_2 is found to be $L_2 = 46.6$ nH.

The value of the coupling factor can be determined from Table 4.1 [or by using (5.32)]. Its approximate value is 0.37.

Because

$$[\omega L_1 L_2 (1 - k^2)]^2 = 462.1 \times 10^{-15}$$

and

$$[L_T R_L]^2 = 400.0 \times 10^{-12}$$

inequality (5.36) does apply, and the tap-point as determined from Table 5.2 will be accurate.

5.7 PARALLEL DOUBLE-TUNED TRANSFORMERS

The narrowband circuits discussed up to this point match the load conjugately to the source at a single frequency only. With the double-tuned transformer shown in Figure 5.9, it is possible to match the source to the load at two different frequencies. If these frequencies are close enough to each other, the ripple in the passband can be designed to be small.

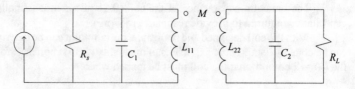

Figure 5.9   A parallel double-tuned transformer.

The transducer power gain (G_T) of the double-tuned transformer is plotted against frequency in Figure 5.10 for different values of the coupling factor. When the coupling factor is smaller than a certain critical value (k_c), the source cannot be matched to the load.

The rejection characteristics of the double-tuned transformer are superior to those of a simple parallel resonant circuit. The difference becomes very pronounced when the circuit Q becomes high.

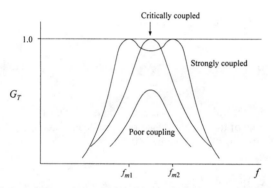

Figure 5.10   Three typical responses of the double-tuned transformer.

The transformer can be analyzed by using the equivalent circuits shown in Figure 5.11.

It is possible to match the load to the source at two different frequencies with the parallel double-tuned circuit by splitting the circuit up, as shown in Figure 5.12 (or Figure 5.13). In Figure 5.12, L_2' and C_2' are transforming elements, while L_{11} and C_1 are compensating elements. The two transforming elements can match the load resistance to R_s at two different frequencies, while the two compensating elements can be designed to cancel the residual reactance at these frequencies.

Equations for the ripple in the passband of the double-tuned transformer (see Figure 5.10) can be established by breaking up the equivalent circuit, as shown in Figure 5.13. The capacitor and inductor of the left-hand matching section cause the series resistance R_L to decrease at high and low frequencies, respectively. R_L is equal to the source resistance R_s at the resonant frequency of the inductor and capacitor. The capacitor in the right-hand section causes the series resistance R_s to decrease monotonically with increasing frequency.

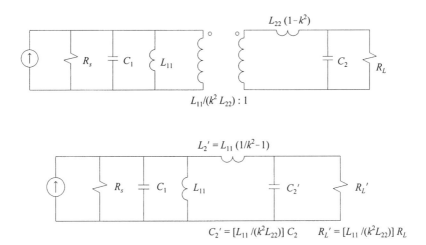

Figure 5.11   Two equivalent circuits for a parallel double-tuned transformer.

If R_L' is small enough, there will be two frequencies where the series resistance of the left-hand section is equal to that of the right-hand section.

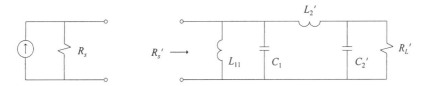

Figure 5.12   An equivalent circuit for determining the frequencies where maximum power is transferred.

Because the transformed resistance varies over the passband, there must inevitably be some ripple in the passband. The size of the ripple is a function of the ratio of the source resistance R_s to the series resistance of the right-hand section at the resonant frequency

$$\omega_0 = 1 / \sqrt{L_{11} C_1}$$

A Design Procedure for a Double-Tuned Transformer

Specifications: The two maximum power transfer frequencies f_{m1} and f_{m2}, the load resistance R_L, the source resistance R_s, and the transducer power gain (G_T) at the center frequency.

1.   As might be expected, the minimum ripple in the passband is a function of

the bandwidth required. It can be shown that, for two given maximum power transfer frequencies, f_{m1} and f_{m2}, the minimum value of the transducer power gain in the passband will always be lower than

$$G_{T,min} = \frac{4 f_{m2} / f_{m1}}{(f_{m2} / f_{m1})^2 + 2 f_{m2} / f_{m1} + 1} \tag{5.41}$$

The first step in the design process, therefore, is to establish whether the specified $G_{T,min}$ can be realized.

2.   Calculate the resistance ratio r ($r = R_s / [R'_L (1 + Q_0^2)]$) for the required passband ripple:

$$r = \frac{1 + |1 - G_T|^{1/2}}{1 - |1 - G_T|^{1/2}} \tag{5.42}$$

3.   Calculate the value of Q_2 (i.e., the second transformation Q of the transforming section in Figure 5.12) at the two desired maximum power transfer frequencies:

$$Q^2_{2\_m1} = r f_{m1} / f_{m2} - 1 \tag{5.43}$$

$$Q^2_{2\_m2} = r f_{m2} / f_{m1} - 1 \tag{5.44}$$

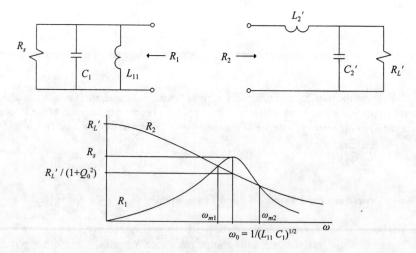

Figure 5.13   The series resistance of the two impedance-matching sections as a function of frequency.

4. Solve the following two equations for the values of L_2' and C_2':

$$-\omega_{m1} L_2' + (1/C_2')/\omega_{m1} = |Q_{2\_m1}| R_s /[1 + Q_{2\_m1}^2] \tag{5.45}$$

$$+\omega_{m2} L_2' - (1/C_2')/\omega_{m2} = |Q_{2\_m2}| R_s /[1 + Q_{2\_m2}^2] \tag{5.46}$$

5. Calculate the value of R_L':

$$R_L' = (1 + Q_{2\_m1}^2)/(\omega_{m1}^2 C_2'^2 R_s) \tag{5.47}$$

4. Calculate the value of the input susceptance (B_{m1}, B_{m2}) of the right-hand section in Figure 5.13 at the maximum power transfer frequencies f_{m1} and f_{m2}.

These susceptances are given by the equations

$$B_{m1} = \text{Imag} \cfrac{1}{j\omega_{m1} L_2' + \cfrac{1}{G_2' + j\omega_{m1} C_2'}} \tag{5.48}$$

$$B_{m2} = \text{Imag} \cfrac{1}{j\omega_{m2} L_2' + \cfrac{1}{G_2' + j\omega_{m2} C_2'}} \tag{5.49}$$

7. Solve the following two equations for the values of L_{11} and C_1:

$$(1/L_{11})/\omega_{m1} - \omega_{m1} C_1 = |B_{m1}| \tag{5.50}$$

$$(1/L_{11})/\omega_{m2} - \omega_{m2} C_1 = -|B_{m2}| \tag{5.51}$$

8. Calculate the values of k, L_{22}, and C_2 by using the following equations:

$$k = 1/(1 + L_2'/L_{11})^{1/2} \tag{5.52}$$

$$L_{22} = (L_{11}/k^2) R_L / R_L' \tag{5.53}$$

$$C_2 = [L_{11}/(k^2 L_{22})] C_2' \tag{5.54}$$

9. If the components of the transformer turn out to be unrealizable, it is often possible to obtain a practical circuit by impedance-scaling the results. L-sections can then be used to transform the load resistance and the source resistance as required. The alternative is to design a series double-tuned transformer.

10.   Check the insertion loss and the frequency response by following the
        procedures outlined in Sections 4.8 and 4.9.

EXAMPLE 5.3   Designing a parallel double-tuned transformer to have a
                         passband ripple of less than 0.5 dB.

The procedure outlined above was followed to design a parallel double-tuned transformer with perfect matching at 9.0 and 11.0 MHz and a passband ripple of less
than 0.5 dB ($G_T = 10^{-0.5/10} = 0.89$). The load resistance is 50Ω, and the source
resistance is 500Ω.
        The results of the calculations in the different steps are repeated here:

1. $G_{T,min} = 0.89$

2. $r = 1.9925$

3. $Q_{2\_9}^2 = 0.630$ ($Q_{2\_9} = 0.739$)

        $Q_{2\_11}^2 = 1.435$ ($Q_{2\_11} = 1.198$)

4. $L_2' = 19.5\,\mu H$

        $C_2' = 13.2\,pF$

5. $R_L' = 5.9\,k\Omega$

6. $B_{m1} = 1.49\,mS$

        $B_{m2} = -2.36\,mS$

7. $C_1 = 157\,pF$

        $L_{11} = 1.71\,\mu H$

8. $k = 0.284$

        $C_2 = 1.55\,nF$

        $L_{22} = 0.18\,\mu H$

9.   The transducer power gain of the double-tuned transformer is given in
        Table 5.3 as a function of the frequency (the components were assumed to
        be lossless).

Table 5.3 The Transducer Power Gain of the Double-Tuned Transformer as a Function of the Frequency

Frequency (MHz)	G_T (dB)
4.00	-23.00
5.00	-19.00
7.50	-5.50
8.00	-2.40
9.00	-0.06
9.95	-0.53
11.00	-0.05
12.00	-3.70
15.00	-19.00

EXAMPLE 5.4   Illustration of the rejection obtainable with a parallel double-tuned transformer.

As an illustration of the good rejection characteristics of a double-tuned transformer, the -30-dB quality factor of a double-tuned transformer with maximum power transfer frequencies at 5.97 and 6.03 MHz (0.1-dB ripple) was calculated.

The -30-dB cutoff frequencies for the designed transformer are approximately 5.915 and 6.082 MHz. The resulting -30-dB Q-factor, therefore, is

$$Q_{-30} = 36$$

If the same results were to be obtained with a simple parallel resonant circuit, the required 3-dB Q-factor would have been 1,026! The highest transformation Q in the double-tuned transformer is 60.

The -3-dB Q-factor for the double-tuned transformer is approximately equal to 7. The ratio of the -3-dB and -30-dB Q-factors of the transformer, therefore, is

$$Q_{-30} / Q = 36 / 7 = 5.1$$

This ratio is 145 times that of the equivalent simple parallel resonant circuit.

5.8   SERIES DOUBLE-TUNED TRANSFORMERS

Although it is not the dual of the parallel double-tuned transformer, the response of the series double-tuned transformer shown in Figure 5.14 is similar to that of the parallel double-tuned transformer.

The insertion loss of the series circuit will be lower than that of the equivalent parallel circuit when the load resistance and source resistance are low.

Impedance transformation can be obtained with the circuit shown in Figure 5.14 by scaling the components on each side of the transformer to the required levels while the coupling factor remains unchanged.

The series-tuned transformer can be designed to match the source conjugately to the load at two specified frequencies by following the procedure outlined below.

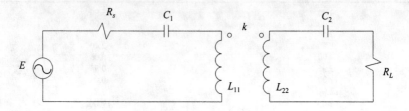

Figure 5.14   A series double-tuned transformer.

A Design Procedure for a Series Double-Tuned Transformer

Specifications: The two maximum power transfer frequencies, f_{m1} and f_{m2}, the load resistance, the source resistance, and the transducer power gain (G_T) at the center frequency.

1.   Calculate the value of the product kQ:

$$kQ = 1/\sqrt{G_T} + \sqrt{\frac{1}{G_T} - 1} \qquad (5.55)$$

k is the required coupling factor and Q is the quality factor of each of the two sides of the transformer when the coupling between them is equal to zero.

2.   Determine the coupling factor required by solving the following equation:

$$k^4 + k^2[(kQ)^4 M^2 - 4(kQ)^2] + [4 - M^2](kQ)^4 = 0 \qquad (5.56)$$

where

$$M = \frac{f_{m1}^2 + f_{m2}^2}{f_{m1} f_{m2}} \qquad (5.57)$$

3.   Determine the required uncoupled Q-factor and the circuit components:

$$Q = (kQ)/k \qquad (5.58)$$

$$f_0 = [f_{m1}f_{m2}(1-k^2)^{1/2}]^{1/2} \tag{5.59}$$

$$C_1 = 1/[\omega_0 R_s Q] \tag{5.60}$$

$$L_{11} = QR_s / \omega_0 \tag{5.61}$$

$$L_{22} = QR_L / \omega_0 \tag{5.62}$$

$$C_2 = 1/[\omega_0 R_L Q] \tag{5.63}$$

If some of these components are unrealizable, the specifications can be changed to be more realistic, or, in some cases, the components can be scaled to more realistic values. In the latter case, L-sections can be used to transform the source and load terminations to those required.

4. The insertion loss and frequency response of the transformer can be determined by following the procedure outlined in Sections 4.8 and 4.9. The 3-dB bandwidth of the series double-tuned transformer can be calculated by using the equation

$$B = \sqrt{b^2 + 2b - 1}\ f_0 / Q \tag{5.64}$$

where

$$b = k / k_c \tag{5.65}$$

and

$$k_c = 1 / Q \tag{5.66}$$

EXAMPLE 5.5   Designing a series double-tuned transformer to transform a 50-Ω load to 20Ω.

As an example of the application of the procedure outlined above, a transformer was designed to have maximum power transfer frequencies at 27 and 28 MHz. The other specifications were $G_T = 0.89$, $R_s = 50\Omega$, and $R_L = 20\Omega$.

The results obtained in the different steps are as follows:

1. $kQ = 1.41156$

2. $M = 2.00132$

    $k^4 + 7.93123\ k^2 - 0.02101 = 0$

$k = 0.05246$

3. $Q = 27.43$
    $f_0 = 27.48 \text{ MHz}$
    $C_1 = 4.2 \text{ pF}$
    $L_{11} = 7.94 \text{ μH}$
    $L_{22} = 3.18 \text{ μH}$
    $C_2 = 10.6 \text{ pF}$

4. If the transformer is assumed to be lossless, the 3-dB bandwidth is $B = 1.97$ MHz. The -3-dB and -30-dB Q-factors of the transformer are

    $Q = 13.98$
    $Q_{-30} = 2.81$

    The ratio of the two Q-factors is

    $Q_{-30} / Q = 0.2$

5.9   MEASUREMENT OF THE COUPLING FACTOR OF A TRANSFORMER

Accurate measurement of the coupling factor of a transformer is sometimes required. Three ways to determine the coupling factor will be discussed here.

In the first method, the open-circuit and short-circuit input inductances of the transformer are measured, and the coupling factor is derived from these values. This method can only be used when the losses in the transformer are negligible.

In the second method, the coupling factor is estimated by measuring the open-circuit voltage gain of the transformer. An oscilloscope or voltmeter with a high input impedance (compared to the leakage reactance of the transformer at the measuring frequency) is required if reliable results are to be obtained with this method.

The Z-parameters of the transformer are used in the last method. It is usually more convenient to measure the S-parameters of the transformer. These parameters can be converted easily to Z-parameters by using (1.148).

5.9.1   Measurement of the Coupling Factor by Short-Circuiting the Secondary Winding

The influence of short-circuiting the secondary winding of two coupled coils can be established by using the equivalent circuit shown in Figure 5.3(b). The resistive losses will be assumed to be negligible.

The input admittance of the transformer, with the secondary winding short-circuited, is given by the equation

$$1/(\omega L_T) = 1/(\omega L_{11}) + 1/[(L_{11}/(k^2 L_{22}))\,\omega L_{22}\,(1-k^2)]$$

$$= 1/(\omega L_{11}) + 1/[\omega L_{11}(1/k^2 - 1)] \tag{5.67}$$

This equation can be rewritten to obtain the value of the coupling factor as a function of the other variables:

$$k = [1 - L_T/L_{11}]^{1/2} \tag{5.68}$$

It is therefore possible to determine the coupling factor of the transformer by measuring the open-circuit and short-circuit impedances of the transformer (i.e., if the losses in the transformer can be ignored).

5.9.2 Measurement of the Coupling Factor by Measuring the Open-Circuit Voltage Gain

When the equivalent circuit of Figure 5.3(b) applies, the open-circuit voltage gain of the transformer is given by the equation

$$V_2 = \pm[j\,\omega\,k(L_{11}L_{22})^{1/2}]V_1/(r_1 + j\omega L_{11}) \tag{5.69}$$

From this equation, the coupling factor can be obtained as

$$k = [(r_1^2 + \omega^2 L_{11}^2)/(\omega^2 L_{11} L_{22})]^{1/2}\, V_2/V_1 \tag{5.70}$$

The coupling factor can, therefore, be determined by measuring the open-circuit voltage gain of the transformer. r_1 and the inductances L_{11} and L_{22} can be determined by measuring the open-circuited primary and secondary input impedances of the transformer.

It is important for the input impedance of the voltmeter or oscilloscope used to be much higher than the leakage reactance $\omega L_{22}(1 - k^2)$. Because of this, it is a good idea to define the low-impedance side of the transformer as the secondary side.

5.9.3 Deriving the Coupling Factor from *S*-Parameter Measurements

If the equipment is available, the *S*-parameters of the transformer can be measured. These parameters can be converted to *Z*-parameters by using standard conversion formulas.

If the equivalent circuit of Figure 5.3(b) applies, the transformer *Z*-parameters are given by the equation

$$Z = \begin{bmatrix} r_1 + j\omega L_{11} & \pm j\omega M \\ \pm j\omega M & r_2 + j\omega L_{22} \end{bmatrix} \tag{5.71}$$

With the Z-parameters known, it is a simple matter to determine the copper losses, as well as the primary and secondary inductances.

The coupling factor can be determined from the mutual inductance M and the magnetizing inductances L_{11} and L_{22} by using (5.7).

QUESTIONS AND PROBLEMS

5.1   Prove that the load of an ideal transformer is transformed by a factor equal to the square of the turn ratio of the transformer.

5.2   Prove that $L_{11}/L_{22} = (n_1/n_2)^2$ if the primary and secondary windings of a transformer are wound around a highly permeable toroidal core.

5.3   The low cutoff frequency (3 dB) of a transformer is to be 2 MHz. The load and source resistance values are 400Ω and 50Ω, respectively. What is the value of the magnetizing reactance required? If the coupling factor can be assumed to be equal to one, what is the required inductance of the secondary winding?

5.4   What is the minimum coupling factor required to match a load of 100Ω to 10Ω over the frequency range 2–30 MHz (that is, a 3-dB bandwidth)?

5.5   Prove that the two circuits shown in Figure 5.3 are equivalent.

5.6   Explain why it is that the coupling factor of a transformer with a core of stacked toroids (or a balun core) is higher than that of an equivalent transformer with a single toroidal core.

5.7   Design a single-tuned transformer to match a load of 150Ω to a source with 50-Ω internal resistance. The Q of the circuit must be 10 and f_0 = 100 MHz.

5.8   Design a tapped-coil resonant circuit to-transform a load of 1,000Ω to 50Ω. The required Q of the circuit is 25 and f_0 = 10 MHz. The insertion losses of the circuit must be as low as practically possible.

5.9   Design the filter shown in Figure 5.15 to have the highest Q practically possible. The resonant frequency must be 10 MHz and the space available for the inductor is 4 cm×4 cm×4 cm. High Q porcelain capacitors may be used. (Note that the Q-factors of these capacitors are not necessarily negligible.)

5.10 Design a double-tuned transformer to match a load of 250Ω conjugately to a source of 1,000Ω at 8.8 MHz and 9.6 MHz. The maximum ripple in the passband must be 1 dB. Determine the -3-dB and -30-dB frequencies of the transformer iteratively.

5.11 Design the double-tuned transformer to match a load of 50Ω to a source with 20-Ω internal resistance. The maximum power transfer frequencies must be 89 MHz and

94 MHz. The maximum ripple in the passband must be less than 0.25 dB.

Determine the -3-dB and -30-dB frequencies of the designed transformer.

5.12 What is the smallest ripple in the passband of the parallel double-tuned transformer when the ratio of the two maximum power transfer frequencies is (a) 1.1, (b) 1.25, (c) 1.50, (d) 1.75, and (e) 2.00?

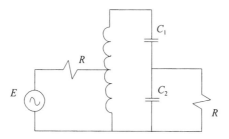

Figure 5.15 A high-Q tapped-coil resonant circuit.

5.13 Determine the insertion losses of the transformers designed in Problems 5.10 and 5.11.

5.14 The transformer in Problem 5.10 is to be replaced with three discrete inductors (refer to Figure 5.5). Calculate the values of the inductances required.

5.15 Find the dual of the series and parallel double-tuned circuits.

5.16 Use the knowledge gained in the previous problem and the standard equations available for transforming a PI-section to an equivalent T-section to design the network shown in Figure 5.16 so as to have maximum power transfer frequencies at 9.75 MHz and 10.25 MHz. $R_L = 250\Omega$; $R_s = 500\Omega$.

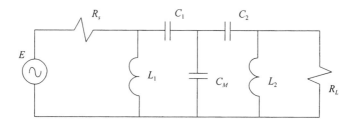

Figure 5.16 Two coupled resonant circuits.

5.17   The Z-parameters of a transformer are

$$\begin{bmatrix} 10 + j50 & -j69 \\ -j69 & 20 + j150 \end{bmatrix}$$

at 25 MHz. Determine the values of L_{11}, L_{22}, r_1, r_2, and k.

5.18   The circuit shown in Figure 5.17 can be used as the output circuit of a vacuum-tube amplifier. The series capacitor is used to tune the circuit. The parallel capacitor is the output capacitance of the tube. The output power of the amplifier is to be as close as possible to 350W over the frequency range 115–150 MHz. The load resistance is 50Ω and the power supply voltage 2 kV. Assuming that the saturation effects and losses in L_{11} can be ignored, find the optimum values for L_{11}, L_{22}, and k. Determine the tuning range of the capacitor C.

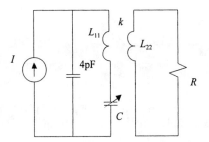

Figure 5.17   An output matching network for a vacuum-tube power amplifier.

REFERENCES

[1]   Skilling, H. H., *Electrical Engineering Circuits*, New York: John Wiley and Sons, 1965.

[2]   Van der Walt, P. W., "A Simple Procedure for Designing Impedance-Matching Networks with Loosely Coupled Transformers," Research Note, University of Stellenbosch, South Africa.

[3]   Penney, R. C., *Linear Algebra: Ideas and Applications*, New York: John Wiley and Sons, 2008.

CHAPTER 6

TRANSMISSION-LINE TRANSFORMERS

6.1 INTRODUCTION

The high-frequency response of a magnetically coupled transformer is limited by the leakage inductance and the parasitic capacitance of the transformer.

The leakage inductance can be decreased significantly if a balun or a stacked core (instead of a toroidal core) is used. It is more difficult, however, to decrease the parasitic capacitance between the two windings and the turns of each winding.

If the outer and center conductor of a coaxial cable are used as the primary and secondary windings of a 1:1 transformer and connected as shown in Figure 6.l(b), a 1:4 impedance transformation can be obtained [similar to that obtained with the auto-transformer shown in Figure 6.1(a)] and the parasitic capacitance between the windings can be controlled.

One would expect the performance of this transmission-line transformer to be optimum when the capacitance between the windings is low, that is, when the characteristic impedance of the line is high. Fortunately, this is not true; rather, there is an optimum characteristic impedance for the line. The high-frequency performance of the transformer is therefore improved by the transmission-line effect.

At low frequencies the transmission-line transformer can be considered to be a conventional transformer with excellent coupling. Because of the transmission-line capacitance, however, this model is not valid at high frequencies. In fact, the magnetic coupling between the windings can be removed completely (by not using a magnetic core and straightening the line), and the high-frequency performance of the transformer will not be influenced at all (i.e., if the losses were negligible).

That this is possible can be appreciated by assuming the currents in the transmission line to be balanced and the line to be short enough for the phase difference between the voltages across the line and the currents in the line to be small. This is illustrated in Figure 6.2.

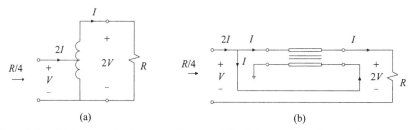

Figure 6.1   (a) A conventional auto-transformer and (b) a 1:4 transmission-line transformer.

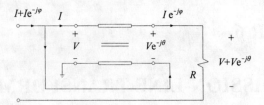

Figure 6.2   The voltages across and the currents in a transmission-line transformer at high frequencies.

It has been assumed in Figure 6.2 that the currents in the transmission line are balanced. If the currents were perfectly balanced, as shown in Figure 6.3(a), the output voltage would have been zero, which is not the case. The currents in the line must be unbalanced for the output voltage to be nonzero.

Because

$$V_o = 2sL_u\ell I_{1u} = 2j\omega L_u\ell I_{1u} \qquad (6.1)$$

the unbalanced current is very small at high frequencies. In (6.1) L_u is the inductance per unit length for the unbalanced current, ℓ the length of the line, and I_{1u} the unbalanced component of the current in each of the two conductors of the line.

Although the high-frequency performance of the transmission-line transformer is not affected by the removal of the magnetic core, the low-frequency response is seriously degraded when this is done. The reason for this is the increase in the unbalanced current. This current increases approximately inversely with frequency. When magnetic material is used, the inductance in (6.1) is increased and the unbalanced current is reduced. If the unbalanced current is small enough, the frequency response of the transformer will be very good at both high and low frequencies.

The bandwidth obtainable with a transmission-line transformer is significantly better than that obtainable with a conventional transformer.

Because of the wide bandwidth, transmission-line transformers are often used to transform resistance. When only one transmission line is used, the only transformation ratios that can be obtained are 1:1 and 1:4. When more than one line is used, it is possible to realize transformers with other transformation ratios.

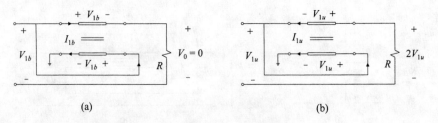

(a)   (b)

Figure 6.3   The output voltage of the 1:4 transmission-line transformer as a function of (a) the balanced and (b) unbalanced currents in the line.

Apart from impedance matching, transmission-line transformers are also used to perform various combining and splitting functions.

The most commonly used configurations will be presented in the next section.

The analyses of the 1:4 transmission-line transformer and other transmission-line transformers will be discussed in detail in Section 6.3. It will be shown in this section that the basic component of any transmission-line transformer is an unbalanced transmission line with increased inductance for the unbalanced currents in the line. This simplifies to a balanced transmission line at high frequencies and a 1:1 transformer with magnetizing inductance at low frequencies.

The design of transmission-line transformers primarily requires that the transformer meets the low-frequency specifications. Compensation at low and/or high frequencies may also be required to extend the bandwidth. The various steps in designing these transformers will be considered in detail in Section 6.4.

Transmission-line transformers are used extensively in RF power amplifiers. The design of impedance-matching networks for these amplifiers will be considered in Section 6.5.

6.2   TRANSMISSION-LINE TRANSFORMER CONFIGURATIONS

Transmission-line transformers are used to change resistance levels in impedance-matching networks and amplifiers, as well as to perform certain splitting and combining functions.

The transformation ratios obtainable with these transformers are limited to those shown in Table 6.1, that is, if less than five transmission lines are used.

The number of lines used in a practical application is limited by the available space. To ensure a good low-frequency response, each line must be wound around a magnetic core.

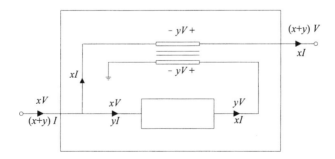

Figure 6.4   The influence of adding an extra line on the impedance transformation ratio of a transmission-line transformer.

More than one line can sometimes be wound around the same core. The polarity of the voltage induced by the flux in the core and the relative size of the voltage across each winding must be taken into account when this is done.

Table 6.1   Transformation Ratios Obtainable with Transmission-Line Transformers as a Function of the Number of Lines Used

Number of lines	1	2	3	4
	$1:1.00 \ (1/1)^2$	$1:2.25 \ (2/3)^2$	$1:1.78 \ (3/4)^2$	$1:1.56 \ (4/5)^2$
	$1:4.00 \ (1/2)^2$	$1:4.00 \ (1/2)^2$	$1:2.78 \ (3/5)^2$	$1:1.96 \ (5/7)^2$
Transformation	—	$1:9.00 \ (1/3)^2$	$1:6.25 \ (2/5)^2$	$1:2.56 \ (5/8)^2$
ratios	—	—	$1:16.0 \ (1/4)^2$	$1:3.06 \ (4/7)^2$
obtainable	—	—	—	$1:5.44 \ (3/7)^2$
	—	—	—	$1:7.11 \ (3/8)^2$
	—	—	—	$1:12.3 \ (2/7)^2$
	—	—	—	$1:25.0 \ (1/5)^2$

The configuration corresponding to a particular transformation ratio can be found by using the technique illustrated in Figure 6.4 [1]. When this technique is applied, the ratio of the input to output voltage changes from x/y to $[x/(x+y)]$. The impedance transformation ratio is changed from $(x/y)^2$ to $[x/(x+y)]^2$. In the simplest case,

$$x = 1 = y$$

and the application of this technique results in the configuration for the 1:4 transmission-line transformer.

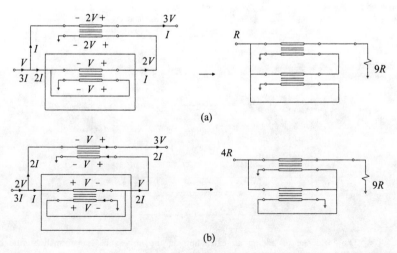

Figure 6.5   Derivation of the configurations for the (a) 1:9 and (b) 4:9 transmission-line transformers.

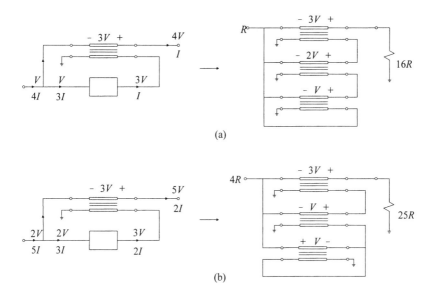

(a)

(b)

Figure 6.6   The configurations of the (a) 1:16 and (b) 4:25 transmission-line transformers.

If the technique is applied to the 1:4 transformer, the 1:9 transformer shown in Figure 6.5(a) is obtained.

By considering the high impedance side of the 1:4 transformer to be the input, the 1:2.25 transformer shown in Figure 6.5(b) is obtained.

The configurations for the $(3/4)^2$, $(1/4)^2$ and the $(2/5)^2$, $(3/5)^2$ transformers can be found by applying the technique to the 1:9 and 9:1 and the 4:9 and 9:4 transformers, respectively. The configurations for the 1:16 and the 4:25 transformers are shown in Figure 6.6.

The transformers most often used in power amplifiers are the 1:4 and 1:9 transmission-line transformers. The high cutoff frequency of the 1:4 transformer can be increased considerably if two lines instead of only one are used, as shown in Figure 6.7 [2].

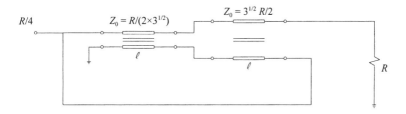

Figure 6.7   The configuration of a 1:4 transmission-line transformer that has no high cutoff frequency (theoretically).

Impedance transformation between a balanced source and a balanced load is often required. The configurations for the balanced 1:4 and 1:9 transmission-line transformers are shown in Figure 6.8(a, b).

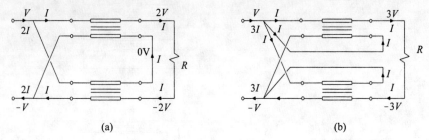

(a)   (b)

Figure 6.8   The configurations for the balanced (a) 1:4 and (b) 1:9 transmission-line transformers.

When either the load or the source is unbalanced, the 1:1 transformer shown in Figure 6.9 can be used to provide the required unbalanced-to-balanced transformation.

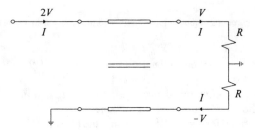

Figure 6.9   The unbalanced-to-balanced 1:1 transmission-line transformer.

At high frequencies, the frequency response of the 1:1 transformer is exactly the same as that of a transmission line terminated in the same load.

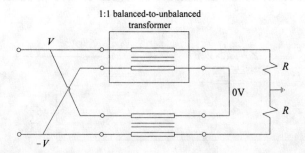

Figure 6.10   Illustration of the equivalence between the 1:4 balanced and the 1:1 balanced-to-unbalanced transmission-line transformers.

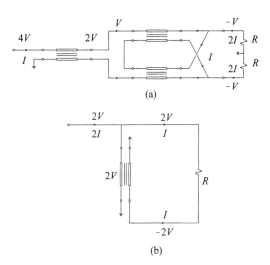

(a)

(b)

Figure 6.11   (a) Combination of a 1:1 transformer and a 4:1 transformer to obtain an unbalanced-to-balanced transformation. (b) A 1:4 unbalanced transmission-line transformer.

Because of the symmetry, the high-frequency response of the 1:4 balanced transformer is identical to that of the 1:1 transformer. The equivalence is illustrated in Figure 6.10.

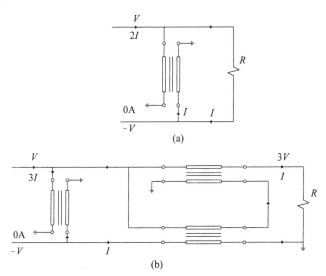

(a)

(b)

Figure 6.12   (a) 1:4 and (b) 1:9 transformers for combining the currents of the transistors in a class B amplifier into a single load.

A 4:1 unbalanced-to-balanced transformation can be obtained by combining the 1:1 and 1:4 balanced transformers, as shown in Figure 6.11(a), or by using the transformer shown in Figure 6.11(b). The latter transformer is less sensitive to nonoptimum impedances than the former, although it has a lower cutoff frequency when the optimum characteristic impedance is used.

The output currents of the two transistors in a push-pull class B amplifier are often combined (at lower frequencies where the conduction angle is 180°) by using either of the 1:4 or 1:9 transformers shown in Figure 6.12.

Although they are used for different purposes, it can be seen that the configurations of the 1:4 transformer shown in Figure 6.12 are similar to that of the 1:4 unbalanced-to-balanced transformer shown in Figure 6.11(b).

By redefining the reference plane of the 1:4 transformer in Figure 6.12(a) (as shown in Figure 6.13), it becomes clear that the frequency responses of both are identical.

The combiner shown in Figure 6.14(a) is often used to combine two in-phase signals at radio frequencies.

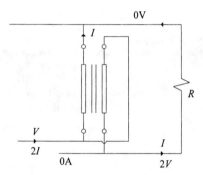

Figure 6.13   The configuration of the 1:4 transformer from Figure 6.12 with a redefined reference plane (ground).

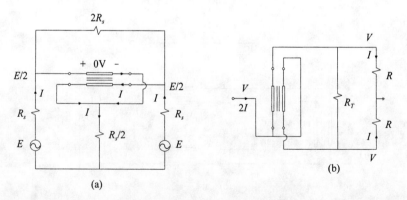

Figure 6.14   (a) A transformer for combining two in-phase signals into the same load; (b) the same transformer used as an in-phase power splitter.

As indicated in the figure, the voltage drop across the 1:1 transformer used in the combiner is equal to zero when the two input signals are equal in amplitude and are in-phase. When the signals are unbalanced, the two sources will be isolated from each other by the transformer. This is illustrated in Figure 6.15 for the case where $E_2 = 0$. If the transformer is assumed to be ideal, and $R_T = 2R_s = 4R_L$, no current will flow in the resistance R_{s2}.

At low frequencies, the isolation obtainable with this transformer is a function of the magnetizing inductance of the 1:1 transformer; that is,

$$S = [4\omega L_{11} / (R_s / 2)]^2 + 1 \qquad (6.2)$$

where S is the ratio of the power dissipated in the load ($R_L = R_s/2$) and the power dissipated in the source resistance (R_{s2}), when $E_2 = 0$.

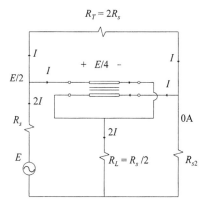

Figure 6.15   Illustration of the isolating action of the hybrid transformer shown in Figure 6.14(a).

The isolation at high frequencies is a function of the electrical length of the line and the characteristic impedance. The combiner can be changed to a splitter by connecting it as shown in Figure 6.14(b).

As in the case of the combiner, the voltage drop across the 1:1 transformer will be zero as long as the loads are balanced. If not, the transformer will cause the power to be distributed more evenly between the two loads than would be the case with a direct connection.

6.3   ANALYSIS OF TRANSMISSION-LINE TRANSFORMERS

The basic building block of a transmission-line transformer is an unbalanced transmission line (see Figure 6.16). The line can be wound around magnetic material or can be shaped as a solenoidal coil. The latter can be done by using semirigid coaxial cable.

The currents in the two conductors of an unbalanced transmission-line are not equal, but are related by the equation

$$I_2(x) = I_1(x) + I_0(x) \tag{6.3}$$

Because the effective current entering the line at any point $[I_{\text{eff}} = I_1(x) - I_2(x) = -I_0(x)]$ must be equal to the current flowing out of the line at any other point further along the line (see Figure 6.17), the unbalanced current $(I_0(x))$ is independent of the distance (x) along the line. Therefore, (6.3) simplifies to

$$I_2(x) = I_1(x) + I_0 \tag{6.4}$$

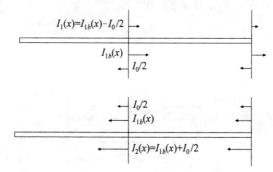

Figure 6.16   The balanced and unbalanced components of the current in a transmission line.

The effect of the magnetic material (or solenoidal shape) is to increase the impedance associated with the unbalanced currents in the line.

Because there is no external magnetic field associated with the balanced currents ($\oint H\,dl = I_{1b} - I_{1b} = 0$), these currents are not influenced by the magnetic material used or by the form in which the line is wound.

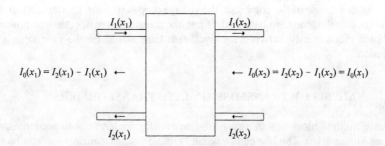

Figure 6.17   The unbalanced current on a transmission line as a function of the distance along the line.

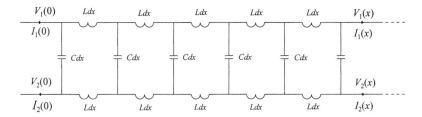

Figure 6.18 The equivalent circuit of an unbalanced transmission-line.

If the influence of the magnetic core or the coil form on the unbalanced currents is ignored initially, the equivalent circuit shown in Figure 6.18 applies and equations for the voltage on and the current in the unbalanced transmission line can be derived in a way similar to the balanced transmission line case (refer to the appendix). The results of the derivation are shown here:

$$I_1(x) = -I_0/2 + Ae^{-\Gamma x} + Be^{+\Gamma x} \tag{6.5}$$

$$I_2(x) = I_0/2 + Ae^{-\Gamma x} + Be^{+\Gamma x} \tag{6.6}$$

$$V_1(x) = V_1(0) - Z_0/2 \cdot (A - B) + Z_0/2 \cdot [Ae^{-\Gamma x} - Be^{\Gamma x}] \\ + sL_u x I_0/2 \tag{6.7}$$

$$V_2(x) = V_2(0) + Z_0/2 \cdot (A - B) - Z_0/2 \cdot [Ae^{-\Gamma x} - Be^{\Gamma x}] \\ + sL_u x I_0/2 \tag{6.8}$$

$$V_{12}(x) = V_1(x) - V_2(x) = Z_0[Ae^{-Gx} - Be^{Gx}] \tag{6.9}$$

where

$$\Gamma = \sqrt{2LC}\,s = j\omega\sqrt{2LC} \tag{6.10}$$

$$Z_0 = \sqrt{2L/C} \tag{6.11}$$

L and C are the inductance and capacitance of the line per unit length, respectively, and x is the position of the point of interest on the line (relative to the LHS). Note that the inductance for the balanced currents (L) and that for the unbalanced currents (L_u) are not the same because of the magnetic coupling between the two conductors of the line.

When magnetic material is used or the line is shaped as a coil, the reactance associated with the unbalanced currents ($I_0/2$) must be changed from $sL_u l$ to

$$X_u = 2sL_{11} \tag{6.12}$$

where L_{11} is the inductance associated with each conductor of the line when the other conductor is open-circuited and l is the length of the transmission line.

The inductance associated with the unbalanced current in each conductor is twice the expected value (L_{11}) because of the excellent coupling between the two conductors of the transmission line. When current is flowing in only one conductor, the voltage across the length of the other conductor will be equal to that of the first, provided that there are no resistive losses in the conductors. The coupling factor, therefore, is very close to unity.

When magnetic material is used or the line is wound as a coil, (6.7) and (6.8) must be changed to

$$V_1(x)=V_1(0)-(Z_0/2)(A-B)+(Z_0/2)[Ae^{-\Gamma x}-Be^{\Gamma x}]$$
$$+s(2L_{11})(x/l)I_0/2$$

and

$$V_2(x)=V_2(0)+(Z_0/2)(A-B)+(Z_0/2)[Ae^{-\Gamma x}-Be^{\Gamma x}]$$
$$+s(2L_{11})(x/l)I_0/2$$

Before (6.5)–(6.9) can be used to determine the voltages on and the currents in any particular transmission-line transformer, the constants A, B, I_0, $V_1(0)$, and $V_2(0)$ must be determined.

These constants can be determined by using the boundary conditions for the transformer under consideration.

When the voltages and currents are known, the power gain and the input and output impedances of the transformer can be determined easily.

Because the transmission line can usually be considered lossless, it is sufficient that the input impedance of a transformer is known (in the lossless case, the magnitudes of the input and output reflection coefficients are equal, and the magnitude of the transducer power gain is only a function of the reflection coefficient). The input impedance of a transmission-line transformer is a function of the frequency, the load impedance, and the length and characteristic impedance of the transmission line used. The expression for the input impedance is therefore usually quite complex.

Although the complexity is not a problem when a computer program is used to analyze a transmission-line transformer, it is possible to simplify the equation for the input impedance at low and high frequencies by making appropriate assumptions. At high frequencies, the reactance associated with each conductor is high compared to the characteristic impedance of the line, and the approximation

$$sL_u l >> Z_0 \qquad (6.13)$$

can be made. Under this approximation, the input impedance of the transformer is only a function of the balanced currents in the line. As far as the impedance is concerned, the transmission line can then be considered balanced.

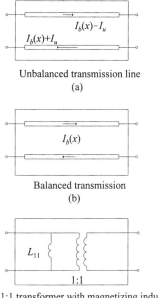

Unbalanced transmission line
(a)

Balanced transmission
(b)

Ideal 1:1 transformer with magnetizing inductance
(c)

Figure 6.19   (a) The basic building block of a transmission-line transformer simplified at (b) high and (c) low frequencies.

At low frequencies, the line is electrically very short and the approximation

$$e^{\pm \Gamma l} = 1 \tag{6.14}$$

can be made and the input impedance of the transformer is independent of the length and the characteristic impedance of the transmission line. The transmission-line transformer can then be considered to be a conventional transformer.

It follows that the basic building block of a transmission-line transformer reduces to a balanced transmission line and a conventional 1:1 transformer with magnetizing inductance L_{11} at high and low frequencies, respectively. This is illustrated in Figure 6.19.

EXAMPLE 6.1   The input impedance of a 1:4 transmission-line transformer.

The input impedance of a 1:4 unbalanced transmission-line transformer (see Figure 6.20) will be determined as an example of the application of (6.5) to (6.11). The boundary conditions for the transformer are as follows:

$$V_2(0) = 0 \tag{6.15}$$

$$V_2(l) = V_1(0) \tag{6.16}$$

$$V_1(l) = Z_L I_1(l) \tag{6.17}$$

These conditions will be used to find two independent equations for the unbalanced current I_0 in terms of A and B. In this way, the relationship between A and B can be established and the input impedance can be found:

$$V_1(0) = V_{12}(0) = Z_0[Ae^{-\Gamma x} - Be^{\Gamma x}] = Z_0(A - B)$$

$$V_2(l) = 0 + 0.5Z_0[A - B] + sL_u l I_0 / 2 - 0.5Z_0[Ae^{-\Gamma l} - Be^{\Gamma l}]$$

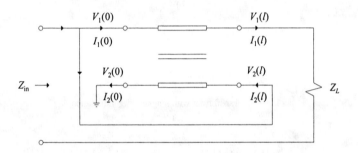

Figure 6.20   The 1:4 unbalanced transmission-line transformer.

Because $V_1(0)$ and $V_2(l)$ are equal, these two equations can be used to obtain an equation for I_0 in terms of A and B. After some manipulation, the following equation is obtained:

$$sL l I_0 = Z_0 / (sL_u l) \cdot [A - B] + Z_0 / (sL_u l) \cdot [Ae^{-\Gamma l} - Be^{\Gamma l}] \tag{6.18}$$

The second equation is established by using the constraint imposed by the load:

$$V_1(l) = V_1(0) - 0.5Z_0[A - B] + sL_u l I_0 / 2 + 0.5Z_0[Ae^{-\Gamma l} - Be^{\Gamma l}]$$

and

$$Z_L I_1(l) = Z_L[-I_0 / 2 + Ae^{-\Gamma l} + Be^{\Gamma l}]$$

By equating these two equations, it follows that

$$[Z_L + sL_u l]I_0 = 2A[Z_0 e^{-\Gamma l} - 0.5 Z_0(e^{-\Gamma l} + 1)]$$
$$+ 2B[Z_0 e^{\Gamma l} + 0.5Z_0(e^{\Gamma l} + 1)] \tag{6.19}$$

The relationship between A and B can now be determined by using (6.18) and (6.19):

$$\frac{B}{A} = \frac{Z_0 E_2[1 + Z_L / (s L_u l)] - 2[Z_L e^{-\Gamma l} - (Z_0 / 2) E_2]}{Z_0 E_1[1 + Z_L / (s L_u l)] + 2[Z_L e^{\Gamma l} + (Z_0 / 2) E_1]} \tag{6.20}$$

where

$$E_1 = 1 + e^{\Gamma l}$$

and

$$E_2 = 1 + e^{-\Gamma l}$$

The input impedance of the transformer is given by the equation

$$Z_{in} = V_1(0) / [I_1(0) + I_2(l)]$$

$$= Z_0 \frac{1 - B / A}{E_2 + (B / A) \cdot E_1} \tag{6.21}$$

If the approximation

$$e^{\pm \Gamma l} \cong 1$$

is used, the equations for the ratio B/A and the input impedance of the transformer simplifies to

$$\frac{B}{A} = \frac{2Z_0 s L_u l + Z_L Z_0 - Z_L s L_u l}{2Z_0 s L_u l + Z_L Z_0 + Z_L s L_u l} \tag{6.22}$$

and

$$Z_{in} = (Z_0 / 2) \frac{1 - B / A}{1 + B / A} = \frac{(Z_L / 4) \cdot s L_u l / 2}{(Z_L / 4) + s L_u l / 2} \tag{6.23}$$

If magnetic material is used, the reactance $s L_u l$ in these equations must be replaced with $(2L_{11})s$. The input impedance is then

$$Z_{in} = \frac{(Z_L / 4) \cdot s L_{11}}{(Z_L / 4) + s L_{11}} \tag{6.24}$$

At high frequencies, the approximation

$$s L_u l \gg Z_0$$

can be made, and the expression for the ratio B/A simplifies to

$$\frac{B}{A} = \frac{Z_0[1 + e^{-\Gamma l}] - Z_L e^{-\Gamma l}}{Z_0[1 + e^{+\Gamma l}] + Z_L e^{+\Gamma l}} \tag{6.25}$$

The impedance is still given by (6.21).

The transducer power gain for the transformer can be determined by using the equation

$$G_T = 1 - \left| \frac{Z_{in} - Z_s^*}{Z_{in} + Z_s} \right|^2 \tag{6.26}$$

where Z_s is the impedance of the source driving the transformer.

EXAMPLE 6.2   The input impedance of a 1:1 balanced-to-unbalanced transformer.

The input impedance of a 1:1 balanced-to-unbalanced transmission-line transformer (see Figure 6.21) can be determined by using the following boundary conditions:

$$V_1(l) = Z_L I_1(l) \tag{6.27}$$

$$V_2(l) = 0 \tag{6.28}$$

$$V_1(0) = -V_2(0) \tag{6.29}$$

By using (6.27), the unbalanced current is found to be

$$I_0/2 = Ae^{-\Gamma l}[1 - Z_0/Z_L] + Be^{\Gamma l}[1 + Z_0/Z_L] \tag{6.30}$$

When (6.28) and (6.29) are used, the second equation necessary for determining the ratio B/A is found to be

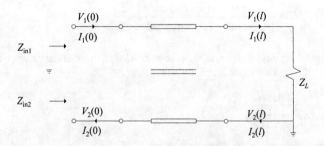

Figure 6.21   The 1:1 balanced-to-unbalanced transmission-line transformer.

$$s L_u l I_0 / 2 = (Z_0 / 2) \cdot [A e^{-\Gamma l} - B e^{\Gamma l}] \tag{6.31}$$

The ratio B/A can now be obtained by using these two equations:

$$\frac{B}{A} = - \frac{e^{-\Gamma l}[1 - Z_0 / Z_L] - [Z_0 / (2s L_u l)] e^{-\Gamma l}}{e^{\Gamma l}[1 + Z_0 / Z_L] + [Z_0 / (2s L_u l)] e^{\Gamma l}} \tag{6.32}$$

When B/A is known, the input impedances Z_{in1} and Z_{in2} can be determined. These impedances are given by the equations.

$$Z_{in1} = \frac{Z_0 [1 - B / A]}{-[Z_0 / (s L_u l)] [e^{-\Gamma l} - (B / A) e^{\Gamma l}] + 2[1 + B / A]} \tag{6.33}$$

$$Z_{in2} = \frac{Z_0 [1 - B / A]}{+[Z_0 / (s L_u l)] [e^{-\Gamma l} - (B / A) e^{\Gamma l}] + 2[1 + B / A]} \tag{6.34}$$

It is clear from the different signs in the denominators of (6.33) and (6.34) that the two input impedances are not equal at low frequencies.

When $s L_u l \gg Z_0$, the two impedances are approximately equal, independent of the characteristic impedance value of the line. Furthermore, the input impedance of the transformer is identical to that of a balanced transmission line terminated in the same load impedance (Z_L).

By using this equivalence, it follows that the input impedance of the 1:1 balanced-to-unbalanced transmission-line transformer will be purely resistive at high frequencies if $Z_0 = R_L$.

Because of the symmetry, the same applies to the 1:4 balanced transmission-line transformer.

6.4   DESIGN OF TRANSMISSION LINE TRANSFORMERS

The design of transmission-line transformers consists of the following:

1.   Determining the characteristic impedance and the diameter of the transmission line to be used;

2.   Determining the minimum value of the magnetizing inductance of the transformer at the lowest passband frequency;

3.   Selecting a suitable magnetic material (if needed);

4.   Determining the type and size of the core to be used;

5.   Calculating the line length and the corresponding high cutoff frequency of the transformer;

6. Compensating the transformer for nonoptimum characteristic impedance, if necessary;

7. Extending the bandwidth by using LC impedance-matching networks, if necessary.

Each of these points will be discussed in detail in the following sections.

6.4.1 Determining the Optimum Characteristic Impedance and Diameter of the Transmission Line to Be Used

At high frequencies, the input impedance of a transmission-line transformer is a function of the characteristic impedance of the transmission line.

The optimum characteristic impedance can be established by taking the ratio of the voltage across one end of the transmission line and the current passing through it. The basic building block of the transformer is then considered to be an ideal 1:1 transformer.

The application of this rule to a 1:4 unbalanced transmission-line transformer is illustrated in Figure 6.22.

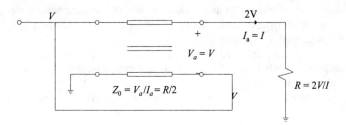

Figure 6.22   Determining the optimum characteristic impedance of an 1:4 unbalanced transmission-line transformer.

If a line with any other characteristic impedance is used, the input reflection coefficient of the transformer will be affected adversely. The effect of the characteristic impedance on the cutoff frequency of the transformer will be discussed later.

When the optimum characteristic impedance is known, the type of line to be used must be chosen. Coaxial cables with 25-Ω and 50-Ω characteristic impedance are freely available. A line with a 12.5-Ω characteristic impedance can be obtained by connecting two 25-Ω lines in parallel, while 100Ω can be obtained by connecting two 50-Ω lines in series (see Figure 3.10).

A wide range of characteristic impedances can be obtained by twisting together pairs of conductors with various diameters. When very low impedances are required (less than 10Ω), many conductors with smaller diameters can be twisted together.

The characteristic impedances of these twisted lines are influenced by the diameter of the wire used, as well as the number of twists per unit length.

Apart from the characteristic impedance, it is also necessary to decide on the diameter of the cable to be used where applicable. This is determined by the losses that can be tolerated and the power to be transmitted through the line.

The attenuation of bifilar or multifilar transmission lines can become a problem at high frequencies, as mentioned in Chapter 3.

6.4.2 Determining the Minimum Value of the Magnetizing Inductance of the Transformer

At low frequencies the transmission-line transformer can be considered to be a conventional 1:1 transformer connected in a special way. When this model is used, the input impedance and power gain versus frequency response at low frequencies can be determined by using Kirchhoff's voltage and current laws on the simplified equivalent circuit. If the load of the transformer consists of a single resistor and the transformer itself is lossless, only the input impedance of the transformer needs to be determined. The power dissipated in the load (and therefore the power gain) can then be found by using the equation

$$P_L = V_{cc}^2 \, G_{\text{eff}} / 2 \tag{6.35}$$

where V_{cc} is the maximum (peak) voltage across the effective parallel input resistance ($R_{\text{eff}} = 1/G_{\text{eff}}$) of the transformer.

When the input impedance and the transfer function are known, the minimum inductance (L_{11}) required to meet the low-frequency specifications can be determined.

The minimum value of the magnetizing inductance of the 1:4 unbalanced and 1:1 unbalanced-to-balanced transformers will be established as examples.

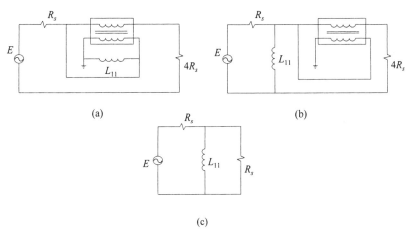

(a)

(b)

(c)

Figure 6.23   (a–c) Simplification of the equivalent circuit of the 1:4 unbalanced transformer at low frequencies.

EXAMPLE 6.3   The magnetizing inductance required in a 1:4 transmission-line transformer.

With the transmission line replaced by a 1:1 transformer with magnetizing inductance, the equivalent circuit of the 1:4 transmission-line transformer can be simplified as shown in Figure 6.23.

If the cutoff frequency is to be the 3-dB cutoff frequency, it is obvious from Figure 6.23(c) that the required magnetizing inductance L_{11} must be such that

$$\omega L_{11} = R_s / 2 \tag{6.36}$$

If this transformer is to be used in a power amplifier, the magnetizing inductance must be high enough for the specified minimum allowable ripple in the passband to be achieved.

Because the power dissipated in the load is given by (6.35), the output power is directly proportional to the effective parallel input resistance.

The efficiency of the amplifier is decreased if the effective load is reactive (refer to Section 2.3.3), that is, if the output impedance of the transistor is assumed to be purely resistive. Specifically, it is decreased by a factor

$$\eta_r = 1 / [1 + (R_{\text{eff}} / X_{\text{eff}})^2]^{1/2} \tag{6.37}$$

where X_{eff} is the effective parallel input reactance of the transformer.

Because R_{eff} is equal to the optimum value (R_s) in this particular problem, the power transmitted through the 1:4 transformer is also equal to the optimum value, that is, at low frequencies.

The relative efficiency is given by

$$\eta_r = 1 / [1 + (R_s / (\omega L_{11}))^2]^{1/2}$$

If $\eta_r = 0.95$ is acceptable, the magnetizing inductance must be such that

$$\omega L_{11} = 3R_s \tag{6.38}$$

(ωL_{11} is often chosen to be equal to $4R_s$).

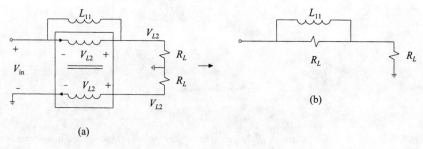

(a)

(b)

Figure 6.24   (a, b) The 1:1 unbalanced-to-balanced transmission-line transformer at low frequencies.

EXAMPLE 6.4   The magnetizing inductance required in a 1:1 transmission-line transformer.

The equivalent circuit for the 1:1 unbalanced-to-balanced transformer is shown in Figure 6.24(a).

By transforming the load on the secondary side of the transformer to the primary side, the equivalent circuit of the 1:1 unbalanced-to-balanced transformer can be simplified to that shown in Figure 6.24(b).

By using this equivalent circuit, the input admittance is found to be

$$
\begin{aligned}
Y_{in} &= \frac{1}{R_L + sL_{11}R_L / [R_L + sL_{11}]} \\
&= \frac{R_L + 2sL_{11}}{R_L^2 + 2sL_{11}R_L} \\
&= \frac{1}{2R_L} \cdot \frac{1 + R_L / (sL_{11})}{1 + R_L / (2sL_{11})}
\end{aligned}
\tag{6.39}
$$

It is clear from this equation that the input resistance will be equal to $2R_L$ if the magnetizing reactance is relatively high.

The relative power dissipation in the two load resistances can be determined by transforming the parallel combination of ωL_{11} and R_L in Figure 6.24(b) to the equivalent series impedance shown in Figure 6.25.

Because the same current flows through the two resistors, the ratio of the power dissipated in each load is equal to the ratio of the resistance of these resistors. If

$$
\omega L_{11} = 4.4R_L
\tag{6.40}
$$

the power dissipated in the two load resistors will differ by 5%. The input power to the transformer will then be 1% higher than the design value, and the relative efficiency will be 0.99.

6.4.3   Determining the Type and Size of the Magnetic Core to Be Used

Toroidal cores are often used in transmission-line transformers. The size of the toroidal core is determined by the inductance required, the maximum flux density in the core (and therefore the allowable losses), and the line length required to meet these specifications.

It was shown in Chapter 3 that if the inductance and flux density specifications are to be met simultaneously, a core with

$$
Al = \frac{\mu_0 \mu_r}{\omega B_{max}^2} \frac{V_{max}^2}{\omega L_{11}}
\tag{6.41}
$$

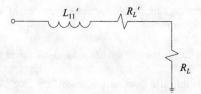

Figure 6.25   The series equivalent of the impedance of the circuit from Figure 6.24(b).

should be used [see (3.33)].

It can be shown that the line length will always increase if a core with an Al-product larger than that given by this equation is used.

If the core size is decreased, it is possible that the line length might be shorter, at least initially.

Whether it will be shorter is a function of the extent to which the inductance must be increased to meet the loss specification (the flux density in the core will be too high if the inductance is not increased), as well as the dimensions of the core.

If the losses in the material increase sharply when the flux density is increased, the optimum core size will always be that given by (3.39).

It is sometimes possible to reduce the line length necessary to provide the required magnetizing inductance by using a number of smaller toroidal cores instead of only one larger core.

The ratio of the line length for a single core to that of N_c stacked cores is given approximately by the equation

$$\frac{l_{e1}}{l_{e2}} = \frac{2w_1 + 2h_1 + 4t}{(A_1 / A_2) \cdot h_2 + (l_2 / l_1) \cdot (4w_2 + 4t)} \tag{6.42}$$

where t is the outer diameter of the transmission line used, l_1 the mean path length of the larger core, l_2 the mean path length of each of the smaller cores, A_1 the effective cross-sectional area of the larger core, and A_2 the effective cross-sectional area of each of the smaller cores. w_1, w_2, h_1, and h_2 are defined in Figure 6.26.

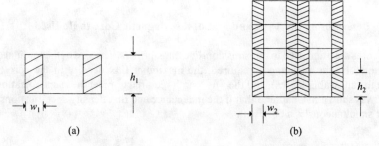

(a)   (b)

Figure 6.26   The cross-section of (a) a single toroidal core and (b) a number of stacked toroidal cores.

Equation (6.42) was derived by assuming the inductance and the flux densities of the two inductors to be equal.

In order to have the same flux density, it is necessary that

$$N_1 / l_1 = N_2 / l_2 \tag{6.43}$$

where N_1 is the number of turns used with the single core and N_2 is the number of turns used with the stacked core.

The inductance of the two inductors will be the same if

$$N_1^2 A_1 / l_1 = N_c N_2^2 A_2 / l_2 \tag{6.44}$$

where N_c is the number of cores used in the stacked-core inductor.

By using (6.43), (6.44) can be changed to

$$A_1 l_1 = N_c \cdot A_2 l_2 \tag{6.45}$$

It follows from this equation that the effective *Al*-products of the single-cored and stacked-cored inductors must be the same.

Equations (6.45) and (6.43) can be used to determine the number of cores and the number of turns required, if using a transformer with stacked cores is worthwhile (i.e., if the core dimensions are known).

If a core with suitable dimensions (comparable to those of the stacked core) is available, a balun core can also be used to decrease the line length of the transformer.

EXAMPLE 6.5   Comparison of the line lengths associated with a stacked core and a single core transmission-line transformer.

As an example of the application of the equations given above, the line lengths of a single toroidal core and a stacked core with $A_2 = 0.5A_1$, $l_2 = 0.5l_1$, $w_1 = h_1$, $w_2 = h_2$, and $t = w_1 / 3$ will be compared.

With $w_1 = h_1$ and $w_2 = h_2$, (6.42) becomes

$$\frac{l_{e1}}{l_{e2}} = \frac{4w_1 + 4t}{2(w_1 / \sqrt{2}) + (1/2) \cdot [4(w_1 / \sqrt{2}) + 4t]}$$

$$= \frac{4w_1 + 4t}{\sqrt{2} \cdot 2w_1 + 2t} \tag{6.46}$$

From this it follows that the ratio of the line length for the two different cores is

$$\frac{l_{e1}}{l_{e2}} \cong \frac{4 + 4/3}{2\sqrt{2} + 2/3} = 1/0.655 = 1.5$$

Because of the reduced line length, the bandwidth of the transformer, therefore, can be increased significantly by using a stacked core or a balun core.

6.4.4   Compensation of Transmission-Line Transformers for Nonoptimum Characteristic Impedances

When a line with the optimum characteristic impedance is not available, it is possible to compensate for the degradation of the high-frequency response of the transformer by using compensating inductors and/or capacitors.

It is usually adequate to use two compensating elements. One element is used in parallel (capacitor) or in series (inductor) with the load to change the input resistance or conductance to the required value at some frequency below the cutoff frequency, while the other is used to remove the reactive part of the input impedance or admittance at the same frequency.

The compensation frequency can be chosen such that the ripple in the output power is equal to a specified value. The optimum compensation frequency can be found iteratively.

The compensation of some frequently used transformers will be considered here as examples.

> **EXAMPLE 6.6**   Compensation of a 1:4 unbalanced transmission-line transformer.

The 1:4 unbalanced transmission-line transformer can be compensated as shown in Figure 6.27, that is, if $Z_0 > 1.35\ Z_{0,opt}$. The values of the two compensation capacitors are given by the equations [3]

$$C_1 = \frac{1 + \cos(\beta l) - \{[1 + \cos(\beta l)]^2 - \sin^2(\beta l) \cdot r^2\}^{1/2}}{\omega_{max}\, r\, R \sin(\beta l)} \tag{6.47}$$

$$C_2 = \frac{2\cos(\beta l) - \{[1 + \cos(\beta l)]^2 - \sin^2(\beta l) \cdot r^2\}^{1/2}}{4\omega_{max}\, r\, R \sin(\beta l)} \tag{6.48}$$

$$r = Z_0 / (2R) \tag{6.49}$$

Equations (6.47) and (6.48) can be derived by setting the real part of the input admittance of the transformer terminated in a resistor ($4R$) in parallel with an unknown capacitor (C_2) equal to $1/R$. C_1 is used to cancel the reactive part of the input admittance.

The compensation frequency [$f_{max} = \omega_{max}/(2\pi)$] is a function of the acceptable variation in the output power and the minimum efficiency required in the passband. It can be found iteratively.

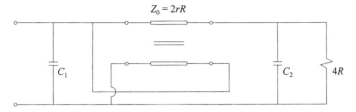

Figure 6.27   Compensation of the unbalanced 1:4 transformer when $Z_0 > 1.35\,Z_{0,opt}$.

The high-frequency response of the transformer can be improved considerably by using this compensation technique. This can be appreciated by comparing the electrical lengths of the line, with and without compensation, at the cutoff frequencies of the transformer.

The electrical lengths of the line at the cutoff frequency, with and without compensation, are compared in Table 6.2. The cutoff frequency was taken to be the frequency at which the output power decreased by more than 17% ($\Delta P > 17\%$) and/or the relative efficiency (η_r) dropped below 95% [refer to (6.17)].

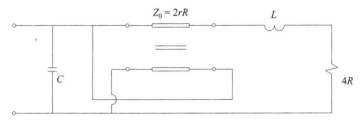

Figure 6.28   Compensation of the unbalanced 1:4 transmission-line transformer when $Z_0 < 1.35\,Z_{0,opt}$.

It can be seen from Table 6.2 that the high cutoff frequency can be increased by more than an octave above its uncompensated value, independent of the characteristic impedance.

The high-frequency response of the transformer can also be improved when

$$Z_0 < 1.35\,Z_{0,opt}$$

This can be done by using an inductor and a capacitor as compensating elements. The inductor is used in series with the high-impedance end of the transformer, while the capacitor is connected in parallel with the low-impedance side, as shown in Figure 6.28.

When $Z_0 = Z_{0,opt}$ the cutoff frequency can be increased by an octave. The required inductance and capacitance can be found by using the following equations [4]:

$$\omega_H L = 0.7558 Z_0 \tag{6.50}$$

$$\omega_H C = 0.8961 Y_0 \tag{6.51}$$

$$\omega_H = \omega_{0.2474\lambda} \tag{6.52}$$

where $\omega_{0.2474\lambda}$ is the radian frequency at which the line length is equal to 0.2474 λ.

Exact equations for the values of the inductance and capacitance can be derived by following the same procedure as before.

Table 6.2 The Electrical Length of the Transmission Line Used in an 1:4 Unbalanced Transmission-Line Transformer at the Compensation and Cutoff Frequencies, with and Without Compensation

$r = Z_0 / (2R)$	Line length without compensation (λ)	Line length at the compensation frequency (λ)	Line length with compensation (λ)
2.0	0.070	0.144	0.160
2.5	0.050	0.116	0.130
3.0	0.036	0.099	0.110
3.5	0.032	0.085	0.095
4.0	0.027	0.075	0.084
4.5	0.024	0.068	0.075
5.0	0.021	0.061	0.068

EXAMPLE 6.7 Compensation of a 1:1 balanced-to-unbalanced transformer.

The 1:1 balanced-to-unbalanced (and 1:4 balanced) transmission-line transformer can be compensated for characteristic impedances that are too high, as shown in Figure 6.29. The capacitance of both capacitors is given by the equation

$$C = \frac{1 - [1 - (r^2 - 1)\tan^2(\beta l)]^{1/2}}{\omega_{max} r R \tan(\beta l)} \tag{6.53}$$

where $r = Z_0 / Z_{0,opt} = Z_0 / R$, and ω_{max} is the compensation frequency.

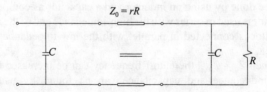

Figure 6.29 Compensation of the 1:1 balanced-to-unbalanced transformer ($r > 1$).

Table 6.3 The Electrical Length of the Transmission Line Used in the 1:1 Balanced-to-Unbalanced Transmission-Line Transformer at the Compensation and Cutoff Frequencies, with and Without Compensation

$r = Z_0 / R$	Line length without compensation (λ)	Line length at compensation frequency (λ)	Line length with compensation (λ)
0.5	0.073	—	—
0.6	0.080	—	—
0.7	0.089	—	—
0.8	0.109	—	—
0.9	0.170	—	—
1.0	∞	—	—
1.1	∞	—	—
1.2	0.119	0.131	0.180
1.3	0.092	0.094	0.155
1.4	0.075	0.075	0.135
1.5	0.065	0.075	0.123
1.6	0.058	0.081	0.116
1.8	0.042	0.081	0.101
2.0	0.041	0.056	0.085
2.5	0.031	0.038	0.066
3.0	0.025	0.044	0.058
3.5	0.021	0.044	0.048
4.0	0.018	0.031	0.043
5.0	0.013	0.031	0.035

The allowable ripple in the output power and the minimum efficiency should be taken into account when the compensation frequency is calculated.

The cutoff frequency ($\Delta P_L \geq 17\%$, $\eta_r \geq 95\%$) of the 1:1 transformer, with and without compensation, is shown in Table 6.3 as a function of the characteristic impedance of the line. It can be seen by inspection of Table 6.3 that the cutoff frequency of the 1:1 transmission-line transformer (and therefore the 1:4 balanced transformer) is more sensitive to deviations in the characteristic impedance than the unbalanced 1:4 transformer.

Compensation of the transformer has a significant effect on the cutoff frequency.

EXAMPLE 6.8 Compensation of a hybrid coupler.

The optimum characteristic impedance of the hybrid transformer shown in Figure 6.30 cannot be determined by using the rule given in Section 6.4.1. The reason for this is that the voltage across the end of the line is equal to zero in the balanced case.

It can be shown (by deriving the exact equations for this transformer) that the optimum characteristic impedance for this transformer is the lowest available characteristic impedance.

Table 6.4   The Electrical Length of the Transmission Line Used in the Hybrid Divider (and Combiner) at the Compensation and Cutoff Frequencies, with and Without Compensation

Z_0 / Z_{L1}	Line length without compensation (λ)	Line length at the compensation frequency (λ)	Line length with compensation (λ)
0.25	0.212	0.283	0.367
0.50	0.121	0.173	0.303
1.00	0.066	0.093	0.139
2.00	0.029	0.048	0.075
3.00	0.020	0.034	0.057

The hybrid transformer can be compensated with an inductor and a capacitor as shown in Figure 6.30. The compensation frequency (as well as the cutoff frequency of the transformer, with and without compensation) is given in Table 6.4 as a function of the characteristic impedance.

The component values are given by the equations

$$\omega_H L = 0.3015\, Z_{L1} / 2 \tag{6.54}$$

$$1 / (\omega_H C) = 1.8005\, Z_{L1} / 2 \tag{6.55}$$

where the compensation frequency (f_H) can be determined by using Table 6.4.

6.4.5   The Design of Highpass LC Networks to Extend the Bandwidth of a Transmission-Line Transformer

The bandwidth of a transmission-line transformer can be extended considerably by using a highpass matching network to compensate for the effect of the magnetizing inductance.

A network that can be used for this purpose is shown in Figure 6.31(a). It is sometimes also possible to use its dual, which is shown in Figure 6.31(b).

The optimum reactance values of the components of the network are given in Table 6.5 as a function of the allowable ripple in the passband. The reactance values are normalized for a load resistance of 1Ω.

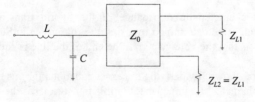

Figure 6.30   Compensation of the hybrid divider shown in Figure 6.14(b).

When this compensation technique is used, the magnetizing reactance required to meet the low-frequency specifications decreases. Because this implies a shorter line, the high cutoff frequency of the transformer will increase.

The exact amount by which the bandwidth will increase is a function of the acceptable ripple in the output power, the magnetic material used, and the losses that can be tolerated.

The new line length can be determined by using the information in Section 6.4.3. It should be noted that the maximum voltage across the line is slightly more than that without compensation. The capacitor in series with the load resistance transforms the resistance slightly upward, and the voltage across the line must therefore be higher than that across the load resistance in order to deliver the same power in the effective resistance in parallel with the magnetizing inductance as in the load. This is illustrated in Figure 6.32.

Table 6.5   The Normalized Values of the Reactance in Figure 6.31(a) at the Lowest Frequency in the Passband as a Function of the Ripple in the Output Power

Ripple (%)	0.5	1.0	2.0	3.0	4.0	5.0
X_C (Ω)	0.30	0.36	0.45	0.50	0.55	0.59
X_L (Ω)	1.80	1.52	1.31	1.20	1.14	1.09
Increase in bandwidth	1.48	1.61	1.73	1.82	1.87	1.90

Because of the decrease in the reactance of the magnetizing inductance and the increase in the effective resistance in parallel with it, the unloaded Q of the magnetizing inductance will also change if the losses are to remain the same as without compensation. The maximum allowable flux density in the core, therefore, will also change (it must be less than before).

Despite the increase in the voltage across the magnetizing inductance and the decrease in the maximum allowable flux density in the core, it is usually worthwhile to use this compensation technique.

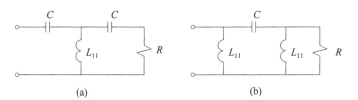

(a)   (b)

Figure 6.31   (a, b) Two low-frequency impedance-matching networks that can be used to extend the bandwidth of a transformer.

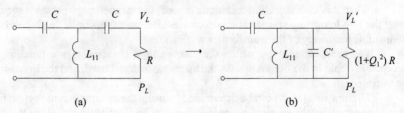

Figure 6.32   Illustration of the increase in the voltage across the magnetizing inductance of the transmission-line transformer with low-frequency compensation.

If the losses in the core can be ignored and the core size remains the same as before compensation, the improvement in the bandwidth will be that given in Table 6.5, as long as the transformer is designed to have the same low cutoff frequency as before.

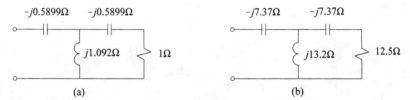

Figure 6.33   The low-frequency impedance-matching network of Figure 6.3l(a) if the passband ripple is 5% and (a) $R_L = 1\Omega$ and (b) $R_L = 12.5\Omega$.

Along with improving the bandwidth, this technique has the added advantage that the required frequency response might be obtained by using an air-cored solenoidal coil instead of magnetic material. If this can be done, the increase in the voltage per turn is not a problem and the flux density consideration does not apply.

EXAMPLE 6.9   Low-frequency compensation of a 1:4 unbalanced transmission-line transformer.

As an example of the application of the low-frequency compensation technique, the required magnetizing inductance and capacitance for a 1:4 unbalanced transmission-line transformer with $R_L = 12.5\Omega$ and $f_L = 2$ MHz will be determined. The passband ripple is to be 5%.

The compensation networks for $R_L = 1\Omega$ and $R_L = 12.5\Omega$ are shown in Figure 6.33. The component values in Figure 6.33(a) were obtained from Table 6.5.

The positions of the compensating capacitors in the 1:4 transformer are shown in Figure 6.34(a). The input impedance of the transformer is the same as that of the equivalent circuit shown in Figure 6.34(b), which is of the same form as the matching network in Figure 6.33(a).

By using this equivalence, it follows that if the required high cutoff frequency of the transformer is low enough, the magnetizing inductance can be realized by using an air-cored solenoidal coil.

6.5 CONSIDERATIONS APPLYING TO RF POWER AMPLIFIERS

The design of RF and microwave power amplifiers differs from that of small-signal amplifiers in the design of the output circuit. While the output circuit in small-signal amplifiers is usually conjugately matched to the load or used to taper the gain response, the load impedance of a power amplifier must be chosen in such a way that the required power can be obtained and that the efficiency is as high as possible.

The output power obtainable from an amplifier stage is limited by the limitations of the transistor used and/or the output circuit designed. The device limitations stem from the finite voltage, current, and power ratings of the transistor and its saturation voltage or saturation resistance. The saturation voltage or resistance of a transistor determines the lowest value of the instantaneous voltage across it. (Ideally the voltage should go down to zero.) Saturation voltages of a few volts for bipolar transistors, and saturation resistance of fractions of an ohm up to a few ohms are typical for FETs. It was shown in Section 2.3.1 that the maximum output power obtainable from a class A or a class B stage at RF frequencies is given by

$$P_{max} = \frac{(V_s - V_{sat})^2}{2(R_L + \alpha R_{sat})} \cdot \frac{R_L}{R_L + \alpha R_{sat}} \tag{6.56}$$

where V_s is the supply voltage, V_{sat} the saturation voltage, and R_{sat} the saturation resistance. α is equal to 2 for class A amplifiers and equal to 1 for class B amplifiers. In deriving this equation (see Figure 6.35), it was assumed that the output susceptance of the transistor was removed by the output matching network and that the output power is voltage limited (i.e., the output voltage clips before the output current).

The saturation resistance of bipolar transistors is usually negligibly small, while the saturation voltage for FETs can be neglected.

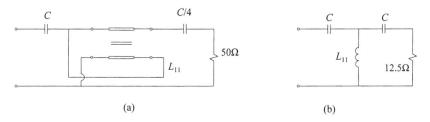

(a)   (b)

Figure 6.34   (a) The positions of the compensating capacitors in the unbalanced 1:4 transmission-line transformer and (b) the equivalent circuit for the input impedance of the transformer.

It follows from (6.56) that, in order to obtain the maximum possible output power from a transistor, it is necessary to use the highest supply voltage possible, and to choose the load resistance as small as possible (the saturation resistance is usually significantly smaller than the load resistance required). The minimum value of the load resistance is determined by the maximum dc and RF currents that can be tolerated through the transistor.

The optimum load for an RF power transistor is usually specified by the manufacturer. When this is not done, the optimum terminations can be determined practically at each frequency of interest by using stub tuners. The optimum terminations can also be estimated by using a large-signal model, if such a model is available. Alternatively, the small-signal S-parameters at the rated current and the dc I/V-curves can be measured, and the power parameter approach outlined in Section 2.3 can be used to generate load-pull contours for the transistor.

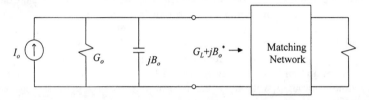

Figure 6.35   The matching problem associated with the output circuit of a power amplifier.

The optimum load is often specified in terms of the equivalent output impedance of the device under the assumption of a conjugate output match. It is important to realize that this impedance is not the same as the actual output impedance of the transistor. The terminations required for the optimum power match are usually not the same as those required for a conjugate match [low output voltage standing wave ratio (VSWR)].

The efficiency of a power amplifier is a function of the class of operation and the effective shunt susceptance in the output circuit (the transistor susceptance included). When the voltage across the output terminals of the transistor is purely sinusoidal, the efficiency will always be less than 50% for class A amplifiers (the conduction angle of the current through the transistor is then 360°), while that for class B amplifiers (180° conduction angle) is constrained to below 78.5%. Higher efficiencies can be obtained with class C amplifiers, but because the same power must be concentrated in a narrower pulse, the peak current though the transistor increases as the efficiency increases. The device specifications for a class C amplifier, therefore, are more severe than those for class A or B amplifiers of the same output power with the same supply voltage. A class C amplifier also cannot be used directly for linear applications.

When the effective load (the output susceptance of the transistor is considered as part of the load) of a power stage is reactive, the efficiency decreases by a factor

$$\eta_r = \frac{1}{\sqrt{1+(B_L/G_L)^2}} \tag{6.57}$$

because of the increase in the supply current caused by the effective shunt susceptance (B_L). In optimizing the efficiency of an amplifier, it is therefore essential to remove the effect of the output susceptance of a transistor.

In order to obtain the required output power, the physical load of a power amplifier (usually 50Ω) must be transformed to a lower value. It will be shown in Chapter 8 that this often can not be done with LC networks (transformation to very low resistance values is usually required and the insertion loss may also be a problem). Consequently, transmission-line transformers are usually used for this purpose at RF frequencies. Combiners and splitters are also required in a balanced or a push-pull configuration or to connect transistors in parallel for higher output power. The cancellation of the output susceptance of a power transistor is carried out with an LC network between these transformers and the transistor.

The gain tapering required in a power amplifier is best done on the input side of each transistor with an RLC matching network. It will be shown in Chapter 8 that these networks can be used to level the gain (the operating power gain in this case) without reactive mismatching. Any reactive mismatching will increase the power required from the driver stages.

It should be noted that ferrites are usually not required in transmission-line transformers at the higher RF frequencies (typically 100 MHz and above).

This section concludes by considering two power amplifier examples. In the first example, designing the output matching network of a narrowband (225–260 MHz) television broadcast amplifier will be considered. The design of a broadband input matching network (2–30 MHz) for a push-pull class B amplifier will be considered in the second example.

EXAMPLE 6.10   Designing an output matching network for a power amplifier.

As an example of the design of an RF power amplifier, an output matching network will be designed for the balanced amplifier shown in Figure 6.36 over the passband 225–260 MHz. The network will be designed for an output power of 165W. The supply voltage will be taken as 28V, and the output capacitance of each transistor as 130 pF.

An approximate value for the required load resistance for each transistor can be obtained from (6.56). The saturation voltage will be taken as 3V, and saturation resistance is assumed to be negligible. Application of (6.56) yields

$$165/2 = P_L = \frac{(28-3)^2}{2R_L}$$

leading to

$$R_L = 3.79\Omega$$

The quarter-wavelength transformers (baluns) in the output circuit and the input circuit are used to transform the load impedance and the source impedance to approximately 6.25Ω ($12.5/2$) for each transistor, and also serves as a combiner for the output power and a splitter for the input power. The actual impedances can be obtained easily by the standard equation for the input impedance of a transmission line (the unbalanced current is very small in this case) and dividing the results by 2 to get the load for each transistor. The impedance thus obtained is the load specification for the output impedance-matching network to be designed.

The source impedance for the output matching network is simply equal to the load resistance required to obtain the specified output power, in parallel with the output capacitance of each transistor. Because a conjugate match to this impedance is required, the transducer power gain required for this matching problem is equal to 1. The specifications for the matching network to be designed are summarized in Table 6.6. This matching problem can be solved by first design-ing an L-section to provide a conjugate match at the highest frequency, after which it can be optimized for the best performance over the passband. The solution shown in Figure 6.37 was synthesized by using the transformation-Q impedance-matching technique described in Section 8.4.3. The deviation from the specified performance is negligibly small.

EXAMPLE 6.11   Designing a input matching network for a power transistor.

Designing a wideband (2–30 MHz) input matching network for a push-pull class B stage [40-W peak envelope power (PEP)] will be considered in this example. No attempt was made to level the gain response in this case. The input impedance of the transistor used (MRF406) is listed in Table 8.1.

The matching network designed is shown in Figure 6.38. The 50-Ω balun on the input side was used to obtain a balanced signal, after which the balanced 4:1 transformer was used to transform the 50-Ω source resistance to 12.5Ω. The LC network was designed to match this resistance to the input impedance of the two transistors (the inputs are effectively connected in series) by using an earlier version of the program LSM provided with this book (refer to Section 8.4.1).

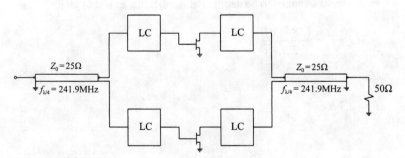

Figure 6.36   The configuration of the push-pull power amplifier in Example 6.10.

Table 6.6 The Specifications for the Output Matching Network of the Power Amplifier of Example 6.10

Frequency (MHz)	Source impedance (Ω)	Load impedance (Ω)	Transducer power gain
225	2.55 - j1.78	6.31 - j1.03	1.0
230	2.51 - j1.79	6.28 - j0.72	1.0
235	2.48 - j1.80	6.27 - j0.43	1.0
240	2.44 - j1.81	6.25 - j0.16	1.0
245	2.41 - j1.82	6.25 + j0.20	1.0
250	2.38 - j1.83	6.27 + j0.49	1.0
255	2.33 - j1.84	6.29 + j0.80	1.0
260	2.30 - j1.85	6.32 + j1.10	1.0

The balanced matching problem can be transformed to a single-ended problem by replacing each of the two capacitors used with two capacitors connected in series and by using the fact that the center points are virtual grounds.

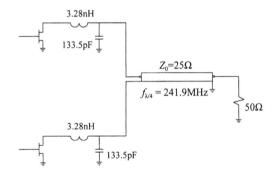

Figure 6.37 The output matching network designed for the power amplifier of Example 6.10.

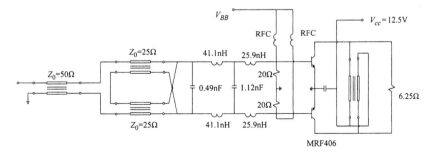

Figure 6.38 The input matching network designed for the push-pull power amplifier of Example 6.11.

The single-ended matching network was designed to match half of the output impedance presented by the 1:4 transformer (approximately 6.25Ω) to the input impedance of a single transistor. The gain (G_T) of the LC network obtained varied between 0.85 and 0.95 over the passband.

With the output power higher than 6W, the input VSWR of the amplifier was measured to be better than 2.6 over the complete passband [4], which is close to the expected value of 2.3.

QUESTIONS AND PROBLEMS

6.1   Is the frequency response of a transmission-line transformer dependent on the coupling between the two conductors of the line at (a) high frequencies or (b) low frequencies?

6.2   Are there any losses in the magnetic core of a transmission-line transformer at high frequencies?

6.3   Verify the transformation ratios given in Table 6.1 for transmission-line transformers with four lines.

6.4   Find the configuration of the 1:2.78 transmission-line transformer by using the technique illustrated in Figure 6.4.

6.5   Show that the high-frequency responses of the balanced-to-unbalanced 1:1 and the balanced 1:4 transmission-line transformers are identical at high frequencies.

6.6   Prove that the unbalanced current in a transmission-line transformer is not a function of the distance along the line.

6.7   Determine the optimum characteristic impedances of the transmission lines used in the transformer of Problem 6.4.

6.8   Determine the optimum characteristic impedances of the transmission lines used in the 1:2.25 transmission-line transformer. Determine the relative lengths of the two lines required if only one core is to be used.

6.9   What is the optimum characteristic impedance for the hybrid divider of Figure 6.14(b)?

6.10   Find the minimum value of the magnetizing inductance of the transmission lines used in the unbalanced-to-balanced 1:4 transmission-line transformer [Figure 6.11(a)]. The input reactance of the transformer must be equal to four times the resistance in parallel with it.

6.11 Determine the minimum value of the magnetizing inductance of a unbalanced-to-balanced 1:1 transmission-line transformer if the output power in the two different loads are to be matched within 2%. $R_{L1} = 12.5\Omega = R_{L2}$.

6.12 Determine the optimum core size for a 12.5:50Ω balanced transmission-line transformer. $P_L = 100\text{W}$; $f = 1.6\,\text{MHz}$. A toroidal core of 4C4 material is to be used; $\mu_r = 120$.

6.13 Can the bandwidth of the transformer in Problem 6.12 be increased if stacked toroidal cores instead of a single toroidal core are used?

6.14 Is it true that the bandwidth of a transmission-line transformer will usually be higher if a balun core instead of a toroidal core is used?

6.15 A 50-Ω coaxial cable is used in a 12.5:50-Ω transmission-line transformer. If the line length required to meet the low-frequency specifications is 25 cm and the velocity on the line is $v = 0.65c$, determine the high cutoff frequency of the transformer.

Determine the cutoff frequency if the transformer is compensated for the nonoptimum characteristic impedance of the transmission line used. Calculate the capacitance values of the compensating capacitors required to compensate the transformer.

6.16 Determine the inductance and capacitance required to improve the high-frequency response of the transmission-line transformer in the previous problem if a transmission line with the optimum characteristic impedance is used.

6.17 Compare the high cutoff frequencies of the balanced and unbalanced 1:4 transmission-line transformers.

6.18 The input impedance of the balanced 9:1 transmission-line transformer shown in Figure 6.39 is given by the equation [1]:

$$Z_{in} = 9R\frac{4 + 5\cos(\beta\ell) + j6r\sin(\beta\ell)}{9\cos(\beta\ell) + j[6/r]\sin(\beta\ell)} \tag{6.58}$$

Determine the cutoff frequency of the transformer if transmission lines with the optimum characteristic impedance are available. Compare the cutoff frequency with those of the 1:4 balanced and unbalanced transmission-line transformers.

6.19 A hybrid divider and combiner are used to obtain 100W from two 50-W power amplifier modules. Two 30-cm lines with $Z_0 = 35\Omega$ are used. If $R_L = R_s = 50\Omega$, determine the high cutoff frequencies of the hybrid transformers.

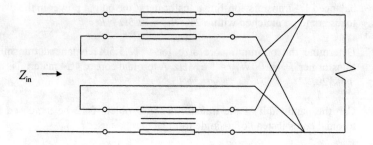

Figure 6.39   The balanced 9:1 transmission-line transformer.

6.20   Determine the positions of the low-frequency compensation capacitors in a balanced
transmission-line transformer. If $R_L = 50\Omega$, calculate the required capacitance of
compensating capacitors.

REFERENCES

[1]   Rotholtz, E., "Transmission-Line Transformers," *IEEE Trans. Microwave Theory Tech.*, V
MTT-29, No. 4, April 1981.

[2]   Dutta Roy, S. C., "A Transmission-Line Transformer Having Frequency Independ
Properties," *Int. J. Circuit Theory App.*, Vol. 8, January 1980, pp. 55–64.

[3]   Hilbers, A. H., "On the Design of HF Wideband Transformers (ECO 6907)," Eindhov
Netherlands: Philips C.A.B. Group, 1969.

[4]   Abrie, P. L. D., "Impedance Matching Networks and Bandwidth Limitations of Clas
Power Amplifiers in the HF and VHF Ranges," Master's Thesis, University of Preto
1982.

SELECTED BIBLIOGRAPHY

Krauss, H. L., and C. W. Allen, "Designing Toroidal Transformers to Optimize Wideb
Performance," *Electronics*, August 16, 1973.

Ruthroff, C. L., "Some Broad-Band Transformers," *Proc. IRE*, August 1959.

Van Nierop, J. H., "Evolution of a 4:1 Impedance Transforming Balun," *IEEE Trans. Anten*
Propag., Vol. AP-30, No. 4, July 1982.

CHAPTER 7

FILM RESISTORS, PARALLEL-PLATE CAPACITORS, INDUCTORS, AND MICROSTRIP DISCONTINUITIES

7.1 INTRODUCTION

The distributed nature of film resistors, parallel-plate capacitors, and inductors cannot be ignored at microwave and millimeter frequencies and will be considered in this chapter.

The behavior of film resistors can be accurately modeled by considering the resistor to be a lossy transmission line. Film resistors will be considered in Section 7.2.

Single-layer parallel-plate capacitors are often used at microwave frequencies. The configurations commonly used in hybrid circuits are shown in Figure 7.1. Metal-insulator-metal (MIM) capacitors (see Figure 7.2) are extensively used in MMICs (monolithic microwave integrated circuits).

At low frequencies these capacitors could be treated as ideal lumped capacitors, but their distributed nature must be taken into account at higher frequencies.

When the capacitor is mounted on a ground plane [bottom plate connected to ground; see Figure 7.1(c)] and the excitation can be taken to be uniform across the width

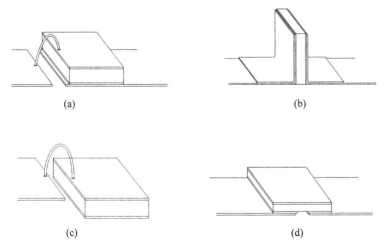

(a)   (b)

(c)   (d)

Figure 7.1   (a) A series connected parallel-plate capacitor, (b) a vertically mounted capacitor, (c) a parallel-plate capacitor mounted on a ground plane, and (d) a gap capacitor.

of the capacitor (narrow width, ribbon, or multiple bond wire cases), the parasitic behavior of the capacitor can be modeled fairly accurately by considering it to be a open-ended transmission line. This case will be considered in Section 7.3.1. The general case (microstrip capacitors) is considered in [1].

Analysis of the vertically mounted parallel-plate capacitor [Figure 7.1(b)] is also straightforward. This capacitor can be considered to be a series connected open-ended stub. The same resonances encountered in an open-ended stub are also encountered in this configuration. Fortunately, the resonant behavior is sharply reduced by any capacitor losses. (This is important when a capacitor is used for wideband coupling or decoupling.)

Analysis of the series configuration shown in Figure 7.1(a) proves to be more challenging. If the capacitance density of the capacitor is high compared to that of the associated microstrip line (which is usually the case) and the behavior at frequencies significantly lower than parallel resonance is considered, these capacitors can be accurately modeled as lumped capacitors cascaded with a transmission line on both sides (line-capacitor-line model) [2]. In this case the transmission-line behavior of the capacitor is essentially that of the microstrip line.

The line-capacitor-line approach is very practical and is adequate in most cases. Modeling of parallel-plate capacitors in this way will be considered in detail in this chapter.

The general case can be handled as described in [2, 3]. The model used for the capacitor in [2] is instructive and is shown in Figure 7.2. Note that the magnetic coupling between the capacitor plates (L_{12}) and capacitance to the ground plane (C_{10}, C_{20}) are included in the model.

Parallel-plate capacitors exhibit series and parallel resonant behavior as the frequency is increased. These effects are very pronounced in high-Q capacitors and are important when designing coupling or decoupling capacitors. The parallel-resonant behavior is not evident in the line-capacitor-line model.

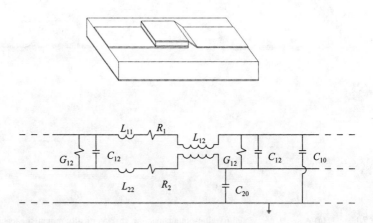

Figure 7.2   The distributed model used for a parallel-plate capacitor in [2].

The basic reason for the parallel resonance in overlay capacitors will be established by considering the parallel-plate capacitor in free space. It will be shown in Section 7.3.3 that a more accurate model for the capacitor would be to use the line-capacitor-line model with a frequency-dependent value for the capacitance. The analysis will be done by considering the series connected parallel-plate capacitor to be an unbalanced transmission line, as was done with transmission-line transformers in the previous chapter. This approach can be extended to handle the microstrip case, too [4]. This section will be concluded by presenting models for MIM series capacitors and overlay capacitors. The parameters in these models can be obtained by curve-fitting the results obtained in EM simulations of capacitors of various sizes for the substrate of interest.

The model for the single-layer parallel-plate capacitor on microstrip can also be used for a multiplate capacitors at microwave frequencies. Good results can be expected up to the first series resonance frequency when this is done.

Lumped and spiral inductors will be considered in Section 7.4. Because of their importance in chip-and-wire modules, calculation of the inductance and the parasitics associated with bond wires will be considered in depth.

Spiral inductors are frequently used in MMICs. Formulas for the inductance and capacitance of spiral inductors are provided in [5–7] and a selection of these will be presented here. A suitable model for these inductors will also be provided [7].

In order to design practical circuits, it is important to appreciate the effect that microstrip discontinuities will have on a circuit. The basic models for the discontinuities normally encountered will be considered in Section 7.5.

7.2   FILM RESISTORS

Thin-film techniques are often used to manufacture resistors at microwave frequencies. By keeping the dimensions of the resistor small, the associated capacitance and inductance can be minimized. The capacitance can be reduced further by depositing the thin film on a substrate with a low dielectric constant.

A film resistor (see Figure 7.3) can be modeled as a lossy transmission line. The relevant equations are as follows

$$r = R_s / W \tag{7.1}$$

$$v_{ph} = c / \sqrt{\varepsilon_{r\_eff}} \tag{7.2}$$

$$C = 1 / (v_{ph} \cdot Z_{0\_LC}) \tag{7.3}$$

$$L = Z_{0\_LC} / v_{ph} \tag{7.4}$$

$$\theta_1 = 0.5[\frac{\pi}{2} + \tan^{-1}(\omega L / r)] \tag{7.5}$$

$$\gamma = \alpha + j\beta = \sqrt{j\omega C \cdot (r + j\omega L)} = \sqrt{\omega C} \sqrt{r^2 + (\omega L)^2} \cdot [\cos\theta_1 + j\sin\theta_1] \tag{7.6}$$

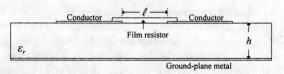

Figure 7.3   A film resistor on microstrip.

$$\theta_2 = -0.5\tan^{-1}[r/(\omega L)] \tag{7.7}$$

$$Z_0 = \sqrt{\frac{r + j\omega L}{j\omega C}} = \sqrt{\frac{\sqrt{r^2 + (\omega L)^2}}{\omega C}} \cdot [\cos\theta_2 + j\sin\theta_2] \tag{7.8}$$

$$\begin{bmatrix} V_I \\ I_I \end{bmatrix} = \begin{bmatrix} \cosh(\gamma l) & Z_0 \sinh(\gamma l) \\ \sinh(\gamma l)/Z_0 & \cosh(\gamma l) \end{bmatrix} \begin{bmatrix} V_L \\ I_L \end{bmatrix} \tag{7.9}$$

where Z_{0_LC} is the characteristic impedance of a lossless line with identical dimensions, W the width (in meters) of the film resistor, l its length (in meters), R_s the resistance per square, r the resistance per unit length (Ohm/m), C the capacitance per unit length (Farad/meter), and L the inductance per unit length (henry/meter). The angles θ_1 and θ_2 are specified in radians.

The influence of the skin effect on the resistance can be incorporated into the resistance per square, R_s. When the resistivity of the film material is high, the skin effect can usually be ignored. Films with resistances of 10Ω–$1,000\Omega$ per square are available.

The transmission matrix equation for a series resistor is defined in (7.9). V_I and I_I in this equation are the input voltage and current, while V_L and I_L are the load voltage and the load current, respectively.

The impedance presented to the circuit by a film resistor (or any transmission line with series losses) connected in series (cascaded) with a load Z_L can be derived from (7.9) and is given by

$$Z_{in} = \frac{Z_L \cosh\gamma\,\ell + Z_0 \sinh\gamma\,\ell}{Z_L \cdot Y_0 \cdot \sinh\gamma\,\ell + \cosh\gamma\,\ell} = Z_0 \cdot \frac{Z_L \cosh\gamma\,\ell + Z_0 \sinh\gamma\,\ell}{Z_L \sinh\gamma\,\ell + Z_0 \cosh\gamma\,\ell}$$

$$= Z_0 \cdot \frac{Z_L + Z_0 \tanh\gamma\,l}{Z_0 + Z_L \tanh\gamma\,l} \tag{7.10}$$

7.3   SINGLE-LAYER PARALLEL-PLATE CAPACITORS

The configurations of single-layer capacitors typically used were considered in Section 7.1. The capacitor can be mounted on the ground plane or on a microstrip line (conductive epoxy is used for this purpose). When mounted on microstrip, the series connection is usually used [see Figure 7.1(a)], but vertical mounting is also an option. Standing axial beam leads are usually used when vertical mounting is required. A gap capacitor [Figure 7.1(d)] has the advantage that no bonding wires or ribbons are required when it is used. This capacitor consists basically of two parallel-plate capacitors connected in cascade.

Parallel-plate capacitors are used for filtering, impedance matching, coupling, and decoupling.

When decoupling to ground is required, the capacitor is usually mounted on the ground plane and connection to the circuit is made with bond wires or a ribbon. The parasitic inductance associated with a ribbon will usually be lower than that associated with bonding wires. Several bonding wires can (and should) be used in parallel, but the inductance will not decrease proportionally with the number of wires used because of the coupling between them.

Discrete parallel-plate capacitors are available in different sizes. Typical widths are 10 mil (D10), 15 mil (D15), 20 mil (D20), and so on.

The capacitance values obtainable from [8] are listed in Table 7.1 as a function of the width (50-V breakdown voltage). Class I materials are used when high-Q capacitors are required (filtering and impedance matching), while class II materials are usually used for resonance-free coupling and decoupling.

Table 7.1   The Capacitance Values (pF) Obtainable as a Function of the Capacitor Width

	Capacitor width					
	D10	D15	D20	D25	D30	D35
Class I	0.05–4.7	0.05–12	0.08–18	0.2–33	0.3–39	0.4–68
Class II	1.8–68	3.3–180	3.9–220	10–470	12–560	20–1000

The length of these capacitors is a function of the dielectric material used and the layer thickness. To provide an idea of the lengths, upper bounds on the lengths are provided in Table 7.2 for various dielectric materials [8] with a dielectric thickness of 4 mil (50-V breakdown voltage). The values were calculated by considering only the plate capacitance and neglecting any fringing capacitance. The dissipation factors [8] for the different materials are also listed in the table. The second group of materials are class II materials.

Accurate information on the exact size of a capacitor can be obtained from the manufacturer.

With the physical dimensions of a capacitor known, the associated characteristic impedance and electrical line length can be determined [vertical mounting is assumed here; see Figure 7.1(b)]. The electrical length can also be estimated by measuring the first parallel resonant frequency (open circuit) of the capacitor.

The first parallel resonant frequency and the characteristic impedance are not independent for a given capacitance value. This follows from the following equations:

$$C_T = (Y_0 / v_{ph}) \cdot \ell = Y_0 \cdot \frac{\sqrt{\varepsilon_{r\_eff}}}{c} \cdot \ell = Y_0 \frac{\sqrt{\varepsilon_{r\_eff}} \cdot \ell}{c} \tag{7.11}$$

$$\Rightarrow$$

$$Y_0 = C_T \cdot \frac{c}{\sqrt{\varepsilon_{r\_eff}} \cdot \ell} \tag{7.12}$$

$$\Delta\theta = 2\pi \cdot \ell / \lambda = 2\pi \cdot \ell / (v_{ph} / f) = \omega \cdot \frac{\sqrt{\varepsilon_{r\_eff}}}{c} \cdot \ell = \omega \cdot \frac{\sqrt{\varepsilon_{r\_eff}} \cdot \ell}{c} \tag{7.13}$$

$$\pi = \omega_0 \frac{\sqrt{\varepsilon_{r\_eff}} \cdot \ell}{c}$$

$$\Rightarrow$$

$$f_0 = \frac{1}{2} \cdot \frac{c}{\sqrt{\varepsilon_{r\_eff}} \cdot \ell} = \frac{1}{2} \cdot \frac{Y_0}{C_T} \tag{7.14}$$

where C_T is the capacitance, Y_0 the characteristic admittance ($Y_0 = 1/Z_0$), ℓ the capacitor length, and f_0 the first parallel resonant frequency.

Table 7.2   Upper Bounds on the Length Required per Picofarad for Different Dielectric Materials (Dielectric Layer Thickness: 4 mils)

Material (DF)	Length per picofarad (mm)					
	D10	D15	D20	D25	D30	D35
CF (0.6%)	2.0616	1.3744	1.0308	0.8246	0.6872	0.5890
CG (0.7%)	0.6479	0.4320	0.3240	0.2592	0.2160	0.1851
NR (0.25%)	0.2926	0.1951	0.1463	0.1170	0.0975	0.0836
NS (0.5%)	0.1463	0.0975	0.0732	0.0585	0.0488	0.0418
NU (1.5%)	0.0756	0.0504	0.0378	0.0302	0.0252	0.0216
NV(1.2%)	0.0454	0.0302	0.0227	0.0181	0.0151	0.0130
BG (2.0%)	0.1134	0.0756	0.0567	0.0454	0.0378	0.0324
BH (2.5%)	0.0181	0.0121	0.0091	0.0073	0.0060	0.0052
BU (2.5%)	0.0082	0.0055	0.0041	0.0033	0.0027	0.0024

It follows from (7.12) and (7.14) that the first parallel resonant frequency and the characteristic impedance associated with a given capacitance value is completely determined by the product $(\varepsilon_{r\_eff})^{0.5} \cdot \ell$. If this product is kept constant, the frequency-dependent behavior of different realizations (different values of ε_r) of the same capacitor value will be identical (i.e., if any difference in the dissipation factors is ignored).

Equations (7.11) and (7.14) can also be combined to give an expression for the capacitance in terms of the Y_0 and f_0:

$$C_T = Y_0 / (2f_0) \tag{7.15}$$

The electrical performance of a parallel-plate capacitor depends on the way it is connected. The different cases will be considered next.

7.3.1 Parallel-Plate Capacitors on a Ground Plane

The equivalent circuit of a capacitor mounted on a ground plane is shown in Figure 7.4. This equivalent circuit is valid if the excitation can be considered to be uniform across the width of the capacitor. This can be ensured by using several bond wires in parallel or by using a ribbon instead of the bond wires.

The bond wire (or ribbon) inductance can and should be minimized by keeping its length as short as possible.

With the equivalent characteristic impedance and the resonant frequency of the parallel-plate capacitor known, the impedance presented by it to the circuit can be calculated. Note that because one side of the capacitor is directly connected to the ground plane, the transmission-line inductance could be reduced by up to one-half compared to the vertically mounted case (this effect will be reduced by coupling effects). A slight change in the capacitance should also be expected because of the difference in the fringing fields.

The equations derived in Section 7.2 for a thin-film resistor (transmission line with series losses) also apply to this case. If the parasitic edge capacitance in Figure 7.4 is ignored, Z_L in (7.10) is an open circuit and (7.10) simplifies to

$$Y_{\text{in}} = Y_0 \tanh \gamma \ell \tag{7.16}$$

$$= Y_0 \frac{\tanh \alpha\ell + \tanh(j\beta\ell)}{1 + \tanh \alpha\ell \cdot \tanh(j\beta\ell)}$$

$$= Y_0 \frac{\tanh \alpha\ell + j \tan \beta\ell}{1 + j \tanh \alpha\ell \cdot \tan \beta\ell} \tag{7.17}$$

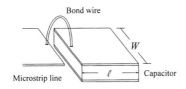

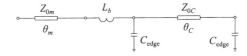

Figure 7.4   The equivalent circuit for a capacitor mounted directly on a ground plane. Z_{0m} is the characteristic impedance of the microstrip line, and L_b is the inductance of the bond wire.

The general case can be handled by using the following equation:

$$Y_{in} = j\omega C_{edge} + Y_0 \cdot \frac{j\omega C_{edge} \cdot \cosh \gamma \ell + Y_0 \sinh \gamma \ell}{j\omega C_{edge} \cdot \sinh \gamma \ell + Y_0 \cosh \gamma \ell} \qquad (7.18)$$

where C_{edge} is the parasitic capacitance at each open end.

If the excitation is at the center of the capacitor instead of at the edge, the capacitor can be considered to consist of two transmission lines connected in parallel. The excitation must be uniform across the width for this to be the case.

7.3.2   Parallel-Plate Capacitors Used as Series Stubs

The equivalent circuit for a parallel-plate capacitor used as a series connected open-ended stub is shown in Figure 7.5. If the fringing capacitance at the open end is ignored, the series admittance presented to the circuit by the capacitor can be calculated by using (7.17).

The insertion loss associated with the capacitor can be calculated by using (7.17) and (1.11):

$$G_T = \left| \frac{y_{21}}{(y_{11} + Y_s)(y_{22} + Y_L) - y_{12}y_{21}} \right|^2 \cdot 4 G_s G_L$$

$$= \left| \frac{-Y_{in}}{(Y_{in} + Y_s)(Y_{in} + Y_L) - (-Y_{in})(-Y_{in})} \right|^2 \cdot 4 G_s G_L$$

$$= \frac{1}{\left| Y_s + Y_L + \dfrac{Y_s Y_L}{Y_{in}} \right|^2} \cdot 4 G_s G_L \qquad (7.19)$$

where Y_s is the admittance to the left of the stub and Y_L is the admittance to the right.

With $Y_s = Y_0 = Y_L$ and $Y_s = G_s$ and $Y_L = G_L$, (7.19) simplifies to

$$G_T = \frac{1}{\left| 1 + \dfrac{1}{2}\dfrac{Y_0}{Y_{in}} \right|^2} \qquad (7.20)$$

Expressed in decibels, this becomes

$$G_T \ (\text{dB}) = -10\log_{10}\left| 1 + \frac{1}{2}\frac{Y_0}{Y_{in}} \right|^2 = -20\log_{10}\left| 1 + \frac{1}{2}\frac{Y_0}{Y_{in}} \right| \qquad (7.21)$$

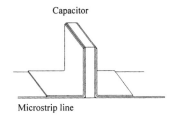

Capacitor

Microstrip line

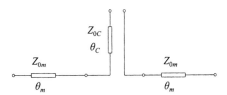

Figure 7.5   The equivalent circuit for a parallel-plate capacitor used as a series stub.

Substitution of the expression for Y_{in} yields that the insertion loss of the capacitor is given by

$$IL = 20\log_{10}\left|1 + \frac{1}{2}\frac{Y_0}{Y_{0C}}\frac{1 + j\tanh\alpha\ell \cdot \tan\beta\ell}{\tanh\alpha\ell + j\tan\beta\ell}\right| \tag{7.22}$$

The insertion loss at the series ($\beta\ell = (2n + 1) \cdot \pi/2$) and the parallel ($\beta\ell = 2n \cdot \pi/2$) resonant frequencies are of interest. Substitution of the relevant values for $\beta\ell$ in (7.22) yields that the insertion loss at the series resonant frequencies is given by

$$IL = 20\log_{10}\left|1 + \frac{1}{2}\frac{Y_0}{Y_{0C}} \cdot \tanh\alpha\ell\right| \tag{7.23}$$

while that at the parallel resonant frequencies is given by

$$IL = 20\log_{10}\left|1 + \frac{1}{2}\frac{Y_0}{Y_{0C}} \cdot \frac{1}{\tanh\alpha\ell}\right| \tag{7.24}$$

Because $\tanh\alpha\ell$ is small when $\alpha\ell$ is small, it follows from (7.23) that the insertion loss will be small at the series resonant frequencies when $\alpha\ell$ is small, as expected. It follows from (7.24) that the insertion loss will be severe at the parallel resonant frequencies when $\alpha\ell$ is small, again as expected.

The attenuation at the parallel resonant frequencies is decreased sharply with increasing $\alpha\ell$. In contrast with this, the attenuation at the series resonant frequencies increases slowly with increasing $\alpha\ell$. It follows that a resonance-free low-impedance connection can be obtained by using a capacitor with significant losses.

It is also clear from (7.23) and (7.24) that the insertion loss at the series and the parallel resonant frequencies will be decreased as the characteristic impedance of the capacitor is decreased. The ideal coupling capacitor, therefore, will have the lowest possible characteristic impedance with sufficient losses to remove any resonance effects.

The characteristic impedance values claimed for the capacitors considered in Section 7.3 [8] range from 0.4Ω to 50Ω (capacitance range: 800–0.05 pF; f_0 range: 1.5–200 GHz; 50-V breakdown voltage).

The ideal capacitor for a filter or an impedance-matching network would be one with negligible losses and with the series resonant frequency ($f_0/2$) far outside the passband. When a coupling capacitor is required, the series resonant frequency ($f_0/2$) should be chosen to be inside the passband, if possible.

7.3.3   Series Connected Parallel-Plate Capacitors

A series connected parallel-plate capacitor [see Figure 7.1(a)] can be considered to be a cascade connection of two transmission lines separated by a lumped capacitor, as explained in Section 7.1. The basic reason for this model will be illustrated in this section by deriving the Y-parameters and the associated model for the capacitor in free space (no ground plane; see Figure 7.6). The results obtained can also be used to refine the line-capacitor-line model by replacing the capacitance value with that obtained in this section for the free-space capacitor. In doing so, the parallel resonant behavior expected is also obtained in the modified model.

An equivalent circuit for the capacitor based on [2] is shown in Figure 7.6. Instead of using this equivalent circuit, the analysis will be done in terms of the balanced and unbalanced currents on the line, as was done for transmission-line transformers in Chapter 5. The effective inductance presented to the balanced and the unbalanced currents will also be different in this case. The relationship can be established by using the equivalent circuit for two coupled coils [see Figure 5.3(a)].

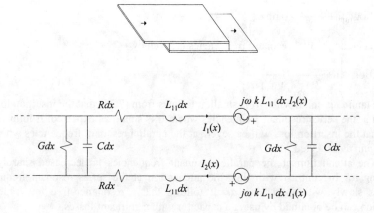

Figure 7.6   An equivalent circuit for the free-space capacitor based on [2].

The effective voltage drop across the inductance and the mutual inductance for an incremental section in the top plate is given by

$$\delta V_1(x) = j\omega L_{11}dx \cdot I_1(x) - j\omega k L_{11}dx \cdot I_2(x)$$

$$= j\omega L_{11}dx \cdot (I_1(x) - k I_2(x))$$

$$= j\omega L_{11}dx \cdot \{[I_b(x) - I_u(x)] - k[I_b(x) + I_u(x)]\}$$

$$= j\omega L_{11} \cdot (1-k) \cdot dx \cdot [I_b(x) - \frac{1+k}{1-k}I_u(x)]$$

$$= j\omega[(1-k)L_{11}]dx \cdot I_b(x) - j\omega[(1+k)L_{11}]dx \cdot I_u(x) \qquad (7.25)$$

while that on the bottom plate is given by

$$\delta V_2(x) = j\omega L_{11}dx \cdot I_2(x) - j\omega L_{11}dx \cdot I_1(x) \cdot k$$

$$= j\omega[(1-k)L_{11}]dx \cdot I_b(x) - j\omega[(1+k)L_{11}]dx \cdot I_u(x) \qquad (7.26)$$

where

$$I_1(x) = I_b(x) - I_u(x) \qquad (7.27)$$

$$I_2(x) = I_b(x) + I_u(x) \qquad (7.28)$$

L_{11} in these equations is the (magnetizing) inductance per unit length of one of the capacitor plates with the other plate open-circuited (zero current).

It follows from (7.25) and (7.26) that the inductance presented to the balanced currents is decreased by a factor $(1-k)$ because of the coupling between the lines, while the inductance presented to the unbalanced currents is increased with a factor $(1+k)$.

The inductance used when the characteristic impedance of a transmission line is calculated is the inductance per unit length associated with the balanced currents $(L_b = [1-k]L_{11})$. The inductance presented to the unbalanced currents is given in terms of this value by

$$L_u = (1+k)L_{11} = \frac{1+k}{1-k}L_b \qquad (7.29)$$

The equivalent circuit shown in Figure 7.6(b) can now be modified as required. The new equivalent circuits are shown in Figure 7.7.

The final equivalent circuit is shown Figure 7.8(b). *Ldx* in this figure should be interpreted as explained earlier.

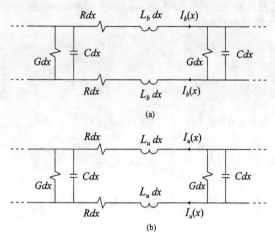

Figure 7.7   The equivalent circuits used to calculate the influence of (a) the balanced and (b) the unbalanced currents on a transmission line.

At this point the free-space capacitor can be analyzed by considering it to be an unbalanced transmission line. The input and output current and voltage of the capacitor will first be established, after which the Y-parameters will be calculated.

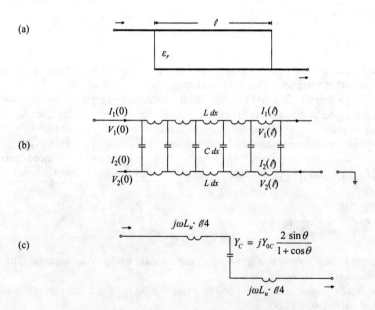

Figure 7.8   (a–c) The equivalent circuit for a single-layer parallel-plate capacitor in free space.

It follows from Figure 7.8(b) that $I_2(0) = 0$ and $I_1(\ell) = 0$. Since the capacitor is in free space and no other path is available for the current, it follows that

$$I_2(\ell) = -I_1(0) \tag{7.30}$$

The current entering the top plate of the capacitor on the left is therefore leaving it at the RHS of the bottom plate.

The currents on the two capacitor plates are given by

$$I_1(x) = -\frac{I_0}{2} + Ae^{-\gamma x} + Be^{\gamma x} \tag{7.31}$$

and

$$I_2(x) = \frac{I_0}{2} + Ae^{-\gamma x} + Be^{\gamma x} \tag{7.32}$$

where $I_1(x)$ is the current on the top plate, $I_2(x)$ the current on the bottom plate, and $I_0/2$ the unbalanced current on the line. Similar to (7.6), γ is given by

$$\gamma = \alpha + j\beta = \sqrt{(G + j\omega C) \cdot 2(r + j\omega L)} \tag{7.33}$$

Note that in this case the total resistance per unit length is $2r$, and the total inductance is $2\omega L$ (see Figure 7.8).

By applying (7.31) at $x = \ell$ and considering that the current at this point is zero, an expression for the unbalanced current is obtained:

$$I_1(\ell) = 0 = -\frac{I_0}{2} + Ae^{-\gamma \ell} + Be^{\gamma \ell}$$

$$\Rightarrow$$

$$\frac{I_0}{2} = Ae^{-\gamma \ell} + Be^{\gamma \ell} \tag{7.34}$$

If (7.32) is applied at $x = 0$, a second expression for I_0 is obtained:

$$I_2(0) = 0 = \frac{I_0}{2} + A + B$$

$$\Rightarrow$$

$$\frac{I_0}{2} = -(A + B) \tag{7.35}$$

A relationship between A and B is obtained by combining (7.34) and (7.35):

$$-(A + B) = Ae^{-\gamma \ell} + Be^{\gamma \ell}$$

$$A(e^{-\gamma \ell} + 1) = -B(e^{\gamma \ell} + 1)$$

$$A = -B \frac{e^{\gamma \ell} + 1}{e^{-\gamma \ell} + 1} \tag{7.36}$$

The expression for the current entering the top plate of the capacitor can be simplified by using this expression:

$$I_1(0) = -\frac{I_0}{2} + A + B \tag{7.37}$$

$$= (A + B) + (A + B) = 2(A + B)$$

$$= 2B \frac{e^{-\gamma \ell} - e^{\gamma \ell}}{e^{-\gamma \ell} + 1} \tag{7.38}$$

Because of the relationship between $I_1(0)$ and $I_2(\ell)$, the expression for $I_2(\ell)$ follows immediately from (7.38):

$$I_2(\ell) = -I_1(0) = -2B \frac{e^{-\gamma \ell} - e^{\gamma \ell}}{e^{-\gamma \ell} + 1} \tag{7.39}$$

With B known, both the input and the output currents are known at this point.

The voltages on the two plates are given by (6.7) and (6.8), while the voltage difference between the two plates is given by (6.9). Since I_0 is known in terms of A and B, and A is known in terms of B, all the voltages are also known in terms of B at this point.

The Y-parameters of the capacitor can now be calculated. In order to derive these parameters, only y_{11} must be calculated ($y_{12} = -y_{11}$, $y_{12} = y_{21}$, and $y_{22} = y_{11}$).

In order to calculate y_{11}, expressions for the input current and the input voltage are required. An expression for the input current has already been derived. Derivation of the expression for the input voltage follows:

$$V_1(\ell) = V_{12}(\ell) = Z_0(A e^{-\gamma \ell} - B e^{\gamma \ell}) \tag{7.40}$$

Substitution of $V_1(\ell)$ in the expression by using (6.7) yields

$$0 = V_1(0) - \frac{Z_0}{2}(A - B) - \frac{Z_0}{2}(A e^{-\gamma \ell} - B e^{\gamma \ell}) + s\ell L_u[-(A + B)]$$

from which it follows that

$$V_1(0) = \frac{Z_0}{2}(A - B) + sL_u\ell(A + B) + \frac{Z_0}{2}(A e^{-\gamma \ell} - B e^{\gamma \ell})$$

$$= A\left[\frac{Z_0}{2}(1+e^{-\gamma\ell}) + sL_u\ell\right] - B\left[\frac{Z_0}{2}(1+e^{\gamma\ell}) - sL_u\ell\right]$$

$$= -B\frac{e^{\gamma\ell}+1}{e^{-\gamma\ell}+1}\left[\frac{Z_0}{2}(1+e^{-\gamma\ell}) + sL_u\ell\right] - B\left[\frac{Z_0}{2}(1+e^{\gamma\ell}) - sL_u\ell\right]$$

$$= -B\,Z_0(e^{\gamma\ell}+1) + sL_u\ell\,B\frac{e^{-\gamma\ell}-e^{\gamma\ell}}{e^{-\gamma\ell}+1} \tag{7.41}$$

With the input current and the input voltage known, the desired expression for y_{11} can now be derived:

$$y_{11} = \frac{2\cdot\dfrac{e^{-\gamma\ell}-e^{\gamma\ell}}{1+e^{-\gamma\ell}}}{-Z_0(e^{\gamma\ell}+1) + sL_u\ell\dfrac{e^{-\gamma\ell}-e^{\gamma\ell}}{1+e^{-\gamma\ell}}}$$

$$= \frac{1}{s\dfrac{L_u\ell}{2} + \dfrac{Z_0}{2}\dfrac{(1+e^{\gamma\ell})(1+e^{-\gamma\ell})}{e^{\gamma\ell}-e^{-\gamma\ell}}} \tag{7.42}$$

If the capacitor is assumed to be lossless, the second term in the denominator of (7.42) can be simplified as follows:

$$X = \frac{Z_0}{2}\frac{(1+e^{\gamma\ell})(1+e^{-\gamma\ell})}{e^{\gamma\ell}-e^{-\gamma\ell}} = \frac{Z_0}{2}\frac{1+e^{-\gamma\ell}+e^{\gamma\ell}+1}{e^{\gamma\ell}-e^{-\gamma\ell}}$$

$$= \frac{Z_0}{2}\frac{e^{-j\beta\ell}+e^{j\beta\ell}+2}{e^{j\beta\ell}-e^{-j\beta\ell}}$$

$$= \frac{Z_0}{2}\frac{2\cos\theta+2}{\cos\theta + j\sin\theta - (\cos\theta - j\sin\theta)}$$

$$= \frac{(\cos\theta+1)/2}{+jY_0\,\sin\theta} \tag{7.43}$$

where $\theta = \beta\ell$.

Substitution of (7.43) in (7.42) yields

$$y_{11} = \cfrac{1}{\cfrac{sL_u\ell}{2} + \cfrac{1}{jY_{0C} \cdot \cfrac{\sin\theta}{(1+\cos\theta)/2}}} \tag{7.44}$$

With y_{11} known, the other Y-parameters can be calculated and the Y-parameter matrix for the parallel-plate capacitor in free space is known.

In the lossless case, these Y-parameters lead directly to the equivalent circuit shown in Figure 7.8(c). It follows that the parallel-plate capacitor in free space can be considered to be purely lumped, with fixed inductance and variable capacitance. As expected, the capacitance at low frequencies reduces to $C\ell$:

$$Y_C = j2Y_{0C}\frac{\sin\beta\ell}{1+\cos\beta\ell} \tag{7.45}$$

$$\cong j2Y_{0C}\frac{\beta\ell}{2}$$

$$= \sqrt{\frac{C}{2L_b}} \cdot \omega\sqrt{(2L_b)\cdot C}\cdot\ell$$

$$= \omega\cdot(C\ell) \tag{7.46}$$

The series inductance obtained in (7.44) is also significant. Since $\omega L_u = \omega L_{11}\cdot(1+k)$, it follows that

$$sL_u\ell/2 = sL_{11}[(1+k)/2]\cdot\ell$$

from which it follows that, if the magnetic coupling between the two capacitor plates is tight, the series inductance will be approximately the same as the uncoupled inductance (zero current in the other capacitor plate) of one of the capacitor plates. Instead of interpreting this inductance as the total inductance of one plate, it would be more accurate to see it as the sum of the inductance of the top plate from the LHS edge to the center and the inductance of the bottom plate from the center to the RHS edge (see Figure 7.8). Note that the inductance of a microstrip line identical to the bottom plate and with the top plate absent would also be $L_{11}\cdot\ell$. The significance of this will be appreciated when the microstrip case is considered.

In order to establish the influence of this inductance on the resonance frequencies of the free-space capacitor, (7.44) must be simplified by replacing L_u in terms of Z_{0C}. This can be done by using (7.29) and (7.15):

$$Z_{0C} = \sqrt{\frac{2L_b}{C}} = \sqrt{\frac{2L_b\cdot\ell}{C\cdot\ell}} = \sqrt{\frac{2L_b\cdot\ell}{C_T}}$$

$\Rightarrow$

$$2 L_b \cdot \ell = Z_0^2 \cdot C_T = Z_0^2 \cdot \frac{Y_0}{2 f_0} = \frac{Z_0}{2 f_0} \tag{7.47}$$

from which it follows [by using (7.15)] that

$$L_u \cdot \ell = \frac{1+k}{1-k} \cdot \frac{Z_0}{4 f_0} \tag{7.48}$$

Substitution of this expression in (7.44) yields that

$$y_{11} = \frac{1}{j\omega \dfrac{1+k}{1-k} \dfrac{Z_{0C}}{8 f_0} + \dfrac{1}{j Y_{0C} \dfrac{\sin\theta}{1+\cos\theta}}} \tag{7.49}$$

$$= j Y_{0C} \frac{2\sin\theta}{(1+\cos\theta) - \dfrac{1+k}{1-k} \cdot \dfrac{\theta}{2} \cdot \sin\theta} \tag{7.50}$$

The effective admittance presented by the free-space capacitor can be calculated by using this expression. In interpreting this expression, it should be kept in mind that the characteristic impedance of the capacitor (Y_{0C}) is not independent of the coupling factor and will approach infinity as the coupling factor approaches unity.

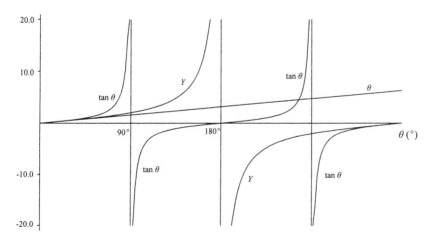

Figure 7.9 Comparison of the tangent function (tan θ) with the functions $Y = \sin\theta \, / \, [(1 + \cos\theta)/2]$, and θ (ωC case). (The variable on the X-axis is the angle in degrees, while the function of interest is displayed on the Y-axis.)

The effective capacitance of the capacitor in the equivalent circuit is determined by the $\sin\theta / [(1+\cos\theta)/2]$ term in (7.44). This function is compared with the tangent function ($\tan\theta$) in Figure 7.9. It is clearly much more linear than the tangent function, and series resonance only occurs when the electrical line length is 180°, not 90° as in the case of the tangent function. Series resonance in the actual capacitor will occur sooner because of the effect of the series inductance. Parallel resonance (open circuit) occurs when the line length is 360°. The parallel resonance frequency is not influenced by the series inductance.

The combined influence of the inductance and the capacitance on the total admittance of the free-space capacitor [as calculated with (7.50)] is shown in Figure 7.10 for a coupling factor of zero and one-half. Series resonance is clearly accelerated drastically by any magnetic coupling between capacitor plates. Fortunately, this problem is eliminated when the capacitor is mounted on a microstrip line.

When the coupling is tight and the capacitor is mounted on a microstrip line with its plates parallel to the substrate surface, the inductance of the bottom plate becomes the inductance of the microstrip line (see Figure 7.11), and this inductance combines with the microstrip capacitance to have a transmission-line effect. The characteristic impedance and phase response of this line section is essentially that of the microstrip line. Similarly, the inductance of the top plate combines with the series combination of the capacitor capacitance and the microstrip capacitance to have a similar line effect (the capacitance of the capacitor usually acts as a short circuit compared to the microstrip capacitance because of the relative difference in the dielectric constants and the thickness of the dielectric layers).

The free-space analysis clearly supports the use of the line-capacitor-line model. When the capacitor is placed with its plates normal to the substrate surface, the model can be enhanced by replacing the fixed capacitance value with a frequency-dependent term based

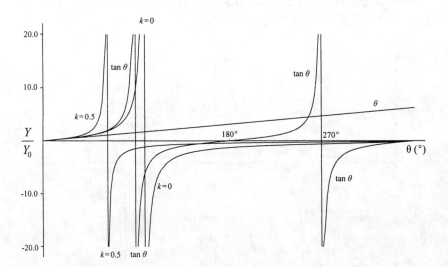

Figure 7.10   Comparison of the tangent function ($Y/Y_0 = \tan\theta$) and the linear case (θ) with the normalized admittance of the free-space capacitor when $k = 0$ and $k = 0.5$ (lossless case).

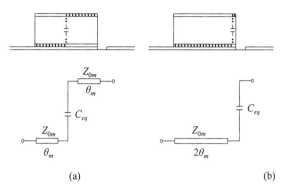

Figure 7.11   (a, b) Transmission-line models for a parallel-plate capacitor mounted on microstrip with the same width as the capacitor. The models are valid below resonance. Limited EM simulations seem to favor the model in (b). Note that C_{eq} is equal to the total capacitance of the capacitor at low frequencies. The electrical length of the microstrip section below the capacitor is $2\theta_m$ at the frequency of interest, and its characteristic impedance is Z_{0m}.

on (7.45). Note that the parasitic capacitance associated with the capacitor plates will not be that of the connecting microstrip line in this case. When the capacitor is mounted with its plates parallel to the substrate surface the resonance behavior will be modified by the extra (asymmetric) capacitance to ground associated with the bottom plate (that is, compared with the free-space case). Also note that in an MIM capacitor the top plate may be smaller than the bottom plate, which will also complicate the resonance behavior.

The model for the single parallel-plate capacitor can be extended easily to obtain a model for gap capacitors too. The gap capacitor model is shown in Figure 7.12. It should be noted that, when possible, the widths of a gap capacitor and the microstrip line should be chosen to be the same. The main reason for this is the parasitic effect of the step discontinuities introduced at the gap when this is not the case.

If the dielectric constant of the capacitor dielectric is much higher than that of the microstrip and the capacitor is thin compared to the microstrip substrate height, the characteristic impedance of the line section associated with the gap can usually also be estimated to be that of the microstrip.

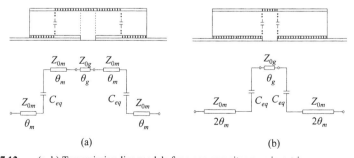

Figure 7.12   (a, b) Transmission-line models for a gap capacitor on microstrip.

Note that for the model to be practical, resistance must be added to it to model the capacitor losses. Extra inductance may also be required to provide a tight fit to measured results. The extended model for a parallel-plate capacitor is shown in Figure 7.13.

The extended model can also be used for multiplate capacitors, at least up to the first series resonance frequency. Note that plates of multiplate chip capacitor can be oriented in parallel with or normal to substrate [9]. When the dielectric constant of the capacitor dielectric is much higher than that of the substrate, the characteristics of the Z_{0m} line will again be close to that of the microstrip line, that is, when the plates are oriented in parallel with the substrate. Z_{0m} should be higher when the plates are oriented normal to the substrate. The best approach in both cases would be to adjust the model parameters to fit the available measurement data as tightly as possible.

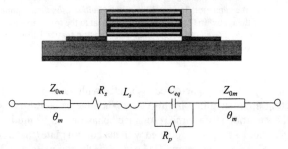

Figure 7.13   The extended transmission-line model for a parallel-plate capacitor mounted on microstrip with the same width as the capacitor. The electrical length of the microstrip section below the capacitor is $2\theta_m$ at the frequency of interest, and its characteristic impedance is Z_{0m}.

The model shown in Figure 7.13 can also be used for MIM capacitors. However, EM simulation of these capacitors on a 100-μm GaAs substrate seems to indicate that a better fit can be obtained at millimeter-wave frequencies if the series capacitance (C_s) is concentrated at the air-bridge connection instead of at the center of the capacitor plates (refer to Figure 7.14). An extra capacitor to ground on the air-bridge side also improved the fit obtained.

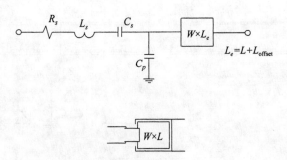

Figure 7.14   An alternative model for a MIM capacitor.

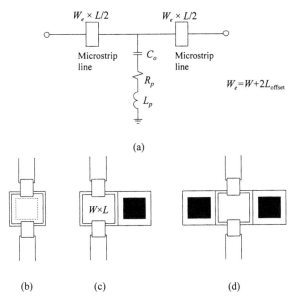

Figure 7.15   (a) An overlay capacitor model for a capacitor with (b) a centered via, (c) one offset via, and (d) two offset vias. L_p is the effective via inductance.

The series capacitance, C_s, in the alternative model was calculated by using the area of the top plate ($A=W{\times}L$) and the capacitance density. Note that the bottom plate was assumed to have an offset, L_{offset}, from the top plate on each side. The reference planes were taken to be at the air-bridge connection to the top plate on the one side, and at the bottom-plate edge on the other side. The air-bridge and the bottom-plate extensions were modeled as standard microstrip lines.

A model for a shunt overlay capacitor is shown in Figure 7.15(a). For minimum inductance to ground a centered via is preferable, but if the process used does not allow this, offset vias can be used as shown in Figure 7.15(c, d). The width of each line in the model is taken to be that of the bottom plate of the capacitor, while the length is half of that of the top plate. EM simulation of overlay capacitors of different sizes indicated that the shunt capacitance, C_o, in the model should be set to be slightly higher than the value calculated by using the area of the top plate ($A=W{\times}L$) and the capacitance density. The shunt inductance was found to be a function of the vias used, as well as the capacitor size (for small capacitors, the inductance increases as the size decreases).

7.4   INDUCTORS

Lumped microwave inductors can be fabricated in different forms. For low inductance values and connections, strip inductors, ribbons, and bonding wires are frequently used, while larger inductance values are realizable with spiral or solenoidal inductors. The basic equations required to design these inductors will be considered here.

7.4.1 Strip Inductors

The inductance of an isolated (no ground plane), flat, ribbon inductor (or strip inductor) is given approximately by [10]

$$L \text{ (nH / mm)} = 0.2 \ \{\ln[l / (w + t)] + 1.193 + 0.2235 \ (w + t) / l\} \tag{7.51}$$

where w is the width of the ribbon, t its thickness, and l its length.

An approximate expression for the Q of a ribbon inductor is [11]

$$Q = 2.15 \times 10^3 \ \frac{L(\text{nH})}{K} \ \frac{w}{l} \left(\frac{\rho(\text{Cu})}{\rho}\right)^{1/2} \left(\frac{f(\text{GHz})}{2}\right)^{1/2} \tag{7.52}$$

where ρ is the resistivity of the material used, and K is a correction factor for the current crowding occurring at the corners of the strip [10]. K is given approximately by the following expression:

$$K = 1.3565 - 0.2319 \ln[w / t] + 0.2386 [\ln(w / t)]^2$$
$$- 0.0536 \ [\ln(w / t)]^3 + 0.0043[\ln(w / t)]^4 \tag{7.53}$$

The inductance of a strip inductor is decreased by the presence of a ground plane. The effective inductance for this case is given in terms of the free-space value by [12, 13]

$$L_{\text{eff}} = [0.570 - 0.145 \ln(W / h)] \cdot L \tag{7.54}$$

7.4.2 Single-Turn Circular Loop

Equations (7.51) and (7.54) can also be used to calculate the inductance of a single-turn circular loop in those cases where the width of the strip (w) is much smaller than the diameter. When the ground plane can be ignored, the following expression [14] can also be used:

$$L(\text{nH / mm}) = 0.2[\ln(l / w + t) - 1.76] \tag{7.55}$$

where l is the average length of the loop and t the conductor thickness. The dimensions must be specified in millimeters.

For (7.55) to apply, the inequality $l >> 2(w + t)$ must be satisfied.

7.4.3 Bond Wire Inductors

Bond wire inductors have the advantage over strip inductors that higher Q-factors can be expected because of the larger surface area. Furthermore, touch-up tuning is possible with bonding wire inductors, while the inductance is fixed for strip inductors. The fixed inductance, however, is an advantage in a first-time-right design.

The inductance associated with a long ($l/d \geq 100$) free-space bonding wire of diameter d and length l can be calculated by using the equation [10]

$$L(\text{nH} / \text{mm}) = 0.20\,[\ln(l/d) + 0.386]$$ (7.56)

The effect of a ground plane can be incorporated by using the equation [10, 12]

$$L(\text{nH} / \text{mm}) = 0.2\,\{\ln\frac{4h}{d} + \ln\frac{l + \sqrt{l^2 + d^2/4}}{1 + \sqrt{l^2 + 4h^2}} +$$

$$\sqrt{1 + \frac{4h^2}{l^2}} - \sqrt{1 + \frac{d^2}{4l^2}} - 2\frac{h}{l} + \frac{d}{2l}\}$$ (7.57)

An approximate expression for the Q of a round wire inductor is [11]

$$Q = 3.38 \times 10^3\,L(\text{nH})\,\frac{d}{l}\left(\frac{\rho(\text{Cu})}{\rho}\right)^{1/2}\left(\frac{f(\text{GHz})}{2}\right)^{1/2}$$ (7.58)

Equations (7.56) and (7.57) are only accurate when $l/d \geq 100$ [9]. When short bond wires are used, the following equation is recommended for the free-space case [15]:

$$L(\text{H}) = [\mu_0 / (2\pi)]\,l\,\{\ln\left[(2l/d) + \sqrt{1 + (2l/d)^2}\right] + d/(2l)$$

$$- \sqrt{1 + (d/(2l))^2} + \mu_r \delta_k\}$$ (7.59)

with

$$\delta_k = 0.25\tanh(4d_s/d)$$ (7.60)

$$d_s = \sqrt{\frac{\rho}{\pi f\,\mu_r \mu_0}}$$ (7.61)

d_s is the skin depth, ρ is the resistivity of the bond wire, and f is the frequency in hertz. When the wire is manufactured with nonmagnetic material, as is usually the case, $\mu_r = 1$.

The $\mu_r\,\delta_k$ term in (7.59) represents the internal inductance of the wire.

The effect of the ground plane is similar to a current image reflection of the inductor. Because of this effect the inductance of the bond wire is decreased when a ground plane is present. The effective inductance is this case is given by [15]

$$L_{\text{eff}}\ (\text{H}) = L - [\mu_0\ /\ (2\pi)] \cdot l \cdot \{\ln\left[l\ /\ (2h) + \sqrt{1 + (l\ /\ (2h))^2}\right]$$
$$- \sqrt{1 + (2h\ /\ l)^2} + 2h\ /\ l\} \tag{7.62}$$

where $2h$ is the center-to-center separation between the wire and its image, and h is the distance from the ground plane.

It is recommended in [15] that h in (7.62) should be replaced by

$$h' = h + 4.6d_s \tag{7.63}$$

to account for the nonperfect ground (finite conductance).

Multiple bond wires are frequently used to make connections in microwave circuits. An equation to calculate the inductance of two parallel bond wires with a center-to-center spacing of s is also given in [15]:

$$L_{\text{pair}} = L_{\text{eff}}\ /\ 2 + M_2\ /\ 2 \tag{7.64}$$

where L_{eff} is given by (7.62), and the mutual inductance M_2 is given by [15]

$$M_2 = [\mu_0\ /\ (2\pi)] \cdot l \cdot \{\ln\left[l\ /\ s + \sqrt{1 + (l\ /\ s)^2}\right] - \sqrt{1 + (s\ /\ l)^2} + s\ /\ l\} \tag{7.65}$$

The expressions given above were based on the assumption that the bond wires were horizontal with the ground plane, which will seldom be the case. This deficiency can be corrected (to a degree) by using the average height of the bond wire(s) above the ground plane. A circular shape for the bond wire was assumed in [15], but a cubic Bézier curve shape based on the angles formed at the two end points of the bond wire will probably be more realistic.

The resistive losses associated with a bond wire is given within 3% by [15]

$$R_s = \begin{cases} R_{dc} \cosh[0.041(d\ /\ d_s)^2]\ \text{for}\ d\ /\ d_s \le 3.394 \\ R_{dc}(0.25d\ /\ d_s + 0.2654)\ \text{for}\ d\ /\ d_s \ge 3.394 \end{cases} \tag{7.66}$$

with

$$R_{dc} = 4\rho\ell\ /\ (\pi d^2) \tag{7.67}$$

When two bond wires are used in parallel, the resistance in each is increased by the proximity effect. The resistance for each bond wire is increased by a factor [15]

$$F_p = 1.0 + 0.8478e^{-0.9435(s/d)} \tag{7.68}$$

Finally, it should be noted that some capacitance is associated with a bond wire. This capacitance can be expected to be of the order of tens of a femtofarad per millimeter of the bond wire length. The capacitance may be small compared to any pad capacitance, but it will have an effect at millimeter-wave frequencies.

A good model for a bond wire would consist of a small capacitor to ground on each side of the bond wire, and the series inductance and resistance estimated for the bond wire. Good results can be expected in practice when such models are based on 3D EM simulations of bond wires.

7.4.4 Single-Layer Solenoidal Air-Cored Inductors

At microwave frequencies, solenoidal inductors are often used as RF chokes in hybrid circuits. When the size is not prohibitively small, they can also be used as inductors.

The inductance of a solenoidal coil is given by

$$L(\text{nH}) = 10.0r^2 N^2 / [2.29l + 2.54r] \tag{7.69}$$

where r is the radius (in millimeters), l is the length (in millimeters), and N is the number of turns of the coil.

In order for its behavior to remain essentially lumped, an inductor must be electrically short. Reasonable results can be expected with shunt inductors when the associated electrical length is shorter than $30°$ (the deviation from the expected linear increase in reactance will then be less than 10%). In the case of a series inductor, the restrictions are more severe because the resistance in series with the inductor will be transformed because of the transmission-line effect too.

It follows from the previous paragraph that the maximum inductance that can be realized with a given conductor length is of interest. By setting the derivative of the inductance [as given by (7.69)] equal to zero, the parameters of the coil with the highest inductance for a specified conductor length can be found. The resulting equations are

$$r_{\text{opt}} = 0.3788\sqrt{l_e \cdot c} \tag{7.70}$$

$$l_{\text{opt}} = 0.4202\sqrt{l_e \cdot c} + c \tag{7.71}$$

$$N = 0.4202\sqrt{l_e / c} \tag{7.72}$$

where c is the wire thickness (in millimeters), r_{opt} the optimum radius, l_e the conductor length, and l_{opt} the optimum coil length.

The wire thickness of the solenoidal coil should be chosen to optimize the Q (refer to Section 3.3.6).

7.4.5 Spiral Inductors

Spiral inductors are often used in MMICs (monolithic microwave integrated circuits; single-chip integrated circuit) to realize the inductance required. Square spiral inductors are usually used, but lower losses can be obtained with circular spiral inductors.

Accurate equations (average errors of less than 2% are claimed) for the inductance and capacitance of square spiral inductors are provided in [5, 7], respectively. The inductance was derived indirectly in [5] by calculating the capacitance associated with a free-space inductor (the inductance is independent of the dielectric constant of the actual substrate to be used) above a ground plane, and then using the duality $LC = \mu_0 \varepsilon_0$. Three fitting factors were introduced in the equations to improve the fit to the inductance obtained by using EM simulations of square spiral inductors. The relevant equations are repeated here:

$$L = 4 \frac{\mu_0 \varepsilon_0}{(C_{1/4far}^n + C_{1/4near}^n)^{1/n}} N^2 (b_1 + b_2)^2 \tag{7.73}$$

where

$$n = 1.3461 - 0.6592\, e^{-0.6139 h/W_s + 0.2918 h/W} \tag{7.74}$$

$$C_{1/4near} = \frac{\varepsilon_0 A_{1/4}}{h} \cdot \frac{W}{W_s} \tag{7.75}$$

$$A_{1/4} = b_2^2 - b_1^2 \tag{7.76}$$

$$b_1 = (d_{in} - S)/2 \tag{7.77}$$

$$b_2 = (d_{out} + S)/2 \tag{7.78}$$

$$c_{f1} = 0.90571 + 0.49425\, e^{-\rho/0.12253} \tag{7.79}$$

$$\rho = \frac{d_{out} - d_{in}}{d_{out} + d_{in}} \tag{7.80}$$

$$W_s = W + S \tag{7.81}$$

$$C_{1/4far} = 1 / \left\{ \left[\frac{1}{c_{f1}\, \varepsilon_0 \sqrt{8\pi A_{1/4}}} + \frac{\Delta p_{11} - \Delta p_{110}}{N\, M} \right] - \frac{1}{4\pi \varepsilon_0 r_0} \right\} \tag{7.82}$$

$$c_f = 0.865 \tag{7.83}$$

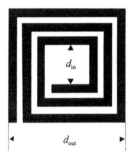

Figure 7.16   An square spiral inductor with three turns.

$$r_0 = \frac{4(b_1^2 + b_1 b_2 + b_2^2)}{3(b_1 + b_2)} \tag{7.84}$$

$$\Delta p_{11} = \frac{1}{c_f \varepsilon_0 \sqrt{8\pi W W_s}} \tag{7.85}$$

$$\Delta p_{110} = \frac{1}{c_f \varepsilon_0 \sqrt{8\pi W_s^2}} \tag{7.86}$$

N is the number of turns, W the turn width (line width), S the spacing between two adjacent line segments, d_{out} the outermost dimension (see Figure 7.16), d_{in} the innermost dimension, h the substrate height, and ε_r the relative dielectric constant of the substrate.

The modified Wheeler formula given in [7] can be used to calculate the inductance of square, hexagonal (6 sides to each turn) and octagonal (eight sides to each turn) planar spiral inductors:

$$L_{mw} = K_1 \mu_0 \frac{N^2 d_{avg}}{1 + K_2 \rho} \tag{7.87}$$

where the shape factor (fill ratio), ρ, is defined in (7.80), N is the number of turns,

$$d_{avg} = 0.5 \left(d_{out} + d_{in} \right) \tag{7.88}$$

and K_1 and K_2 are defined in Table 7.3.

Table 7.3   Coefficients for the Modified Wheeler Expression [7]

Layout	K_1	K_2
Square	2.34	2.75
Hexagonal	2.33	3.82
Octagonal	2.25	3.55

An equation that can be used for circular spiral inductors too is also given in [7]:

$$L_{gmd} = \frac{\mu_0 N^2 d_{avg} c_1}{2}[\ln(c_2 / \rho) + c_3\rho + c_4\rho^2] \qquad (7.89)$$

where the c_i constants are given in Table 7.4. The accuracy claimed for this equation is 8% when $S \le 3W$.

Table 7.4   Coefficients for the Current Sheet Expression [7]

Layout	c_1	c_2	c_3	c_4
Square	1.27	2.07	0.18	0.13
Hexagonal	1.09	2.23	0.00	0.17
Octagonal	1.07	2.29	0.00	0.19
Circular	1.00	2.46	0.00	0.20

A lumped-element model for a spiral inductor [7] on a lossy substrate is shown in Figure 7.17. R_{sub} can be set to zero on a substrate with low losses. The transmission-line model [4] shown in Figure 7.18 is also useful, and is suitable for spiral inductors realized on low-loss substrates.

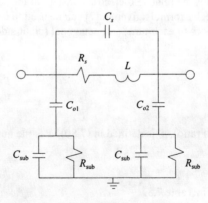

Figure 7.17   A lumped element model for a planar spiral inductor.

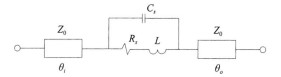

Figure 7.18 A transmission-line model for a spiral inductor on a low-loss substrate.

It should be noted that a spiral inductor is actually not a symmetrical component, and the model should also reflect this. A good fit has been obtained for inductors on a GaAs substrate with the model in Figure 7.18 when the same characteristic impedance with different line lengths were used.

7.5 MICROSTRIP DISCONTINUITY EFFECTS AT THE LOWER MICRO-WAVE FREQUENCIES

Microstrip discontinuity effects associated with bends, curves, changes in the line width, T-junctions, crosses, and open-ended stubs add undesirable inductance and capacitance to designed circuits. The magnitude of these parasitics at the lower microwave frequencies (below X-band) [12] will be considered in this section.

7.5.1 Open-Ended Stubs

The effect of the fringing capacitance associated with the open end of an open-ended stub is similar to extending the length of the line slightly. The equivalent additional line length is given empirically by [16]. The expression for the phase shift (in degrees) is

$$\Delta\theta_{oc} = 4.944 \times 10^{-7} h f \sqrt{\varepsilon_{r\_eff}} \frac{\varepsilon_{r\_eff} + 0.300}{\varepsilon_{r\_eff} - 0.258} \frac{W/h + 0.264}{W/h + 0.800} \tag{7.90}$$

where h is the thickness of the substrate (in meters) and f the frequency (in hertz).

The maximum relative error in (7.90) as compared to the more accurate expression of Silvester and Benedek [17] is less than 4% for $W/h \geq 0.2$ and $2 \leq \varepsilon_r \leq 50$ [18].

As an illustration of the magnitude of the open-end parasitic, the parasitic electrical line length (at 10 GHz) associated with different width-to-height ratios and dielectric constants ($\varepsilon_r = 2.5$ and $\varepsilon_r = 10.2$) are tabulated in Table 7.5 for a substrate with $h = 0.635$ mm. It is clear from these results that the parasitic influence of an open end cannot be neglected at the higher frequencies and that this effect is more pronounced with higher dielectric constants and low impedance lines.

Table 7.5   The Increase in Electrical Line Length Caused by an Open End as a Function of the Dielectric Constant and the Width-to-Height Ratio of the Line ($f = 10$ GHz; $h = 0.635$ mm)

ε_r	Z_0 (Ω)	W/h	θ ($°$)
	25	7.20	5.6
	50	2.80	5.0
2.5	75	1.35	4.4
	100	0.70	3.7
	125	0.38	3.2
	15	6.90	9.3
	25	3.35	8.4
10.2	50	0.90	6.2
	75	0.30	4.5

The simplest way to compensate for the increase in line length is to reduce the length of the designed line by the correct amount. A distance of at least the equivalent line length should be allowed between the end of an open-ended stub and the substrate edge.

7.5.2   Steps in Width

The parasitic effect of a step junction is similar to that of an open end. The effect of the fringing capacitance associated with the wider line of the step discontinuity is similar to an increase in the length of that line. The change in the electrical length (in degrees) can be estimated by using the equation [18]

$$\Delta\theta_{step} = \Delta\theta_{oc}[1 - W_2 / W_1] \tag{7.91}$$

where $\Delta\theta_{oc}$ can be calculated by using (7.90).

An alternative and more accurate approach to characterizing a step discontinuity is to use the equivalent circuit shown in Figure 7.19(b). An approximate expression for the inductance $L_s = L_1 + L_2$ is ($\pm 5\%$ for $W_1 / W_2 \leq 5.0$ and $W_2 / h = 1.0$) [16]

$$\frac{L_s}{h}(\text{nH} / \text{m}) = 40.5\left(\frac{W_1}{W_2} - 1.0\right) - 75\log\left(\frac{W_1}{W_2}\right) + 0.2\left(\frac{W_1}{W_2} - 1\right)^2 \tag{7.92}$$

The individual inductances are given by [16]

$$L_1 = L_{w1} / [L_{w1} + L_{w2}] \cdot L_s \tag{7.93}$$

and

$$L_2 = L_{w2} / [L_{w1} + L_{w2}] \cdot L_s \tag{7.94}$$

where L_{w1} and L_{w2} are the inductances associated with the characteristic impedances of the two lines.

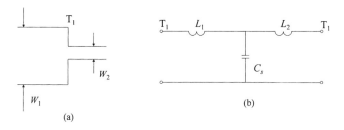

Figure 7.19 (a, b) The equivalent circuit of a step discontinuity.

An approximate closed form expression for the capacitance C_s in Figure 7.19(b) is ($\pm$10% for $\varepsilon_r \leq 10$ and $1.5 \leq W_1 / W_2 \leq 3.5$) [16]

$$\frac{C_s}{\sqrt{W_1 W_2}} (\text{pF} / \text{m}) = [10.1 \log \varepsilon_r + 2.33] \frac{W_1}{W_2} - 12.6 \log \varepsilon_r - 3.17 \tag{7.95}$$

An idea of the magnitude of the parasitic effects associated with step discontinuities can be obtained from the extensions in line length resulting from an open-ended line as given in Table 7.5 and (7.91).

A first-order compensation technique for a step discontinuity would be to decrease the length of the wider line by the appropriate amount. The phase shift associated with a step discontinuity will always be less than that caused by an open end in a line with the lower characteristic impedance.

7.5.3   Microstrip Bends and Curves

The equivalent circuit for a microstrip bend with lines of equal width is shown in Figure 7.20. Closed-form expressions for the right-angled bend discontinuity capacitance and inductance are [12]

$$\frac{C_b}{W} (\text{pF} / \text{m}) = \begin{vmatrix} \dfrac{(14\varepsilon_r + 12.5)W / h - (1.83\varepsilon_r - 2.25)}{\sqrt{W / h}} + \dfrac{0.02\varepsilon_r}{W / h} & (W / h \leq 1) \\[4mm] (9.5\varepsilon_r + 1.25)W / h + 5.2\varepsilon_r + 7.0 & (W / h \geq 1) \end{vmatrix} \tag{7.96}$$

$$L_b / h \ (\text{nH} / \text{m}) = 100[4\sqrt{W / h} - 4.21] \tag{7.97}$$

Equation (9.67) is accurate to within 5% for $2.5 \leq \varepsilon_r \leq 15$ and $0.1 \leq W/h \leq 5$. The accuracy of (7.97) is about 3% for $0.5 \leq W/h \leq 2.0$ [16].

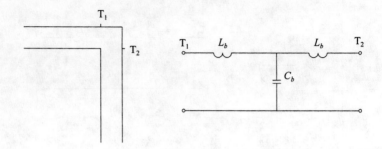

Figure 7.20   Equivalent circuit for a microstrip bend.

An idea of the magnitude of the parasitics associated with a bend discontinuity can be obtained from Table 7.6. The theoretical VSWRs associated with two bends are shown in this table as a function of the frequency.

Although the effect of a single bend may be small at the lower microwave frequencies, it should be kept in mind that it will increase with frequency, the number of bends used in cascade, and the line width.

Table 7.6   The VSWR (Theoretical) Associated with an 90° Bend (Not Chamfered) in a 75-Ω (ε_r= 2.5) Line and a 50-Ω (ε_r= 10.2) Line as a Function of the Frequency (h = 0.508 mm)

ε_r	Z_0 (Ω)	f(GHz)	VSWR
2.5	75	2	1.03
		4	1.06
		8	1.12
		10	1.15
10.2	50	2	1.06
		4	1.13
		8	1.28
		10	1.36

The parasitic effects of bend discontinuities are usually reduced by mitering the bend as shown in Figure 7.21. The optimum value of L in this figure is about 1.8W for 50-Ω lines on alumina and rexolite substrates, and it seems to be independent of the bend angle [12].

When $W/h \geq 0.25$ and $\varepsilon_r \leq 25$, the length L can be calculated by using the following equation:

$$L / W = \sqrt{2}\,[1.04 + 1.3\,e^{-1.35W/h}] \tag{7.98}$$

The equivalent electrical line length of the miter (l) can be estimated by using the equation

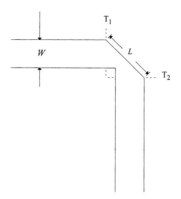

Figure 7.21   Compensation of a microstrip bend.

$$l = L / \sqrt{2} \tag{7.99}$$

When the line is too thin ($W/h \le 0.25$), the optimum miter cannot be used.

Curving a line is frequently a better option than mitering it. When the curving radius is larger than twice the width of the line, the main parasitic effect is a change in the effective line length. The effective length $[l_{\text{eff}} = (2\pi R_{\text{eff}}) \times \theta_c/(2\pi)]$ of the curve ($3 < R/W < 7$) can be estimated by assuming the effective radius to be [19]

$$R_{\text{eff}} = R_{\text{inner}} + 0.3W \tag{7.100}$$

This is illustrated in Figure 7.22.

Curving a line also has the advantage that the direction of the line can be changed with any arbitrary angle (θ_c).

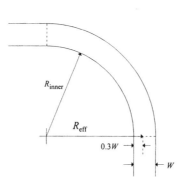

Figure 7.22   The effective curving radius (R_{eff}) when a line is curved.

7.5.4   T-Junctions and Crosses

Hammerstad's approach [18] to characterizing the parasitic effects of a T-junction with constant main-line width is illustrated in Figure 7.23. The different parameters are defined by the following equations:

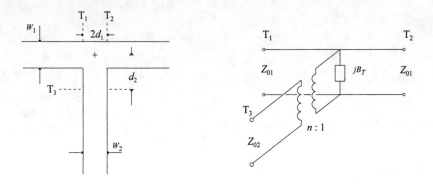

Figure 7.23   The equivalent circuit for a microstrip T-junction.

$$D_1 = 120\pi h / Z_{01}(\text{air}) \qquad D_2 = 120\pi h / Z_{02}(\text{air}) \tag{7.101}$$

$$d_2' = \frac{D_1}{2} - d_2 \tag{7.102}$$

$$n^2 = \left\{ \frac{\sin\left(\dfrac{\pi}{2}\dfrac{2D_1}{\lambda_m}\dfrac{Z_{01}}{Z_{02}}\right)}{\dfrac{\pi}{2}\dfrac{2D_1}{\lambda_m}\dfrac{Z_{01}}{Z_{02}}} \right\}^2 \left\{ 1 - \left(\pi \frac{2D_1}{\lambda_m}\frac{d_2'}{D_1}\right)^2 \right\} \tag{7.103}$$

$$d_1 / D_2 = 0.05 n^2 Z_{01} / Z_{02} \tag{7.104}$$

$$d_2' / D_1 = \{0.076 + 0.2(2D_1 / \lambda_m)^2 + 0.663\,e^{-1.71 Z_{01}/Z_{02}} \\ - 0.172\ln(Z_{01} / Z_{02})\} Z_{01} / Z_{02} \tag{7.105}$$

$$\frac{B_T \lambda_m}{Y_{01} D_1} = \begin{vmatrix} -[1 - 2D_1 / \lambda_m] Z_{01} / Z_{02} & Z_{01} / Z_{02} \le 0.5 \\[2mm] [1 - 2D_1 / \lambda_m][3Z_{01} / Z_{02} - 2] & Z_{01} / Z_{02} \ge 0.5 \end{vmatrix} \tag{7.106}$$

$$\lambda_m = \lambda_0 / \sqrt{\varepsilon_{r\_eff}} \qquad (7.107)$$

When $Z_{01} / Z_{02} \geq 2$, the calculated value of d_2 / D_1 is too high. In this range, a better value for d_2 can be obtained by replacing Z_{01} / Z_{02} in (7.105) with its inverse [18].

T-junctions can be compensated easily for the reference plane offsets by simply adjusting the lengths of the different lines. The offset in the main line is usually very small, and the main effect is on the length of the stub.

The best solution to the transformer effect is to keep the width of the stub narrow enough for the transforming effect to be negligible (n should be close to unity).

Because of the approximate nature of the equation for the loading susceptance at the junction, no compensation for this effect is recommended. As with the transformer effect, the best option is again to limit the stub width to values for which the loading susceptance will be negligible. If this cannot be done, a better model for the junction should be obtained or physical compensation of the junction should be considered [20].

As a first order approximation, a cross can be considered to be two T-junctions in parallel.

7.5.5 Via Holes

The best option when short circuits (connections to the ground plane) are required is to use via holes. The parasitic effect of a via hole is usually not severe and the same performance can be expected every time (repeatability).

Via holes are made by drilling holes in the substrate at the appropriate positions before the tracks are etched (a drill file is usually created for this purpose), after which the substrate is treated chemically and metal is deposited electrolytically or by sputtering on the cylindrical surface of these holes. Manufacturing reliable via holes in teflon substrates is not a simple matter and is best left to experts.

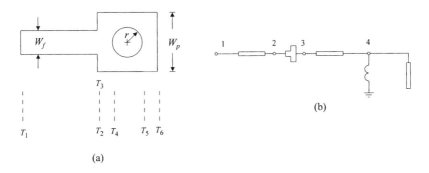

Figure 7.24   (a) A single via hole and (b) an equivalent circuit for it.

The main parasitic associated with the hole itself is the inductance to ground. Depending on the diameter and the substrate thickness, this inductance is usually very small. The via hole inductance (cylindrical via hole) can be estimated by using the following equation [21]:

$$L_{via} = \frac{\mu_0}{2\pi}\left[h\cdot\ln\left(\frac{h+\sqrt{r^2+h^2}}{r}\right)+\frac{3}{2}(r-\sqrt{r^2+h^2})\right]\tag{7.108}$$

where h is the substrate height and r is the radius of the hole.

The parasitics associated with the via hole pad and any step associated with the feeding line will often dominate the effect of the via and will usually increase the effective inductance of the via significantly. An equivalent circuit for the via hole is shown in Figure 7.24(b). Note that the length of the open stub used is determined by the pad section to the right of the via hole (T_5 to T_6).

QUESTIONS AND PROBLEMS

7.1   Assuming that a lossless material with a very high dielectric constant is available, can a single layer plate capacitor be designed to provide a perfect short at very high frequencies?

7.2   Explain why a capacitor mounted as shown in Figure 7.1(b) would have parallel and series resonance frequencies.

7.3   If two capacitors with the same capacitance but different unloaded Qs are available, which one would you choose as (a) a broadband coupling capacitor or (b) a coupling capacitor in the input matching network of a narrowband low-noise amplifier (LNA)?

7.4   A capacitor is mounted on the ground plane for decoupling. For the lowest impedance to ground, would you connect a bonding wire to the center of the capacitor or to the edge? Assume the bond wire lengths to be the same.

7.5   Design a 50-Ω thin-film resistor on a microstrip substrate with ε_r=9.98, h=100 µm and t=17.5 µm, where h is the substrate height and t the copper thickness. Assume that the thin-film thickness is very small (30 nm). Fit a model to the resistor and compare it with the ideal case. Repeat the exercise for 10-Ω, 220-Ω, and 1,000-Ω resistors.

7.6   Use (7.14) to calculate the characteristic impedance of a parallel-plate capacitor of 12 pF if its first parallel-resonant frequency is at 13.5 GHz. Use the information provided in Table 7.2 to estimate the length if a D25 capacitor with NS material is used. Estimate the dielectric constant by using the calculated length and (7.14).

7.7 Explain why one would expect a series connected single-layer parallel-plate capacitor (refer to Figure 7.6) to have series and parallel resonance frequencies. Consider the free-space case for simplicity.

7.8 Does it matter whether a multiplate chip capacitor is mounted with its plates parallel or normal to the substrate surface?

7.9 Set up an EM experiment to simulate the behavior of the capacitor in Problem 7.6 if it is mounted flat on alumina microstrip line (substrate thickness 100 μm) of the same width. Set the capacitor dielectric thickness to 4 mils and the metal thickness to 3 μm. Connect to the top plate of the capacitor with a line of the same width at the same height. Fit a model to the simulated S-parameters.

7.10 Calculate the dimensions of a 1-nH square spiral inductor on a GaAs substrate (h=100 μm) if the line width is 5 μm and the spacing between the turns is 5 μm. Use three turns (12 segments). Use the different equations provided for this purpose in Section 7.4.

7.11 Setup an EM experiment for the inductor designed in Problem 7.10 (use an airbridge to connect to the inner port) and fit a model to the simulated S-parameters. Compare the inductance with the design value.

7.12 Calculate the expected Q-factor of a bonding-wire inductor with a 50-μm diameter and length of 2 mm at (a) 2 GHz and (b) 10 GHz.

7.13 Determine the dimensions of a circular 2-nH single-loop inductor. The conductor thickness is 35 μm.

7.14 Calculate the increase in line length associated with an open-ended 25.0-Ω line on a substrate with $\varepsilon_r = 10.0$, $h = 0.635$ mm, $t = 17.5$ μm at (a) 2 GHz and (b) 10 GHz.

7.15 Calculate the parasitic change in the lengths of a 75.0–50.0-Ω step junction on the substrate used in Problem 7.14 at the same frequencies. Determine the equivalent circuit of the step junction.

REFERENCES

[1] Wolff, I., and N. Knoppik, "Rectangular and Circular Microstrip Disk Capacitors and Resonators," *IEEE Trans. Microwave Theory Tech.*, Vol. MTT-22, No. 10, October 1974.

[2] Mondal, J. P., "An Experimental Verification of a Simple Distributed Model of MIM Capacitors for MMIC Applications," *IEEE Trans. Microwave Theory Tech.*, Vol. MTT-35, No. 4, April 1987.

[3] Giancarlo, B., et al., "MIM Capacitor Modeling: A Planar Approach," *IEEE Trans. Microwave Theory Tech.*, Vol. MTT-43, No. 4, April 1995.

[4] *MultiMatch RF and Microwave Impedance-Matching, Amplifier and Oscillator Synthesis Software*, Stellenbosch, South Africa: Ampsa (PTY) Ltd; http://www.ampsa.com., 2008.

[5] Tang, W. C., and Y. L. Chow, "Simple CAD Formula for Inductance Calculation of Square Spiral Inductors with Grounded Substrate by Duality and Synthetic Asymptote," *Microwave and Optical Technology Letters*, Vol. 34, No. 2, July 20, 2002.

[6] Tang, W., et al., "CAD Formula of the Capacitance to Ground of Square Spiral Inductors with One- or Two-Layer Substrate by Synthetic Asymptote," *Microwave and Optical Technology Letters*, Vol. 48, No. 5, May, 2006.

[7] Mohan, S. S., et al., "Simple Accurate Expressions for Planar Spiral Inductances," *IEEE Journal of Solid-State Circuits*, Vol. 34, No. 10, October 1999.

[8] "Di-Cap Microwave Ceramic Capacitors," Cazenovia, NY: Dielectric Laboratories Inc., 1998.

[9] Ingalls, M., and G. Kent, "Monolithic Capacitors as Transmission Lines," *IEEE Trans. Microwave Theory Tech.*, Vol. MTT-35, No. 11, November 1987.

[10] Terman, F. E., *Radio Engineers Handbook*, New York: McGraw Hill,1943.

[11] Young, L., *Advances in Microwaves*, New York: Academic Press, 1977.

[12] Gupta, K. C., R. Garg, and R. Chadha, *Computer-Aided Design of Microwave Circuits*, Dedham, MA: Artech House, 1981.

[13] Chaddock, R. E., "The Application of Lumped Element Techniques to High Frequency Hybrid Integrated Circuits," *Radio and Electronics Eng.* (GB), Vol. 44, 1974, pp. 414–420.

[14] Dukes, J. M. C., *Printed Circuits, Their Design, and Application*, London: MacDonald, 1961, pp. 120–135.

[15] March, S. L., "Simple Equations Characterize Bond Wires," *Microwaves & RF*, November 1991, pp. 105–110.

[16] Garg, R., and I. J. Bahl, "Microstrip Discontinuities," *Int. J. Electronics*, Vol. 45, July 1978.

[17] Silvester, P., and P. Benedek, "Equivalent Capacitance of Microstrip Open Circuits,"

IEEE Trans. Microwave Theory Tech., Vol. MTT-20, 1972, pp. 511–516.

[18] Hammerstad, E. O., "Equations for Microstrip Circuit Design," *Conference Proceedings 5th European Microwave Conference,* September 1975, Hamburg.

[19] *Foundry Manual*, GaAs Foundry Services, Texas Instruments, Inc., January 1991.

[20] Chadha, R., and K. C. Gupta, "Compensation of Discontinuities in Planar Transmission Lines," *IEEE Trans. Microwave Theory Tech.*, Vol. MTT-30, No. 12, December 1982, pp. 2151–2156.

[21] Goldfarb, M. E., and R. A. Pucel, "Modeling Via Hole Grounds in Microstrip," *IEEE Microwave and Guided Wave Letters*, Vol. 1, No. 6, June 1991.

SELECTED BIBLIOGRAPHY

Bahl, I., *Lumped Elements for RF and Microwave Circuits*, Norwood, MA: Artech House, 2003.

Gupta K. C., et al., *Microstrip Lines and Slotlines*, Second Edition, Norwood, MA: Artech House, 1996.

CHAPTER 8

THE DESIGN OF WIDEBAND IMPEDANCE-MATCHING NETWORKS

8.1   INTRODUCTION

An impedance-matching network usually matches a load to a source inside the passband and may also be used to attenuate unwanted signals outside it.

When the load impedance and the source impedance are purely resistive, inductor-capacitor (LC) networks can be designed relatively easily to fulfill the filter specifications in wideband matching networks. It is difficult, however, if not impossible, to scale impedances over a wideband by using only a limited number of inductors and capacitors. This transformation function can only be done with transformers when the bandwidth-transformation product becomes large. If the required bandwidth is relatively small, however, it is possible to transform resistance over large distances with LC networks.

When the load impedance or the source impedance is reactive, part of the impedance transformation function of the matching network is to remove this reactiveness. The extent to which this can be done is a function of the load impedance itself, as well as the transducer power gain versus the frequency response required.

The limitations of a specific load impedance for a specific frequency response can be determined in at least three ways. Fano's set of integral equations can be used to determine these constraints [1], while Youla formulated the constraints in terms of Laurent series expansions [2]. Carlin advanced an iterative procedure for this purpose [3]. Because of its relative simplicity, only the iterative technique developed by Carlin will be presented here.

While the underlying theory will not be considered here, the integral constraints on simple resistor-inductor (RL) and resistor-capacitor (RC) networks lead to simple and useful upper limits on the gain [4]. These gain limits will be considered in Section 8.3.3, along with the Youla gain-bandwidth constraints associated with a parallel RC or a series RL load (Chebyshev response).

With the limitations of a particular load (or source) known, a network that will provide the required power gain versus frequency response can be designed by using direct synthesis or iterative techniques. Both of these approaches will be discussed in this chapter.

Networks for matching a complex load to a complex source are often required. A theoretical approach to solving this class of problems was developed by Chen and Satyana-rayana [5], and more recently an alternative and simplified theory was introduced by Carlin and Yarman [6]. Carlin and Yarman also developed iterative techniques for matching a complex load to a complex source [6, 7]. Because of its relative simplicity and its superior results [8], only iterative techniques for matching a complex load to a complex source will be considered.

271

It is often possible to design matching networks for complex terminations by initially assuming the terminations to be purely resistive. The reactances are then absorbed parasitically into the network when the design is completed. Because the effort required to design a network in this way is minimal when it can be done, this approach will also be considered in this chapter.

Impedance-matching networks are often required to provide a transducer power gain versus frequency response with a positive slope in the passband. LC networks can be designed either interactively or directly to fulfill this requirement. There is, however, the disadvantage that the source will inevitably be mismatched at the lower frequencies in the passband. LC networks with gain slopes also tend to be sensitive to changes in the component values. Because this does not necessarily apply to RLC impedance-matching networks, the design of these networks will also be examined in this chapter.

8.2   FITTING AN IMPEDANCE OR ADMITTANCE FUNCTION TO A SET OF IMPEDANCE VERSUS FREQUENCY COORDINATES

When impedance-matching networks are designed, impedance (or admittance) functions that will approximate a set of discrete impedance versus frequency coordinates are often required. The set of coordinates might be the measured input impedance of a transistor or antenna, or it could be the output or input impedance (admittance) of a network to be designed.

It is sometimes possible to approximate the measured impedance of a device with simple RC, RL, or RLC equivalent circuits. This can usually be done when the resistive part of the impedance or admittance is more or less constant over the frequency range of interest. The components of such an equivalent circuit can be determined by setting up an equation for the input impedance or admittance of the network chosen, and equating its real and imaginary parts to the measured values. Although this technique can be used, more sophisticated techniques are often required.

A major problem in finding an impedance function that will fit a given set of coordinates is its realizability. The function obtained must be positive-real.

A technique that usually gives good results is based on the fact that the reactance (susceptance) of a minimum-impedance (admittance) function can be determined when the resistance (conductance) is known [9]. The equivalent circuit of a minimum-impedance function that has a parallel capacitor or inductor as the last element is shown in Figure 8.1.

Because the reactance can be determined when the resistance is known, it follows that the impedance itself is known when its resistive part is known. In terms of equations [9], if

$$R(\omega) = [\sum_{2n} C_j / (\omega - \omega_j) + R_0 / 2] + [\sum_{2n} C_j^* / (\omega - \omega_j^*) + R_0 / 2] \qquad (8.1)$$

then

$$Z(j\omega) = \sum_{2n} 2C_j / (\omega - \omega_j) + R_0 \qquad (8.2)$$

The poles ω_1, ω_2, ..., ω_{2n} are the first and second quadrant poles of the resistance function (which is an even function), while C_1, C_2, ..., C_{2n} are the residues of these poles. The poles, ω_1^*, ω_2^*, ..., ω_{2n}^* and the residues, C_1^*, C_2^*,..., C_{2n}^* are the conjugates of the first and second quadrant poles and the residues, respectively. R_0 is a real-valued (positive) constant.

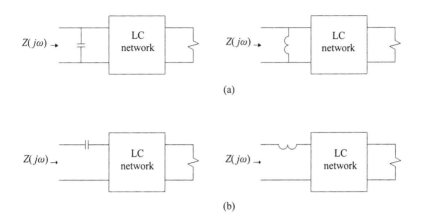

(a)

(b)

Figure 8.1   Equivalent circuits for (a) minimum-impedance and (b) minimum-admittance functions.

If a rational function for the resistive part of the minimum impedance function can be found, the impedance function itself can be found by using (8.1) and (8.2).

If the resistance function is assumed to be of the form

$$R(\omega) = \omega^{2p} / (a_{2m}\omega^{2m} + a_{2m-2}\omega^{2m-2} + ... + a_0) \tag{8.3}$$

the unknown coefficients in the function can be found by using the equations available for fitting a polynomial to a given set of coordinates with least-square error.

Equation (8.3) can be changed to have the following form

$$T(\omega) = \omega^{2p} / R(\omega)$$

$$= a_{2m}(\omega^2)^m + a_{2m-2}(\omega^2)^{m-1} + ... + a_0$$

$$= b_m x^m + b_{m-1} x^{m-1} + ... + b_0 \tag{8.4}$$

where

$$x = \omega^2.$$

Because $R(\omega)$ and the number of zeros at the origin are known (the designer must choose the number of zeros), the function $T(\omega)$ is also known at discrete frequencies.

The coefficients b_0, b_1, ..., b_m can now be determined by solving the following set of equations:

$$s_0 b_0 + s_1 b_1 + ... + s_m b_m = t_0$$

$$s_1 b_0 + s_2 b_1 + ... + s_{m+1} b_m = t_1$$

.

.

$$s_{m+1} b_0 + s_{m+2} b_1 + ... + s_{2m} b_m = t_m \qquad (8.5)$$

where

$$s_0 = h$$

$$s_1 = \sum x_i$$

$$s_{2m} = \sum x_i^{2m} \qquad (8.6)$$

and

$$t_0 = \sum T(\omega_i)$$

$$t_1 = \sum T(\omega_i) x_i$$

$$t_m = \sum T(\omega_i) x_i^m \qquad (8.7)$$

where h is the number of coordinates $[x_i, T(\omega_i)]$.

With these equations solved, the coefficients in (8.4) and, therefore, those in (8.3) are known. The minimum-impedance function itself can now also be determined by using (8.1) and (8.2). In order to use these equations, the poles of the resistance function must be determined first. The impedance function fitted will be positive-real if care is taken to ensure that the approximation function has no real zeros in the ω-plane.

It is possible that the input impedance, as given by the synthesized minimum impedance function, will deviate slightly from the initial set of coordinates. This situation can be improved by adding a pole at the origin or infinity to the impedance function. Alternatively, either the minimum-admittance function corresponding to the set of impedance coordinates can be determined or a computer optimization program can be used to improve the match between the two sets of impedances.

With the resistance or conductance known at a number of frequencies, a polynomial approximation function can be found for it, as described, by using the program LSM provided on the CD accompanying this book. The minimum-impedance or minimum-admittance function associated with the resistance or conductance function is also calculated, and the cascade network associated with this impedance or admittance function can also be extracted by continued fractionation.

It happens occasionally that the polynomial, $T(\omega)$, determined by the program is not positive-real. There are two reasons for this.

First, the number of coordinates specified in areas where $T(\omega)$ approaches zero may be insufficient. This can be remedied easily by specifying more coordinates in these areas.

Second, the increase in $T(\omega)$ may be too slow at high frequencies. In such a case the polynomial will have a zero on the real axis of the ω-plane. This, in turn, implies a real pole in the resistance function $R(\omega)$, and therefore a pole on the $j\omega$-axis for the function $R(s)$, which is, of course, not allowable.

In order to overcome this problem, the option to add an extra data point at a frequency one and a half times the highest frequency specified is provided. The initial value specified for the resistance or conductance at this extra frequency should be incremented until the zero problem is resolved. The initial value must be specified relative to the value specified at the highest frequency.

The initial value specified at the extra frequency will be incremented for a specified number of times. If the zero problem still persists, more iterations and/or a larger initial value should be used.

EXAMPLE 8.1   Fitting a function to a set of resistance coordinates.

As an example of using the program LSM, an equivalent circuit for the input impedance of the Motorola MFR406 power transistor will be determined.

The input impedance of the MFR406, as specified by the manufacturer, is tabulated in Table 8.1. It can be seen from this data that the resistance approaches a constant at low frequencies. The approximation function for the resistance, therefore, should be of lowpass form; that is, the function must be of the form

$$R(\omega) = 1 / [a_{2m}\omega^{2m} + a_{2m-2}\omega^{2m-2} + ... + a_0] \qquad (8.8)$$

In polynomial form, this becomes

$$T(\omega) = 1 / R(\omega) = b_m(\omega^2)^m + b_{m-1}(\omega^2)^{m-1} + ... + b_0 \qquad (8.9)$$

The data can be entered directly into the program, or the reflection coefficients of the associated impedances or admittances can be imported from a Touchstone one-port S-parameter data file (file type: .s1p). A maximum of 120 data sets are allowed.

The data for this problem is available in the file "MRF406ZinModel.lsm" which can be found in directory "\artechLSM\Examples" after running the LSM installation program (LSM_V3setup.exe). Use the *File | Open* command to open this file. The resistance values specified are shown in Table 8.1. The values specified in the program can be viewed by using the *Edit | Rout/Gout for Polynomial Fit* command and are displayed in Figure 8.2.

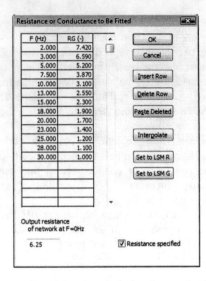

Figure 8.2　　The resistance for which an impedance function was fitted in Example 8.1.

Note the check box for indicating whether values for the resistance or conductance were specified in Figure 8.2. Also note the edit box provided for specifying the resistance of conductance at dc.

Table 8.1　　The Input Impedance of the MRF406 as a Function of the Frequency

Frequency (MHz)	Input impedance (Ω)
2	$7.5 - j2.6$
5	$5.2 - j2.3$
10	$3.1 - j1.9$
15	$2.3 - j1.8$
20	$1.7 - j1.7$
25	$1.2 - j1.4$
30	$1.0 - j1.0$

With the specifications made, the *Synthesize | Network to Fit Resistance/Conductance* command should be used next. The specifications made are displayed first by the synthesis wizard, after which the required specifications can be made. The number of elements (n) to be used in the network to be synthesized, the number of transmission zeros in the network (n_z; $0 \le n_z \le n$), and the normalization resistance to be used for the reflection coefficient to be calculated must be specified (refer to Figure 8.3).

There is no guarantee that the polynomial fitted to the specified conductance or resistance will pass through the dc value specified. This condition can be enforced by selecting the *Force DC Resistance* option provided (see Figure 8.3). The impedance or admittance function obtained will be scaled to provide the resistance desired when this option is selected. This will obviously upset the fit at the other frequencies. The alternative is to add more data points around dc.

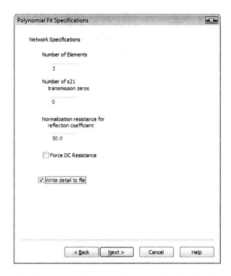

Figure 8.3 The specifications page for the resistance to be fitted in Example 8.1.

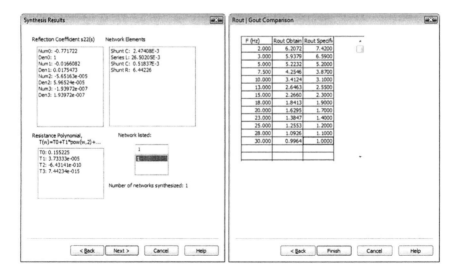

Figure 8.4 The results obtained in Example 8.1.

The output information provided by the program can also be obtained in text format when the option to write the detail to a file is selected (see Figure 8.3). Note that more information is supplied in this file than is displayed on the screen. The name of this file is the same as that of the project, but with the file extension (".lsm") replaced with "output.txt".

When zeros on the real axis are required in the function fitted, the option to add an extra data point (as explained above) will be provided.

The polynomial obtained from the program is (see Figure 8.4)

$$T(\omega) = 0.1552 + 0.3733 \times 10^{-4} \, \omega^2 - 0.6431 \times 10^{-9} \, \omega^4 + 7.4423 \times 10^{-15} \, \omega^6$$

The resistance function obtained is

$$R(\omega) = 1/[0.1552 + 0.3733 \times 10^{-4} \, \omega^2 - 0.6431 \times 10^{-9} \, \omega^4 + 7.4423 \times 10^{-15} \, \omega^6]$$

The poles of this function are

$$\omega = \pm243.324 \pm j118.555$$

$$000.000 \pm j62.338$$

There are four poles in the complex ω-plane and two poles on the imaginary axis. The poles are listed in the text file associated with the project.

After multiplication of the numerator and the denominator by $-j$, the minimum-impedance function obtained from the program is

$$Z(j\omega) = \frac{29.4214 \times 10^6 + j121.0335 \times 10^3 \, \omega - 404.1906\omega^2}{4.5669 \times 10^6 + j88.042 \times 10^3 \, \omega - 299.446\omega^2 - j\omega^3}$$

The impedance function is given as a function of s by the equation

$$Z(s) = \frac{29.4214 \times 10^6 + 121.0335 \times 10^3 \, s + 404.1906s^2}{4.5669 \times 10^6 + 88.042 \times 10^3 \, s + 299.446s^2 + s^3} \tag{8.10}$$

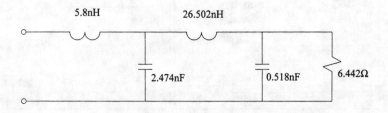

Figure 8.5 An equivalent circuit for the input impedance of the MRF406 transistor.

Table 8.2   The Input Impedance of the MRF406 Compared to the Impedance as Given by (8.10), as Well as the Input Impedance of the Equivalent Circuit Shown in Figure 8.5

Frequency (MHz)	Input impedance of transistor (Ω)	Impedance as given by (8.10) (Ω)	Input impedance of the equivalent circuit shown in Figure 8.2 (Ω)
2.0	7.4 − j2.6	6.2 − j1.2	6.6 − j1.3
5.0	5.2 − j2.3	5.2 − j2.5	5.2 − j2.3
10.0	3.1 − j1.9	3.4 − j3.1	3.4 − j2.7
15.0	2.3 − j1.8	2.3 − j2.9	2.3 − j2.4
20.0	1.7 − j1.7	1.6 − j2.6	1.6 − j1.9
25.0	1.2 − j1.4	1.3 − j2.3	1.3 − j0.9
30.0	1.0 − j1.0	1.0 − j2.1	1.0 − j1.0

The resistance as given by this equation is compared to the measured resistance in Figure 8.4(b), while the impedances are compared in Table 8.2. The equivalent circuit associated with the impedance function is shown in Figure 8.5. The network (without the 5.8-nH series inductor) was extracted by continued fractionation from the impedance function (Cauer development).

It can be seen from Table 8.2 that the impedance given by (8.10) correlates well with the input impedance of the transistor. The match can be improved, however, by adding a pole at infinity (series inductor) to the impedance function. The resulting network is shown in Figure 8.5.

If necessary, the correlation between the impedance of the transistor and that given by the new equation can be improved by entering the network into an optimization program. Alternatives are to fit a minimum-admittance function with four elements to the measured data, or to design a network to provide matching to the conjugate of the input impedance of the transistor. Both of these alternatives can be explored with LSM. Note the *Set LSM R* and *Set LSM G* options in Figure 8.2. These options can be used to set the resistance or conductance to be fitted to that associated with the single-matching problem specified in the program.

8.3   THE ANALYTICAL APPROACH TO IMPEDANCE MATCHING

Impedance-matching networks can be designed directly (analytically) or iteratively. The direct approach will be discussed in this section.

In its simplest form the load impedance and the source impedance of an impedance-matching problem are purely resistive and equal. In this case the matching network has only a filtering function.

When an explicit function for the required transducer power gain versus frequency response is specified, the network required to meet the filtering specifications can be designed easily with the well-known Darlington synthesis technique. Darlington synthesis will be discussed in Section 8.3.1.

When a bandpass network is designed by using Darlington synthesis, the source or load resistance of the network synthesized often does not have the required value. When the bandwidth is relatively narrow (less than two octaves) and the network contains L-sections consisting only of inductors or capacitors, it is sometimes possible to transform the source or the load to have the required value by using LC transformers. The design of these networks will be discussed in Section 8.3.2.

The only solution possible when the required bandwidth is large is to use conventional or transmission-line transformers. Although transformation over a wide bandwidth is possible when a transmission-line transformer is used, the transformation ratios are limited (1/4, 1/9, ...).

At microwave frequencies, wideband transformation of resistance is possible when tapered or stepped-impedance lines are used.

It is often possible to eliminate the need for resistance transformation by designing a network with a semi-lowpass or semi-highpass transducer power gain versus frequency response (i.e., matching networks without transmission zeros at the origin or infinity, respectively). This technique can also be used to provide matching between unequal load and source terminations when the transformation bandwidth is small enough.

It should be noted that the transformation distance obtainable over a given bandwidth with LC impedance-matching networks is limited. This follows from the fact that a high transformation Q is required when the transformation distance is large. A high transformation Q, in turn, implies a high network Q and therefore a narrow bandwidth.

The gain-bandwidth products of impedance-matching networks are also limited when the load or source impedance is reactive. The extent to which this reactiveness can be removed is limited because negative inductors and capacitors do not exist.

When only the load (or source) impedance is reactive, the gain-bandwidth constraints imposed by the load (or source) on a particular transducer power gain versus frequency response can be determined by using the integral constraints formulated by Fano, the Laurent series constraints of Youla, or the iterative approach of Carlin.

The constraints on several types of loads, usually for Chebyshev responses, are available in the literature in explicit form. Only the limitations imposed by simple RC and RL loads will be considered here, in Section 8.3.3.

The constraints imposed by any other load can be determined by using Carlin's iterative approach, which will be discussed in Section 8.4. The procedure to be described has been implemented in the LSM program.

The analytical design of networks for matching a complex load to a purely resistive source is discussed in Section 8.3.4. Two analytical approaches to solving impedance-matching problems belonging to this class will be considered.

When the technique discussed in Section 8.3.4.1 is used, the complexity of the load is immaterial. As long as the specified transducer power gain versus frequency response is realizable, any load can be matched to a resistive source.

The parasitic absorption approach discussed in Section 8.3.4.2 can only be used when the terminations can be modeled with simple equivalent circuits.

When the load is parasitically absorbed into an impedance-matching network, it is initially assumed to be purely resistive. A network with a suitable topology is then designed and, if the gain-bandwidth constraints imposed by the reactive load on the transducer power gain versus frequency response chosen were taken into account, it will be possible to absorb the reactive part of the load into the designed network.

Although it is limited to simpler problems, this technique has the advantage that less effort is required in designing the network. Parasitic absorption can also be used to solve simple problems where both the source and the load terminations are reactive.

The principle of parasitic absorption is illustrated in Figure 8.6.

Although it is also possible to match a complex load to a complex source analytically [5, 6], the relevant theory will not be considered here, because much better results can be obtained with considerably less effort by using iterative techniques [8].

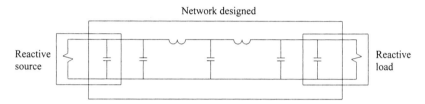

Figure 8.6   Illustration of the principle of parasitic absorption.

The additional theory necessary to design commensurate distributed networks will be covered in Section 8.3.6. Richards' transformation, Kuroda and Norton's identities, and unit elements and their extraction will be considered.

Under Richards' transformation all of the theory applicable to the design of lumped element networks also apply to commensurate distributed networks.

8.3.1   Darlington Synthesis of Impedance-Matching Networks

When a resistive load is matched to a resistive source (see Figure 8.7), a network that will provide the required transducer power gain versus frequency response can be designed by following the procedure outlined here.

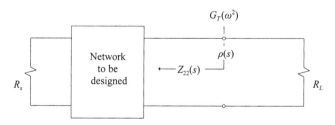

Figure 8.7   The design of an impedance-matching network when the terminations are purely resistive.

It should be noted that the source (or load) resistance of the network designed by following this procedure will often not be equal to the specified value. When the network designed contains bandpass L-sections, it is sometimes possible to use LC transformers to adjust the resistance level. Transformers can be used to change the impedance levels in wideband designs.

If a network without transmission zeros at $\omega = 0$ or $\omega \to \infty$ is designed and the gain-bandwidth limitations are not a problem, the source resistance will always have the required value.

Darlington Synthesis

Specifications: The transducer power gain versus frequency response required and the values of the load resistance and the source resistance.

1.   Replace all the ω^2 terms in the specified transducer power gain function $G_T(\omega^2)$, with $-s^2$ terms.

Determine the product $\rho(s)\rho(-s)$, where $\rho(s)$ is the reflection coefficient $s_{22}(s)$ corresponding to the specified transducer power gain function:

$$\rho(s)\rho(-s) = 1 - G_T(-s^2) \tag{8.11}$$

2.   The next step is to determine $\rho(s)$.

Assign all the left-hand plane (LHP) poles of $\rho(s)\rho(-s)$ to $\rho(s)$. It is not necessary to assign only LHP zeros to $\rho(s)$. Any combination of zeros can be assigned to it, as long as they are assigned in conjugate pairs and the relationship between $\rho(s)$ and $\rho(-s)$ is kept in mind.

When a minimum-phase network (i.e., a network with minimum phase variation in the passband) is required, all the LHP zeros must be assigned to $\rho(s)$.

When the parasitic absorption approach is followed, the right-hand plane (RHP) zeros must be assigned to $\rho(s)$ if the source is reactive. When both the load and the source impedances are reactive, it is usually best to try all possible combinations.

The sign assigned to $\rho(s)$ is often important. When lowpass networks are designed and the load and the effective source resistance, as viewed from the load terminals at $\omega = 0$, are not equal, the sign of $\rho(s)$ is determined by its value at the origin. The sign must be such that

$$\rho(0) = \frac{Z_{22}(0) - R_L}{Z_{22}(0) + R_L} \tag{8.12}$$

When the two resistance values are equal, a plus or a minus sign may be assigned to $\rho(s)$.

When the value of $\rho(s)$ is plus one (open-circuit) at infinity, the first element of the network (as viewed from the load terminals) will be a series inductor. When the sign is negative (short-circuit), the first element will be a parallel capacitor. The sign of $\rho(s)$ must be such that

$$\rho\left(\infty\right) = \frac{Z_{22}\left(\infty\right) - R_{L}}{Z_{22}\left(\infty\right) + R_{L}} \tag{8.13}$$

Where the load resistance is equal to the source resistance and a network with a series capacitor as first element is required, the sign of $\rho(s)$ must be such that

$$\rho(0) = +1$$

When

$$\rho(0) = -1$$

the first element will be a parallel inductor.

The relationship between the value of the reflection coefficient at zero or infinity and the topology of the network is summarized in Table 8.3 for lowpass, highpass, and bandpass networks.

The information in this table is useful when networks are designed to absorb the reactive part of the load impedance parasitically.

3. Find the impedance function corresponding to the reflection coefficient $\rho(s)$ by using the equation

$$\frac{Z_{22}\left(s\right)}{R_{L}} = \frac{1 + \rho\left(s\right)}{1 - \rho\left(s\right)} \tag{8.14}$$

4. Synthesize the required network by using standard filter theory. If the topology is important, the transmission zeros and the poles at the origin and infinity must be extracted in the proper sequence.

5. If the source resistance of the network does not have the required value, transformers or LC transformers can be used to change the impedance level as required.

EXAMPLE 8.2   Darlington synthesis of a Butterworth network.

A third-order Butterworth network will be synthesized as an example of the application of this procedure. $R_L = 1\Omega = R_s$ and the required 3-dB cutoff frequency is 1 rad/s. A lowpass network with an inductor as the first element is required.

1. The transducer power gain function for the third-order Butterworth characteristic is

Table 8.3   The Relationship Between the Reactive Part of the Output Impedance of a Network and the Value of the Corresponding Reflection Coefficient $\rho_{22}(s)$ at the Origin and Infinity

Type of network	$\rho(s)$	$Z_{22}(s)$ at $\omega = 0$	$Z_{22}(s)$ at $\omega \to \infty$	Example
Lowpass	$\rho(0) = \dfrac{R_s - R_L}{R_s + R_L}$	Resistive		
	$\rho(\infty) = 1$		Inductive	
	$\rho(0) = \dfrac{R_s - R_L}{R_s + R_L}$	Resistive		
	$\rho(\infty) = -1$		Capacitive	
Highpass	$\rho(0) = 1$	Capacitive		
	$\rho(\infty) = \dfrac{R_s - R_L}{R_s + R_L}$		Resistive	
	$\rho(0) = -1$	Inductive		
	$\rho(\infty) = \dfrac{R_s - R_L}{R_s + R_L}$		Resistive	
	$\rho(0) = 1$	Capacitive		
	$\rho(\infty) = 1$		Inductive	
Bandpass	$\rho(0) = -1$	Inductive		
	$\rho(\infty) = 1$		Inductive	
	$\rho(0) = 1$	Capacitive		
	$\rho(\infty) = -1$		Capacitive	
	$\rho(0) = -1$	Inductive		
	$\rho(\infty) = -1$		Capacitive	

$$G_T(\omega^2) = 1/[1+\omega^6]$$

By substituting each ω^2 term with $-s^2$, this becomes

$$G_T(-s^2) = 1/[1+(-s^2)^3] = 1/[1-s^6]$$

The product $\rho(s)\,\rho(-s)$ can now be determined:

$$\rho(s)\rho(-s) = 1 - G_T(-s^2)$$

$$= \frac{s^6}{s^6 - 1}$$

$$= \frac{\pm s^3}{s^3 + 2s^2 + 2s + 1} \quad \frac{\pm s^3}{-s^3 + 2s^2 - 2s + 1}$$

2.   After calculation of the pole positions and the zero positions, all the LHP poles and half of the $j\omega$-axis zeros are assigned to $\rho(s)$. Because $R_L = 1\Omega = R_s$, a positive or a negative sign can be assigned to $\rho(s)$ if the topology was not important. Because an inductor is required as the first element in this example, the output impedance of the network will be high at high frequencies and therefore a positive sign must be assigned to $\rho(s)$:

$$\rho(s) = s^3 / (s^3 + 2s^2 + 2s + 1)$$

3.   The output impedance of the network to be designed is given by (8.14):

$$\frac{Z_{22}(s)}{R_L} = \frac{1 + \rho(s)}{1 - \rho(s)}$$

$$= \frac{2s^3 + 2s^2 + 2s + 1}{2s^2 + 2s + 1}$$

4.   The network can now be synthesized by continued fractionation (Cauer development) of the impedance function:

$$\frac{Z_{22}(s)}{R_L} = s + \cfrac{1}{2s + \cfrac{1}{s + 1/1}} = sL_1 + \cfrac{1}{sC_2 + \cfrac{1}{sL_3 + 1/G_4}}$$

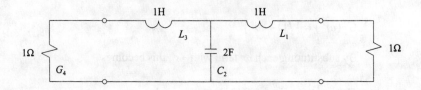

Figure 8.8   A network with a third-order Butterworth response.

Because the source resistance has the required value, no transformers or LC transformers are required in this particular case.

The network designed by this example is shown in Figure 8.8.

8.3.2   LC Transformers

The impedance level in a bandpass network containing an L-section consisting of capacitors or inductors only can be changed by replacing the L-section with a suitable T- or PI-section (see Figures 8.9 and 8.10). Similar to the L-sections discussed in Chapter 4, the output impedance will be transformed downward when the element of the L-section to the left is a parallel element and will be transformed upward when it is a series element. The T- and PI-section equivalents for the bandpass L-sections are shown in Figures 8.6 and 8.7.

The maximum transformation distance of these sections (n^2) is limited by the ratio of the series reactance and the parallel reactance of the original section, as can be seen by inspection of (8.15)–(8.18).

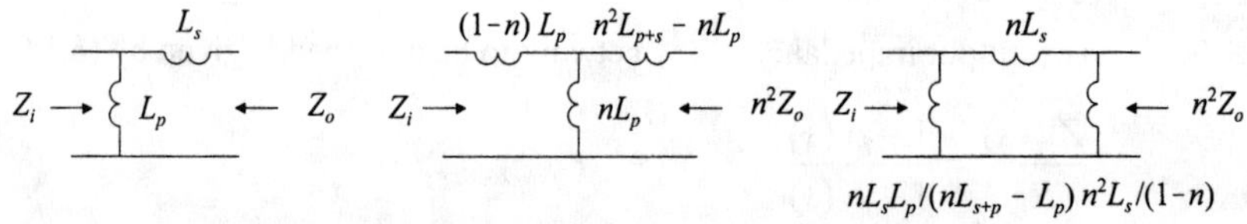

$$(1+\frac{L_s}{L_p})^{-1} < n < 1 \tag{8.15}$$

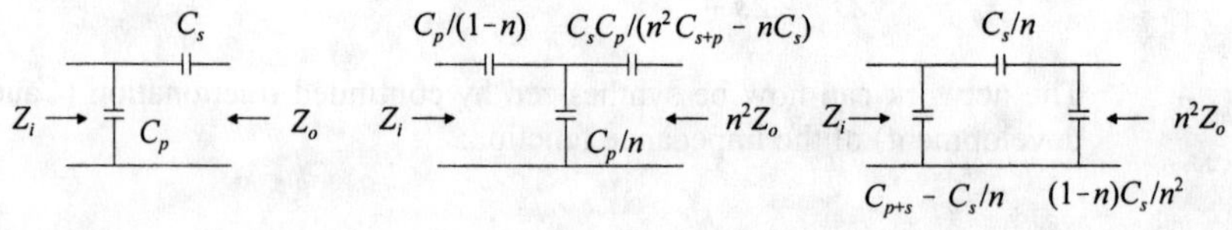

$$(1+\frac{C_p}{C_s})^{-1} < n < 1 \tag{8.16}$$

Figure 8.9   LC transformers yielding a downward transformation.

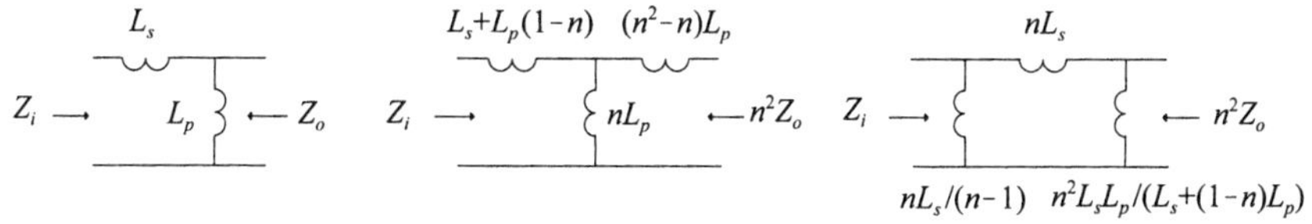

$$L_s$$

$$L_s + L_p(1-n) \quad (n^2 - n)L_p$$

$$nL_s$$

$$Z_i \longrightarrow \quad L_p \quad \longleftarrow Z_o \qquad Z_i \longrightarrow \quad nL_p \quad \longleftarrow n^2 Z_o \qquad Z_i \longrightarrow \qquad \longleftarrow n^2 Z_o$$

$$nL_s/(n-1) \quad n^2 L_s L_p/(L_s + (1-n)L_p)$$

$$1 < n < 1 + \frac{L_s}{L_p} \tag{8.17}$$

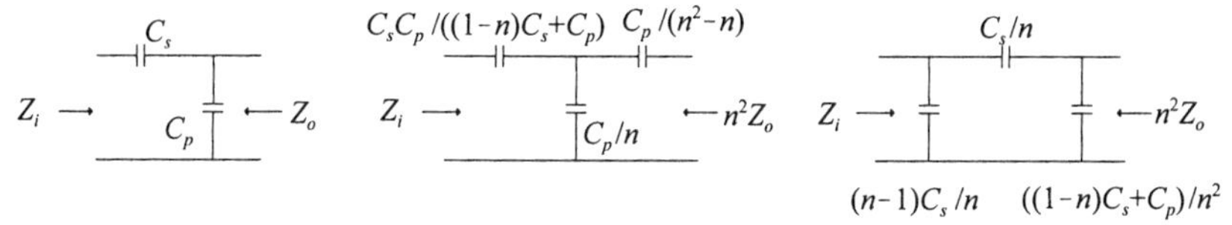

$$C_s$$

$$C_s C_p /((1-n)C_s + C_p) \quad C_p /(n^2 - n)$$

$$C_s/n$$

$$Z_i \longrightarrow \quad C_p \quad \longleftarrow Z_o \qquad Z_i \longrightarrow \quad C_p/n \quad \longleftarrow n^2 Z_o \qquad Z_i \longrightarrow \qquad \longleftarrow n^2 Z_o$$

$$(n-1)C_s/n \quad ((1-n)C_s + C_p)/n^2$$

$$1 < n < 1 + \frac{C_p}{C_s} \tag{8.18}$$

Figure 8.10   LC transformers yielding an upward transformation.

The LSM program can be used to calculate the components required *(Functions | Design an LC Transformer* command).

EXAMPLE 8.3   The transformation properties of a second-order Chebyshev bandpass network.

A bandpass network with a second-order Chebyshev response is shown in Figure 8.11. The ripple in the passband is 0.5 dB, the center frequency is f_0 and the bandwidth is B.

The input impedance of the network can be transformed downward (or the output impedance can be transformed upward) by replacing either of the two bandpass L-sections with an equivalent T- or PI-section (the second L-section is obtained by moving the series capacitor to the left and the shunt capacitor to the right).

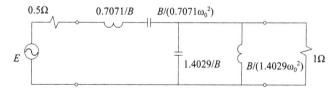

Figure 8.11   A bandpass network with a second-order Chebyshev response (center frequency: ω_0 rad/s; bandwidth: B rad/s; passband ripple: 0.5 dB).

Table 8.4   The Maximum Transformation Ratio for an LC Transformer in the Network Shown in Figure 8.11 as a Function of the Relative Bandwidth (ω_H / ω_L)

Relative bandwidth	Transformation ratio of the LC transformer
2	9.0
3	3.0
5	1.7
10	1.2

The maximum transformation ratio (n^2) possible by replacing either of the two bandpass L-sections in Figure 8.11 with its equivalent T- or PI-section can be determined by using (8.15)–(8.18):

$$n^2_{max} = 1 / \left[1 + \frac{1.4209}{B} / \frac{B}{0.7071\omega_0^2} \right]^2$$

$$= \left[\frac{1}{1 + (\omega_0 / B)^2} \right]^2 \tag{8.19}$$

The transformation ratio obtainable, therefore, is a function of the ratio of the center frequency and the bandwidth of the network. (This ratio can be defined as the Q-factor of the network.)

The maximum transformation ratio for different values of the relative bandwidth (ω_H / ω_L, where $\omega_0^2 = \omega_H \times \omega_L$ and B=$\omega_H - \omega_L$) is shown in Table 8.4. The transformation ratio obtainable is large when the bandwidth is narrow and is small when the bandwidth is wide.

The transformed network for $\omega_0 = 1.732$ rad/s, $B = 2$ rad/s ($\omega_H / \omega_L = 3$), $R_L = 1\Omega$, and $R_s = 1/6\ \Omega$ is shown in Figure 8.12. Note that the original inductor on the input side and the LC transformer itself were impedance-scaled by a factor of 1/3. This was required because the load resistance targeted remained constant (1Ω), while the transformer increased the output impedance by a factor of 3 [see Figure 8.10(b)].

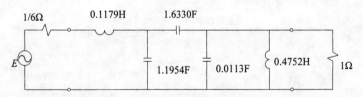

Figure 8.12   A network for matching a 1Ω load to a source with 1/6Ω internal resistance (Chebyshev response, 1/2-dB ripple, $\omega_0 = 1.732$ rad/s, $B = 3$ rad/s).

8.3.3   The Gain-Bandwidth Constraints Imposed by Simple RC and RL Loads

The constraints imposed by a parallel RC load on the gain-bandwidth product of a lossless network with reflection coefficient $\rho(s)$ can be expressed in the form [1]

$$\int_0^\infty \ln\frac{1}{|\rho(j\omega)|}\,d\omega \le \frac{\pi}{RC} \tag{8.20}$$

If the RC time-constant in (8.20) is replaced with L/R, the gain-bandwidth constraints associated with a series RL load can also be determined by using (8.20).

The gain-bandwidth expression for a series RC load is [1]

$$\int_0^\infty \omega^{-2}\ln\frac{1}{|\rho(j\omega)|}\,d\omega \le \pi RC \tag{8.21}$$

The constraints associated with a parallel RL load also follow from (8.21) by replacing RC with L/R.

Because the value of $\ln(1/|\rho(j\omega)|)$ is normally high inside the passband and close to zero outside it, these integral equations clearly illustrate the trade-off possible between the gain and the bandwidth of the matching network. It follows that any increase in the bandwidth will bring about a decrease in the gain when the gain-bandwidth product exceeds the limit imposed by the load.

Assuming that the matching network has an ideal response ($G_{T,\max}$ inside the passband and zero outside it), (8.20) and (8.21) can be manipulated to obtain the absolute maximum transducer power gain associated with a bandwidth B (Hz).

The upper limit resulting from (8.20) is

$$G_{T,\max} \le 1 - e^{-1/(RCB)} \tag{8.22}$$

A similar expression can be derived for the series RC case. Because this derivation is more involved, the details are shown below:

$$\int_0^\infty \omega^{-2}\ln\left|\frac{1}{\rho(j\omega)}\right|\partial\omega = \int_0^{\omega_L}\omega^{-2}\ln\left|\frac{1}{\rho(j\omega)}\right|\partial\omega + \int_{\omega_L}^{\omega_H}\omega^{-2}\ln\left|\frac{1}{\rho(j\omega)}\right|\partial\omega$$

$$+ \int_{\omega_H}^\infty \omega^{-2}\ln\left|\frac{1}{\rho(j\omega)}\right|\partial\omega$$

$$= \int_0^{\omega_L}\omega^{-2}\ln\left|\frac{1}{1}\right|\partial\omega + \int_{\omega_L}^{\omega_H}\omega^{-2}\ln\left|\frac{1}{\rho_{\min}(j\omega)}\right|\partial\omega + \int_{\omega_H}^\infty \omega^{-2}\cdot 0\,\partial\omega$$

$$= -\ln\left|\frac{1}{\rho_{\min}}\right|\omega^{-1}\Big|_{\omega_L}^{\omega_H} = \ln|\rho_{\min}|\left(\frac{1}{\omega_H} - \frac{1}{\omega_L}\right) = \pi RC$$

It follows from this result that

$$|\rho_{min}| = e^{-\frac{\pi RC\omega_L\omega_H}{\omega_H-\omega_L}}$$

(8.23)

Because $G_T = 1 - |\rho|^2$, the required expression for the gain can now be obtained:

$$G_{T,min} = 1 - e^{-\frac{2\pi RC\omega_L\omega_H}{\omega_H-\omega_L}}$$

(8.24)

It has been shown in [4] that these gain expressions can be simplified to

$$G_{T,min} = 1 - e^{-2\pi Q_c/Q_L}$$

(8.25)

in all four cases.

Q_c in (8.25) is the circuit Q and is defined by

$$Q_c = \frac{\sqrt{\omega_H\omega_L}}{\omega_H-\omega_L} = \frac{\omega_0}{B}$$

(8.26)

(as usual), while Q_L is the Q-factor of the load at the center frequency ($\omega_0 C/R$ in the parallel RC case, $1/(\omega_0 CR)$ in the series RC case, $\omega_0 L/R$ in the series RL case, and so on).

The expressions derived by using the gain-bandwidth theory developed by Youla [2] are generally more convenient when a specific transducer power gain versus frequency response is of interest. If the parallel RC load is considered and the transducer power gain function is a lowpass Chebyshev function with ripple factor ε and cutoff frequency ω_c (rad/s), the maximum gain in the passband (K_n) can be determined by using the equation [10]

$$[1 - K_n]^{1/2} = \varepsilon \sinh\left\{n\sinh^{-1}\left[\sinh\left(\frac{1}{n}\sin^{-1}\frac{1}{\varepsilon}\right) - \frac{2\sin\frac{\pi}{2n}}{RC\omega_c}\right]\right\}$$

(8.27)

where n is the order of the network.

Curves illustrating the relationship between the maximum realizable gain (K_n) as a function of the RCf_c product ($\omega_c = 2\pi f_c$) are given in Figure 8.13 for a Chebyshev response with 0.5-dB ripple in the passband [10].

It can be seen from these curves that the maximum power gain will be less than 1 when the RCf_c product is greater than approximately 0.3. When the RCf_c product increases above this value, the maximum realizable gain drops rapidly.

If more elements (n) are used in the impedance-matching network, the gain-bandwidth product increases. The improvement, however, is small when more than four elements are used.

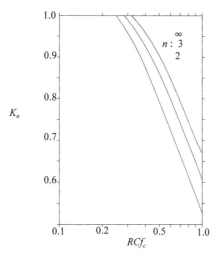

Figure 8.13   The maximum realizable power gain (K_n) of a parallel RC load as a function of the RCf_c product of the load (lowpass Chebyshev response with 0.5-dB ripple in the passband and cutoff frequency f_c) (after [10]).

8.3.4   Direct Synthesis of Impedance-Matching Networks When the Load (or Source) Is Reactive

If the specified transducer power gain can be realized, a network for matching a reactive load to a resistive source can be designed by following the procedure outlined here [10].

Similar to Darlington synthesis, the source resistance obtained will often not be equal to that specified. The required transformation can usually be obtained by using transformers or LC transformers.

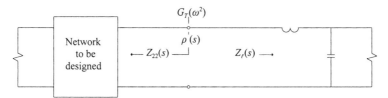

Figure 8.14   Illustration of the design of impedance-matching networks between a resistive source and a reactive load.

Design Procedure

Specifications: The source resistance, the load impedance [$Z_t(s)$], and the transducer power gain versus frequency response required (see Figure 8.14).

1. Replace all ω^2 terms in the specified transducer power gain versus frequency function $[G_T(\omega^2)]$ with $-s^2$ terms.
    Determine the product $\rho(s)\rho(-s)$ by using (8.11):

$$\rho(s)\rho(-s) = 1 - G_T(-s^2)$$

2. The next step is to determine $\rho(s)$. Assign all LHP poles to $\rho(s)$. If a minimum phase solution is required, assign all the LHP zeros to $\rho(s)$. If not, any combination of the zeros can be assigned to it, as long as they are assigned in conjugate pairs, and the relationship between $\rho(s)$ and $\rho(-s)$ is kept in mind.

    When the load resistance differs from the source resistance and a lowpass or highpass network is designed, the sign of $\rho(s)$ can be determined by determining the value of $\rho(s)$ as defined by (8.28) at $\omega = 0$ and when $\omega \to \infty$, respectively.

$$\rho(s) = \frac{Z_{22}(s) - Z_l(-s)}{Z_{22}(s) + Z_l(s)} A(s) \tag{8.28}$$

$$A(s) = \prod_i [s - s_i] / [s + s_i] \tag{8.29}$$

    where $A(s)$ is an all-pass function with poles (s_i) equal to the open LHP poles (LHP poles with the $j\omega$-axis poles excluded) of the load impedance function $Z_l(s)$. $A(s)$ ensures that all the poles of the reflection coefficient, $\rho(s)$, will lie in the LHP by canceling the RHP poles caused by $Z_l(-s)$ in (8.28).

    $Z_{22}(s)$ is the output impedance of the network to be designed.

    When the load and source impedances are purely resistive and bandpass networks are designed, the sign of $\rho(s)$ is indeterminate and either sign can be used, that is, unless a specific topology is required. This does not always apply when the load impedance is reactive.

    Whether there are any constraints on the sign of $\rho(s)$ can be determined by considering (8.28) at $\omega = 0$ and when $\omega \to \infty$.

3. Determine the output impedance of the impedance-matching network as seen from the load terminals. This can be done by using the following equations:

$$r_l(s) = 0.5\,[Z_l(s) + Z_l(-s)] \tag{8.30}$$

    where $r_l(s)$ is the even (resistive) part of the load impedance function $Z_l(s)$

$$F(s) = 2r_l(s)A(s) \tag{8.31}$$

$$Z_{22}(s) = \frac{F(s)}{A(s) - \rho(s)} - Z_l(s) \tag{8.32}$$

4. Synthesize the required network by using standard filter theory. When the topology is important, the elements of the network must be extracted in the proper sequence.

EXAMPLE 8.4   Darlington synthesis when the load is complex.

As an example of the application of this procedure, a network will be designed to match a load consisting of a 1-Ω resistor in parallel with a 1.39-F capacitor to a source with 0.5-Ω internal resistance. The transducer power gain versus frequency function is to be a second-order lowpass Chebyshev function with 0.5-dB ripple in the passband.

$\omega_c = 1$ rad/s; $K_n = 1$

The specified transducer power gain function is

$$G_T(\omega^2) = \frac{K_n}{1 + \varepsilon^2 C_n^2(\omega / \omega_c)}$$

$$= \frac{1}{1 + 0.12202(4\omega^4 - 4\omega^2 + 1)}$$

(The ripple factor and the polynomial $C_2(\omega)$ were obtained from standard filter tables.)

Because $RCf_c = 0.221$, it is clear from Figure 8.13 ($n = 2$) that the gain function specified is realizable.

Step 1

$$G_T(-s^2) = \frac{1}{1 + 0.12202(4s^4 + 4s^2 + 1)}$$

$$\rho(s)\rho(-s) = 1 - G_T(-s^2)$$

$$= \frac{s^4 + s^2 + 0.2500}{s^4 + s^2 + 2.2988}$$

Step 2

By assigning the LHP poles of the product $\rho(s)\rho(-s)$ to $\rho(s)$, its denominator is found to be

$$\rho(s) = s^2 + 1.4257s + 1.5126$$

By assigning the LHP zeros to $\rho(s)$, the numerator is found to be

$$q(s) = s^2 + 0.5000$$

The reflection coefficient $\rho(s)$ is therefore

$$\rho(s) = \pm \frac{s^2 + 0.5000}{s^2 + 1.4256s + 1.5162}$$

Since, by using (8.29),

$$A(s) = \prod_i [s - s_i] / [s + s_i]$$

$$= \frac{s - 0.7914}{s + 0.7914}$$

and, by using (8.28),

$$\rho(0) = \frac{Z_{22}(0) - Z_l(0)}{Z_{22}(0) + Z_l(0)} A(s)$$

$$= \frac{0.5 - 1.0}{0.5 + 1.0} \frac{0 - 0.7194}{0 + 0.7194}$$

$$= \frac{0.5}{1.5}$$

it follows that a positive sign must be assigned to $\rho(s)$.

Step 3

$$Z_l(s) = 1 / [1 + 1.39s]$$

$$r_l(s) = 0.5 [Z_l(s) + Z_l(-s)]$$

$$= -0.5176 / [s^2 - 0.5175]$$

$$F(s) = -1.0351 / [s + 0.7194]^2$$

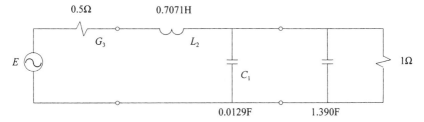

Figure 8.15   A network for matching a capacitive load to a source with 0.5-Ω internal resistance (0.5-dB
ripple, second-order Chebyshev response).

$$Z_{22}(s) = F(s) / [A(s) - \rho(s)] - Z_l(s)$$

$$= \frac{1.4297s}{0.0185s^2 + 0.0129s + 2.0360}$$

Step 4

$$Z_{22}(s) = \cfrac{1}{0.0129s + \cfrac{1}{0.7071s + \cfrac{1}{2.0}}} = \cfrac{1}{sC_1 + \cfrac{1}{sL_2 + \cfrac{1}{G_3}}}$$

Step 5

The source resistance is equal to the specified value and, therefore, no transformer
is required.

The designed network is shown in Figure 8.15.

8.3.5   Synthesis of Networks for Matching a Reactive Load to a Purely Resistive or a Reactive Source by Using the Principle of Parasitic Absorption

When the load impedance or the source impedance can be approximated with simple RC,
RL, or RLC networks, impedance-matching networks for the reactive terminations can be
designed by at first ignoring the reactiveness. If the gain-bandwidth constraints are taken
into account and a network with a suitable topology is designed, it will be possible to
absorb the reactive parts of the terminations into the designed network.

The topology of the network is a function of the order of the gain function chosen,
its transmission zeros, and the sign of $\rho(s)$, as was explained in Section 8.3.1.

When only the load or source impedance is reactive and it can be approximated with a parallel RC or series RL network, the maximum gain in the passband (K_n) can be determined for a Chebyshev transducer power gain versus frequency response with a specified ripple factor ε, by using (8.27).

Although this equation gives the optimum K_n corresponding to a specified ripple factor, it gives no indication as to which ripple factor will cause the lowest insertion loss in the passband; in other words, the optimum ripple factor is not known.

The optimum ripple factors corresponding to some values of the load or source quality factors at the highest frequency in the passband ($2\pi RCf_c$) were determined iteratively by substituting various values for the ripple factor into (8.27). The corresponding values for the highest gain and the lowest gain in the passband [K_n; $K_n/(1+\varepsilon^2)$] are tabulated for different values of the Q-factor and the number of elements used in the network in Table 8.5.

It follows from the table that the insertion loss will be approximately 0.5 dB when $Q = 2.25$ and four matching elements are used (an ideal transformer will also be required if the source or load resistance differs from that required).

When both the source impedance and the load impedance are reactive and can be approximated with parallel RC or series RL networks, the optimum values for the maximum gain in the passband (K_n) and the ripple factor (ε) can be determined by using the following set of equations [11]:

$$X = [1/Q_2 + 1/Q_1]\sin\frac{\pi}{2n} \tag{8.33}$$

$$Y = [1/Q_2 - 1/Q_1]\sin\frac{\pi}{2n} \tag{8.34}$$

$$A = \sinh^{-1} X$$

$$= \ln[X + \sqrt{X^2 + 1}] \tag{8.35}$$

$$B = \ln[Y + \sqrt{Y^2 + 1}] \tag{8.36}$$

$$= 1/\sinh[nA] \tag{8.37}$$

$$C = 0.5\frac{\sinh^2[nB]}{\sinh^2[nA]} + 0.5\sqrt{\frac{\sinh^4[nB]}{\sinh^4[nA]} + \frac{4}{\sinh^2[nA]}} \tag{8.38}$$

$$K_n = 1/[C^2 \sinh^2(nA)] \tag{8.39}$$

$$G_T(\omega^2) = K_n/[1 + \varepsilon^2 C_n^2(\omega)] \tag{8.40}$$

Q_1 and Q_2 are, respectively, the source and load Q-factors at the highest frequency in the passband.

Table 8.5 The Values of the Highest and the Lowest Transducer Power Gain in the Passband [K_n; $K_n/(1 + \varepsilon^2)$] of the Optimum Lowpass Chebyshev Function as a Function of the Load or the Source Q-Factor at the Highest Frequency in the Passband and the Number of Elements Used

Q	K_n; $K_n/(1+\varepsilon^2)$			
	$n = 2$	$n = 3$	$n = 4$	$n = 5$
	1.0000	1.0000	1.0000	1.0000
0.25	0.9998	1.0000	1.0000	1.0000
	0.9997	0.9999	1.0000	1.0000
0.50	0.9969	0.9994	0.9998	1.0000
	0.9997	0.9989	0.9992	0.9994
0.75	0.9876	0.9961	0.9981	0.9988
	0.9929	0.9951	0.9962	0.9968
1.00	0.9703	0.9876	0.9925	0.9946
	0.9814	0.9875	0.9894	0.9905
1.25	0.9459	0.9729	0.9816	0.9856
	0.9685	0.9749	0.9789	0.9805
1.50	0.9165	0.9527	0.9655	0.9715
	0.9515	0.9589	0.9626	0.9665
1.75	0.8839	0.9284	0.9451	0.9533
	0.9319	0.9419	0.9453	0.9492
2.00	0.8499	0.9016	0.9216	0.9319
	0.9111	0.9206	0.9256	0.9294
2.25	0.8157	0.8731	0.8963	0.9083
	0.8877	0.9004	0.9046	0.9082
2.50	0.7821	0.8442	0.8699	0.8834
	0.8666	0.8776	0.8828	0.8861
2.75	0.7496	0.8152	0.8431	0.8580
	0.8440	0.8548	0.8609	0.8638
3.00	0.7184	0.7868	0.8165	0.8325
	0.8223	0.8344	0.8391	0.8415
3.25	0.6887	0.7592	0.7903	0.8073
	0.7997	0.8126	0.8177	0.8195
3.50	0.6606	0.7326	0.7648	0.7827
	0.7800	0.7915	0.7968	0.8000
3.75	0.6340	0.7071	0.7402	0.7587
	0.7597	0.7695	0.7749	0.7791
4.00	0.6090	0.6827	0.7165	0.7355

When the load or source impedance can be approximated with series or parallel RLC resonant circuits, the inductance or capacitance can be increased to cause resonance at the center frequency of the passband (ω_0), and the bandpass problem can be transformed to an equivalent lowpass problem ($\omega_c = 1$ rad/s) by using the standard transformation formulas repeated in Figure 8.16. The optimum Chebyshev gain function can then be determined as described above.

The lowpass Q-factors corresponding to the bandpass Q (at the center frequency) can be determined by using the equation

$$Q_L = Q_B / (\omega_0 / B) \tag{8.41}$$

This equation can be derived easily by using (8.44) and (8.45).

EXAMPLE 8.5   A gain-bandwidth example based on Table 8.5.

The optimum values of K_n and ε (two-element Chebyshev matching network) will be determined for the load of Example 8.4 ($1\Omega \parallel 1.39F$).
       Since

$$Q_L = \omega_c R C = 1.39$$

it follows from the first column of Table 8.5 and linear interpolation of the data that

$$K_n \cong 0.9814 - \frac{1.39 - 1.25}{1.50 - 1.25}[0.9814 - 0.9685]$$

$$= 0.9742$$

and

$$\frac{K_n}{1 + \varepsilon^2} \cong 0.9459 - \frac{1.39 - 1.25}{1.50 - 1.25}[0.9459 - 0.9165] = 0.9294$$

It follows by manipulation of the last equation that the optimum value of the ripple factor is

$$\varepsilon = 0.2196$$

Therefore, the optimum two-element gain function is

$$G_T(\omega^2) = \frac{0.97}{1 + 0.0482 C_2^2(\omega)}$$

The maximum value of the insertion loss is 0.32 dB.

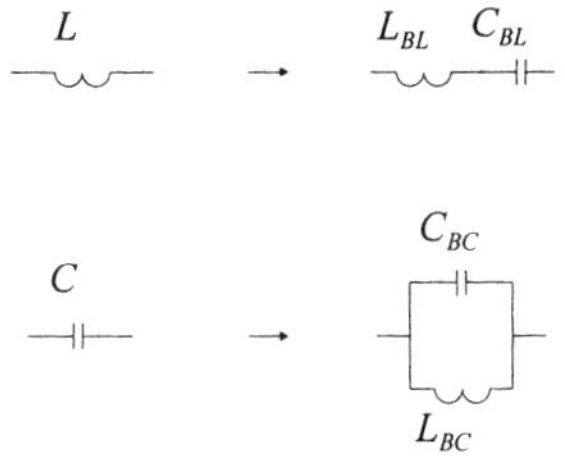

$$L \quad \rightarrow \quad L_{BL} \quad C_{BL}$$

$$L_{BL} = L / B \tag{8.42}$$

$$C_{BL} = B / (\omega_0^2 L_{BL}) \tag{8.43}$$

$$C \quad \rightarrow \quad \begin{matrix} C_{BC} \\ \\ L_{BC} \end{matrix}$$

$$C_{BC} = C / B \tag{8.44}$$

$$L_{BC} = B / (\omega_0^2 C_{BC}) \tag{8.45}$$

Figure 8.16 Formula for transforming a lowpass network ($\omega_c = 1$ rad/s) to a bandpass network with center frequency ω_0 and bandwidth B (rad/s).

8.3.6   The Analytical Approach to Designing Commensurate Distributed Impedance-Matching Networks

By using Richards' transformation [12], the analytical theory applicable to the design of lumped-element networks also applies to commensurate distributed networks (i.e., distributed networks in which the line lengths are all equal). Open-ended lines are transformed to lumped capacitors, and short-circuited lines to lumped inductors under this transformation.

Unlike short-circuited and open-ended stubs, the series transmission lines used in distributed designs have no lumped equivalents under Richards' transformation. The influence of these series lines (unit elements) on the gain function and their extraction from an impedance function when a network is synthesized will be discussed in Section 8.3.6.1, together with Richards' transformation.

The series short-circuited stubs, which are often found in a network designed by the use of Richards' transformation, are not realizable in planar form. When the designed network is to be realized in planar form, these unwanted stubs can be removed by using Kuroda's lowpass identities.

As in the case of lumped networks, impedance scaling is often required for a designed impedance-matching network. This impedance scaling function can be performed by using Kuroda's highpass identities and Norton's bandpass identities.

Kuroda's and Norton's identities will be discussed in Section 8.3.6.2.

8.3.6.1   Richards' Transformation

By using Richards' transformation [12]

$$S = j\Omega = j \tan[\beta l] = j \tan\left[\frac{\pi}{2} \frac{\omega}{\omega_0}\right] \tag{8.46}$$

open-ended and short-circuited stubs are mapped to capacitors and inductors in the S-plane.

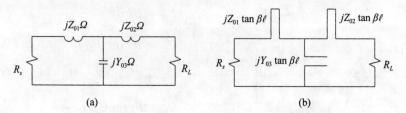

(a)   (b)

Figure 8.17   (a) A lumped-element equivalent for the distributed network in (b) under Richards' transformation.

The inductance and capacitance of the lumped equivalents are respectively equal to the characteristic impedance and admittance of the short-circuited and open-ended lines in the distributed network. This is illustrated in Figure 8.17.

The frequency response of a commensurate distributed network is compared with that of its lumped equivalent in Figure 8.18. Note that the response of the distributed network is periodic (βl versus $\tan \beta l$ characteristic) and that the gain at the even harmonics (including $\omega = 0$) and the uneven harmonics of ω_0 is equal to that of the lumped equivalent at $\omega = 0$ and $\omega \rightarrow \infty$, respectively. The distributed response is simply a compressed, periodic version of its lumped equivalent.

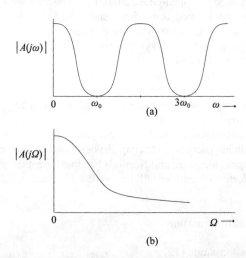

(a)

(b)

Figure 8.18   (a) The change in the frequency response of (b) a lowpass network with Richards' transformation [see (8.46)].

Series transmission lines are often used in distributed designs. The transmission matrix for a unit element (series transmission line in a commensurate distributed network) is

$$T = \begin{bmatrix} \cos\left(\dfrac{\pi}{2}\dfrac{\omega}{\omega_0}\right) & jZ_0 \sin\left(\dfrac{\pi}{2}\dfrac{\omega}{\omega_0}\right) \\ jY_0 \sin\left(\dfrac{\pi}{2}\dfrac{\omega}{\omega_0}\right) & \cos\left(\dfrac{\pi}{2}\dfrac{\omega}{\omega_0}\right) \end{bmatrix}$$ (8.47)

By using Richards' transformation, the transmission matrix becomes

$$T = \frac{1}{\sqrt{1-S^2}} \begin{bmatrix} 1 & Z_0 S \\ Y_0 S & 1 \end{bmatrix}$$ (8.48)

The transducer power gain of a commensurate distributed cascade network with N_u unit elements, N_h highpass elements, and order N is given by an expression of the form [13]

$$s_{21}(S)s_{21}(-S) = K_0 \frac{S^{2N_h}[1-S^2]^{N_u}}{G_N(S^2)}$$ (8.49)

where $G_N(S^2)$ is an Nth degree polynomial in S^2. Each unit element, therefore, contributes a factor $(1-S^2)$ to the numerator of the transducer power gain function.

With the gain function chosen, the input impedance of the corresponding network can be determined by Darlington synthesis, as described previously for lumped impedance-matching networks. With the input impedance established, the network can be synthesized. This can be done as before except for the extraction of unit elements.

A unit element can be extracted from the impedance function when the even part of the input impedance [$R_{in}(S)$; see (8.30)] at $S = 1$ is equal to zero. The characteristic impedance of the unit element is given by

$$Z_0 = Z_{in}(S)\Big|_{S=1}$$ (8.50)

With the unit element extracted, the remaining input impedance can be determined by using the expression [12]

$$Z'_{in} = Z_{in}(1) \frac{S Z_{in}(1) - Z_{in}(S)}{S Z_{in}(S) - Z_{in}(1)}$$ (8.51)

This impedance function will always have a common factor $S^2 - 1$ in its denominator and numerator which can be canceled.

With its lumped-element equivalent known, the design of the required distributed matching network is completed, that is, if impedance scaling is not required.

EXAMPLE 8.6   Extraction of a unit element.

The extraction procedure for a unit element will be illustrated by synthesizing the network with input impedance.

$$Z_{in}(S) = \frac{75S^2 + 125S}{1.5S^2 + 1.5S + 1.0}$$

$$= \frac{[75S^2] + 125S}{[1.5S^2 + 1.0] + 1.5S}$$

Because the numerator (N) of the even part of $Z_{in}(S)$ at $S = 1$ is given by

$$N\left\{ Z_{even} \right\}\Big|_{S=1} = N\left\{0.5\left[Z(S) + Z(-S)\right]\right\}\Big|_{S=1}$$

$$= [75S^2][1.5S^2 + 1.0] - [125S][1.5S]\Big|_{S=1} = 0$$

and

$$Z_{in}(1) = 50$$

a unit element of 50Ω can be extracted.

     The input impedance with the unit element removed can be determined by applying (8.51):

$$Z'_{in} = Z_{in}(1)\frac{S Z_{in}(1) - Z_{in}(S)}{S Z_{in}(S) - Z_{in}(1)}$$

$$= 50\frac{75S^3 - 75S^2}{75S^3 + 50S^2 - 75S - 50}$$

$$= 50\frac{[S^2 - 1]75S}{[S^2 - 1][75S + 50]}$$

$$= \frac{1}{1/[75S] + 1/50}$$

The synthesized network is shown in Figure 8.19.

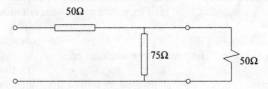

Figure 8.19   The network synthesized in Example 8.6.

8.3.6.2 Kuroda and Norton's Identities

Kuroda's lowpass identities are shown in Figure 8.20. These identities can sometimes be used to transform unrealistic impedances to more realistic levels, but they are more frequently used to remove unwanted series short-circuited stubs from planar designs.

Kuroda's highpass identities are shown in Figure 8.21. These identities can be used to change the impedance level in a matching network, as illustrated. The impedance level to the right of the transformed components is scaled with a factor k, while the input impedance (Z_{in}) remains unchanged.

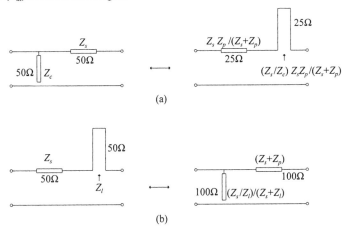

Figure 8.20 (a, b) Kuroda's lowpass identities for commensurate networks.

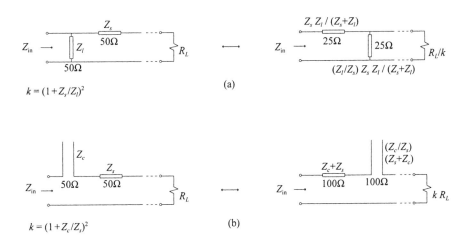

Figure 8.21 (a, b) Kuroda's highpass identities for commensurate networks.

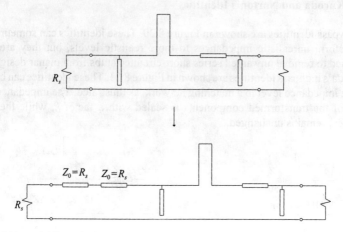

Figure 8.22   Adding unit elements to a commensurate network with a resistive source.

In applying Kuroda's identities it is useful to know that when the load or source impedance is purely resistive any number of unit elements (with $Z_0 = R_s$ or $Z_0 = R_L$) can be added in series with it without changing the amplitude response. This is illustrated in Figure 8.22.

The impedance level in a network can also be changed by using Norton's identities. Unlike Kuroda's identities in which unit elements are always involved, Norton's identities are applied to L-sections consisting only of open-ended or short-circuited stubs. These identities are shown in Figure 8.23.

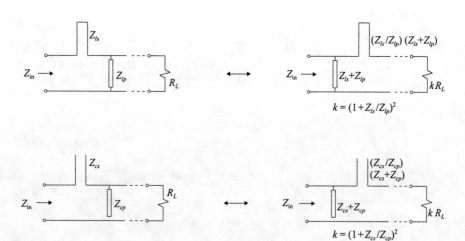

Figure 8.23   Norton's bandpass identities for commensurate networks.

8.4   THE ITERATIVE DESIGN OF IMPEDANCE-MATCHING NETWORKS

Instead of following the analytical approach, impedance-matching networks can also be designed iteratively. The "real-frequency" iterative techniques considered here have a major advantage over analytical and other techniques in that equivalent circuits for the load or source impedances, or an analytical expression for the transducer power gain versus frequency response, is not required. The networks synthesized by using iterative techniques are generally simpler in form with superior gain properties [6–8].

The first "real-frequency" technique was introduced by Carlin [3]. A complex load can be matched to a purely resistive source by using this technique. In this technique a piece-wise linear approximation of the output resistance or conductance of the network to be designed is optimized by using a least-square optimization routine. The technique has been implemented in the LSM program provided with this book.

The convergence properties of this technique are very good, and it can be used generally to estimate the gain-bandwidth constraints imposed by any complex load.

It has the disadvantage, however, that the response outside the passband is usually unnecessarily constrained, and occasionally it will not give the best results obtainable without considerable experimentation with the response outside the passband.

When bandpass networks are designed, the reactance of the network may not approximate the expected reactance well because of the difficulty of detecting and approximating the narrow spikes that can occur in the resistance function of a bandpass network. In these cases, the actual response will be poorer than that expected from the line-segment results [14].

Despite these disadvantages, the networks synthesized by using this technique are superior to those obtainable by direct application of analytical theory. This technique will be discussed in detail in Section 8.4.1.

Apart from matching a complex load to a purely resistive source, the "real-frequency" technique introduced by Yarman and Carlin in [7] can also be used to match a complex load to a complex source. In this technique, the numerator coefficients of the input reflection parameter (s_{11}) of the network is optimized. s_{11} is calculated with respect to purely resistive (and constant) normalization resistance values, and not in terms of the actual complex terminations. This technique has also been implemented in the LSM program.

Compared to the line-segment technique (where only one of the terminations is complex), the reflection coefficient technique has the advantage that it has no approximation step.

Initialization of the reflection coefficient procedure is not as simple as it is for the line-segment technique where the unknown output impedance of the network to be synthesized is taken to be equal to the resistive part of the known reactive load. However, excellent results can usually be achieved if the results obtained from the line-segment technique are used for initialization.

Note that when the line segment results are used to initialize the reflection coefficient procedure, the normalization resistance used to calculate the reflection coefficients should be the same.

Although the solution achieved with the reflection coefficient procedure may not necessarily be the best solution obtainable, it is, as a rule, much better than anything obtainable by direct application of analytical theory.

The reflection coefficient technique will be discussed in detail in Section 8.4.2.

In another technique proposed by Yarman and Carlin to solve double-matching problems (i.e., problems in which the load and the source are complex) [6], the output resistance of the matching network terminated at the input in a purely resistive load is optimized. Because this procedure has no significant advantage over the reflection coefficient technique, it will not be covered here. The interested reader is referred to [6].

The double-matching problem can also be solved very effectively by doing a synthesis-based systematic search on the transformation Q-factors (introduced in Chapter 3) to obtain initial solutions, which can then be optimized. Instead of optimizing only the best solution obtained in the search, it is a good idea to store a number of the best solutions obtained (10–25 are usually adequate) and then optimize all of these.

This approach has the distinct advantage that initial solutions are generated by the software and are not required from the designer.

If the systematic search is done thoroughly enough and enough solutions are stored for optimization, the probability of finding the optimum solution to any impedance-matching problem is very high.

Other major advantages of this approach are that many solutions are obtained (not just one) and that transformers are never required in the solutions synthesized. This technique is also very robust and can easily be extended to incorporate a great variety of constraints (topology constraints, constraints on the element values, and so forth).

Synthesis with this technique is over all topologies (if required), as is the case with the "real-frequency" techniques introduced by Carlin et al.

The transformation-Q approach can also be extended to form the basis of a algorithm for the design of distributed matching networks. This can be done without resorting to Richards' transformation, being restricted to commensurate solutions, and having to deal with any short-circuited stubs in the main line. The distributed matching techniques implemented in [15] are based on the T- and PI-section lumped-element equivalents introduced for a series transmission line in Section 8.4.3.2, as well as the open-ended and short-circuited stub equivalents considered in Section 8.4.3.1.

When commensurate networks are designed by following this approach, the variables are the characteristic impedances and the line lengths are fixed. The synthesis procedure can also be generalized so that the different lengths are used for the different types of lines (main-line sections, open-ended stubs, and shorted stubs), as was done in [15].

By choosing the different line lengths in a generalized commensurate network carefully, matching networks can be synthesized to control the performance at the harmonic frequencies, as well as the fundamental frequency. The short-circuits and open circuits required with the different classes of amplifiers (class B, E, F; refer to Chapter 2) can also be realized in this way. When the harmonic terminations must be presented to the intrinsic transistor, the power parameters introduced in Chapter 2 can be used to set up the associated matching problem.

When they are not used to control the harmonic performance, the line lengths used for the stubs can also be constrained to be electrically short to allow replacing them with lumped elements (mixed lumped/distributed solutions) [15].

Instead of synthesizing matching networks with fixed line lengths, the characteristic impedances to be used for the different line types can be fixed and the optimum lengths can be determined [15]. To approximate the results obtainable with lumped solutions, the characteristic impedance assigned to the main-line sections and any shorted stubs should be as high as possible, while that used for open-ended stubs should be as low as possible.

The characteristic impedances should not be set arbitrarily high, but should be set with line losses, as well as the impedance levels required in mind. In microstrip solutions, the lowest characteristic impedance to be used is usually determined by the T-junction parasitics introduced by wide junctions. In general, the transformer effect and the loading effect at each junction should be kept small [15].

The lengths of the lines used can be reduced by using lumped elements, if space is a problem. Part of the line to be replaced should be retained as a pad for the lumped component. Again the transformation-Q technique can be extended easily to achieve this [15].

The basic transformation-Q technique will be considered in detail in Section 8.4.3.

The advantages and disadvantages of the three techniques considered in this section are compared in Table 8.6.

8.4.1   The Line-Segment Approach to Matching a Complex Load to a Resistive Source

The gain-bandwidth constraints imposed by a reactive load (or source) can be determined iteratively by assuming that the output impedance (or admittance) of the network is a minimum-impedance (admittance) function (see Figure 8.24). When this is done, the output reactance of the network is known when the resistance is known. The optimum resistance can then be determined by minimizing the mean-square deviation between the desired and the actual transducer power gain.

If the resistance is approximated with line segments, the problem is well-behaved and the optimization can be done with a simple least-square optimization routine.

A detailed description of the procedure [3, 16] and the mathematics involved follows. This technique is implemented in the program LSM, which is provided with this book. The Fortran source code for the programs which preceded the latest version of LSM (LSMfortran, RCDMfortran, PLNMfortran, ZVRfortran, CFRfortran and LCTRANSfortran) is also provided. Note that these programs were integrated and refined in the latest version of LSM. LSM is a Visual C++ 2008 Unicode MFC program compiled under Windows Vista. The program will also run on Windows XP.

The Line-Segment Approach

1. Assume that the output impedance/admittance of the optimum network is a minimum-impedance/minimum-admittance function.

2. Assume as a first approximation that the resistive part $[R(\omega)]$ of the output impedance of the optimum network is equal to the resistive part of the measured load impedance (Z_L); that is,

$$R(\omega_i) = R_L(\omega_i) \tag{8.52}$$

Table 8.6   Comparison of the Different Iterative Impedance-Matching Techniques Considered Here

Line-segment technique	Reflection coefficient technique	Transformation-Q technique
Single-matching technique.	Double-matching technique.	Double-matching technique.
Topology independent, except for the number of transmission zeros at the origin which must be specified.	Topology independent, except for the number of transmission zeros at the origin which must be specified.	Topology independent.
Topology control only through the number of transmission zeros at the origin and the sign of the reflection coefficient.	Topology control only through the number of transmission zeros at the origin. and the sign of the reflection coefficient.	Topology control can be implemented easily.
Initialization by setting the output resistance equal to the required load resistance.	Initialization by using the results of the line-segment technique.	Initialization by synthesis-based systematic searches.
The results obtained are usually degraded in the approximation step present in this procedure.	No approximation step. Can be used to optimize the solution obtained with the line-segment approach.	No approximation step.
Excellent convergence properties, for lowpass or highpass solutions, but no guarantee of finding the global optimum.	Strongly dependent on initial solutions. No guarantee of finding the global optimum.	Dependent on the search range and density used. If the systematic search is done densely enough, it is highly likely that the global optimum will be obtained.
Ideal transformers may be required.	Ideal transformers may be required in bandpass solutions.	Ideal transformers are never required.
Single solution for each set of specifications.	Single solution for each set of specifications.	Many solutions to each matching problem.
Limited to lumped or commensurate distributed networks.	Limited to lumped or commensurate distributed networks.	Can be generalized easily to synthesize lumped solutions, commensurate solutions (without any series stubs), noncommensurate distributed solutions, and mixed lumped/distributed solutions.
Easy to implement.	Easy to implement.	More involved.

3.   Approximate the rational output resistance of the network with a piece-wise linear function, as illustrated in Figure 8.25.

    Enough increment frequencies (the frequencies at which the slope of the linear function changes) must be chosen to ensure a reasonable approximation of the unknown resistance. This can usually be done by choosing the frequencies to ensure a good approximation of the measured load resistance.

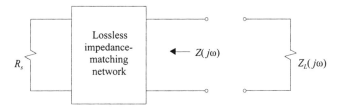

Figure 8.24 The impedance-matching problem under consideration.

For the sake of simplicity, the resistance $R(\omega)$ is assumed to equal zero at frequencies greater than the last increment frequency (ω_n).

The linear resistance function can be considered to be the sum of the semi-infinite functions $a_1(\omega)$, $a_2(\omega)$, ..., $a_n(\omega)$ shown in Figure 8.26, each with an appropriate weight factor r_k:

$$R(\omega) = r_0 + \sum_k r_k a_k(\omega) \tag{8.53}$$

where

$$a_k(\omega) = \begin{vmatrix} 0 & \text{if } \omega < \omega_{k-1} \\ \dfrac{\omega - \omega_{k-1}}{\omega_k - \omega_{k-1}} & \text{if } \omega_{k-1} < \omega_k \\ 1 & \text{if } \omega_k < \omega \end{vmatrix} \tag{8.54}$$

and

$$r_0 = Z_s(0) \tag{8.55}$$

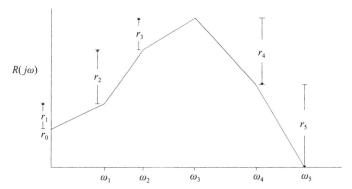

Figure 8.25 Approximation of the output resistance of the matching network with a piece-wise linear function.

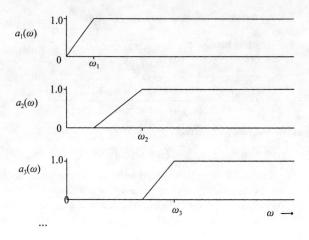

Figure 8.26   Illustration of the semi-infinite functions used in the line-segment approach.

When the optimum lowpass network is determined,

$$r_0 = R_s \tag{8.56}$$

where R_s is the source resistance as shown in Figure 8.24. In all other cases r_0 is equal to zero.

Because the resistance $R(j\omega)$ is equal to zero when the frequency is greater than the last increment frequency (ω_n), the increment factor (weight factor) r_n is not independent of the other increment factors. The following equation applies:

$$r_n = -[r_0 + r_1 + r_2 + ... + r_{n-1}] \tag{8.57}$$

When this value for r_n is substituted into (8.53), it changes to

$$R(\omega) = [1 - a_n(\omega)]r_0 + \sum_k r_k [a_k(\omega) - a_n(\omega)] \tag{8.58}$$

In vector form this becomes

$$R(\omega) = [1 - a_n(\omega)]r_0 + \overline{a}^T(\omega)\overline{r}' \tag{8.59}$$

where

$$\overline{a}^T = [a_1(\omega) - a_n(\omega), a_2(\omega) - a_n(\omega), ..., a_{n-1}(\omega) - a_n(\omega)] \tag{8.60}$$

and

$$\bar{r}' = \begin{bmatrix} r_1 \\ r_2 \\ \cdot \\ r_{n-1} \end{bmatrix}$$

(8.61)

The resistance value at any particular frequency can be calculated by using (8.59).

Because the impedance function was assumed to be a minimum-impedance function, the reactance associated with the resistance $R(j\omega)$ is known. It can be determined by using the equation

$$X(\omega) = \bar{b}^T(\omega)\,\bar{r}$$

(8.62)

where

$$\bar{b}^T(\omega) = [b_1(\omega), b_2(\omega), \ldots, b_n(\omega)]$$

(8.63)

and

$$b_k(\omega) = \frac{1}{\pi[\omega_k - \omega_{k-1}]} \int_{\omega_{k-1}}^{\omega_k} \ln\left|\frac{y+\omega}{y-\omega}\right| dy$$

(8.64)

The value of the integral in the last equation is high when the relevant increment frequencies are close to ω, and the value decreases as the increment frequencies deviate from ω. It follows from this and (8.62), that the reactance associated with the resistance at a particular frequency will be high when the resistance changes rapidly at nearby frequencies.

The integral in (8.64) has a simple closed form evaluation [9], which is useful in determining its value.

If the dependence of r_n on the other increment factors is taken into account, (8.62) becomes

$$X(\omega) = [0 - b_n(\omega)]r_0 + \bar{b}'^T(\omega)\bar{r}'$$

(8.65)

where

$$\bar{b}'(\omega) = \begin{bmatrix} b_1(\omega) - b_n(\omega) \\ b_2(\omega) - b_n(\omega) \\ \cdot \\ b_{n-1}(\omega) - b_n(\omega) \end{bmatrix}$$

(8.66)

The resistance and reactance corresponding to a particular

increment vector, $\bar{r}'$ can be determined at any particular frequency by using (8.59) and (8.66).

4.   Calculate the transducer power gain associated with the initial value ($\bar{r}_0'$) of the increment vector, $\bar{r}'$, at the various frequencies of interest. This can be done by using the equation

$$G_T(\omega) = 1 - \left|s_{11}\right|^2$$

$$= 1 - \left|\frac{Z_L(\omega) - Z^*(\omega)}{Z_L(\omega) + Z(\omega)}\right|^2$$

$$= 1 - \left|\frac{[R_L(\omega) + jX_L(\omega)] - [R(\omega) + jX(\omega)]^*}{[R_L(\omega) + jX_L(\omega)] + [R(\omega) + jX(\omega)]}\right|^2$$

$$= \frac{4\, R_L(\omega)\, R(\omega)}{[R_L(\omega) + R(\omega)]^2 + [X_L(\omega) + X(\omega)]^2} \qquad (8.67)$$

where $Z_L(\omega)$ is the measured load impedance.

5.   Determine the optimum value of the increment vector $\bar{r}'$ iteratively by minimizing the sum of the relative difference in the actual transducer power gain $[G_T(\omega)]$ and the desired transducer power gain $[G_I(\omega)]$ squared at the different frequencies of interest. This can be done by using a least-square optimization routine.

The relevant equations are as follows:

$$E = \sum_j e^2(\bar{r}', \omega_j) \qquad (8.68)$$

$$= \sum_j \left[\frac{G_T(\bar{r}', \omega_j)}{G_I(\omega_j)} - 1\right]^2 \qquad (8.69)$$

$$\bar{f}' = \frac{\partial e(\bar{r}', \omega)}{\partial \bar{r}_0'} \qquad (8.70)$$

$$= \frac{\partial e[R(\omega), X(\omega)]}{\partial R(\omega)}\frac{\partial R(\omega)}{\partial \bar{r}_0'} + \frac{\partial e[R(\omega), X(\omega)]}{\partial X(\omega)}\frac{\partial X(\omega)}{\partial \bar{r}_0'} \qquad (8.71)$$

$$= \frac{\partial e[R(\omega), X(\omega)]}{\partial R(\omega)} \, \overline{a}(\omega) + \frac{\partial e[R(\omega), X(\omega)]}{\partial X(\omega)} \, \overline{b}(\omega) \tag{8.72}$$

where $\overline{r}_0'$ is the current initial value of the increment vector, and $\overline{f}'(\omega)$ is the gradient vector associated with the error function $e\,(\overline{r}', \omega)$, where

$$e[R(\omega), X(\omega)] = G_T[R(\omega), X(\omega)] / G_l(\omega) - 1 \tag{8.73}$$

with $G_T[R(\omega), X(\omega)]$ as defined in (8.67).

$$e(\overline{r}', \omega_j) = e(\overline{r}_0', \omega_j) + \overline{f}'^T(\omega_j) \cdot \overline{\delta} \tag{8.74}$$

with $\overline{\delta}$ defined by the equation

$$\overline{r}' = \overline{r}_0' + \overline{\delta} \tag{8.75}$$

where $\overline{r}'$ is the new initial value of the increment vector.

$$\frac{\partial \sum_j e^2(\overline{r}', \omega_j)}{\partial \overline{\delta}} = 0 \tag{8.76}$$

$$\Rightarrow$$

$$\sum_j \overline{f}'(\omega_j) \overline{f}'^T(\omega_j) \, \overline{\delta} = -\sum_j e(\overline{r}_0', \omega_j) \overline{f}'(\omega_j) \tag{8.77}$$

With the optimum increment vector known, the gain-bandwidth constraints imposed by the load, as well as the output impedance (admittance) of the optimum network, are known.

6.   The next step is to determine the optimum network.

Since the optimum increment vector is known, the output resistance of the network is known at any particular frequency. A rational approximation function and the corresponding minimum-impedance (admittance) function can be obtained by following the procedure outlined in Section 8.2.

The order of the network and the number of zeros at the origin are variables in the approximation stage.

With the minimum-impedance and minimum-admittance functions known, the optimum network can be synthesized easily.

Whether a minimum-impedance or minimum-admittance function will be the best solution to a particular problem is usually not known at the outset. If good results are not obtained by using one, the other can be tried.

When the load resistance is higher than the source resistance, a minimum-impedance solution often yields better results.

When the gain-bandwidth product in a problem is a limiting factor, it will be found that the results are dependent on the position of the last increment frequency. Some experimentation with this frequency will then be necessary.

When bandpass networks are designed, both the first and the last increment frequencies have a significant influence on the results when the gain-bandwidth product is limited.

As stated previously, the line-segment technique is implemented in the LSM program. This program can be used to solve single-matching problems (load complex, source purely resistive), as well as double-matching problems.

When single-matching problems are solved, the optimum increment vector and the gain-bandwidth constraints associated with any reactive load are established by the program, after which a resistance or conductance function is fitted to the numerical data, and a network is extracted from the associated minimum-impedance or minimum-admittance function.

The complex load impedances and the transducer power gain required at each frequency ($0 < G_T \leq 1.0$) must be specified at a number of frequencies over the passband of interest. The data can be specified manually or can be imported from a Touchstone one-port S-parameter data file (file type .s1p).

Note that S-parameter data should only be supplied at the frequencies of interest when the .s1p option is used. Only the reflection coefficients associated with the load terminations to be used can be specified in this file. The gain will be assumed to be unity (that is, the assumption is that a conjugate match is desired), but can be edited in the program.

The normalization used for the load reflection coefficients specified is usually 50Ω, but a different value can be specified by in the # line of the .s1p file. An example of a typical .s1p file is provided in the LSM Examples Folder.

Data can also be imported from ASCII text files when the format used in the previous version of LSM is used. Example files (.txt file extensions) that can be used as templates are provided in the LSM Examples folder. Note that the file type used for the data files associated with the latest version is .lsm. These files cannot be edited outside LSM.

The input data for the program consist of the following:

1.   The impedance to be matched and the required transducer power gain at the different frequencies of interest (f, R, X, G_T).

2.   The increment frequencies and the initial values for the output resistance or conductance of the network to be synthesized (f_j, r_j). The output resistance or conductance of the network at dc must also be specified.

4.   The number of iterations to be done.

5.   The amount by which the last increment frequency must be incremented (in hertz) and the number of times this must be done.

6.   The number of elements to be used in the matching network, as well as the

number of transmission zeros $[s_{21}(s)]$. Note that the transmission zeros must be consistent with the resistance or conductance specified at dc. When the value specified is not zero, the network cannot have transmission zeros at dc.

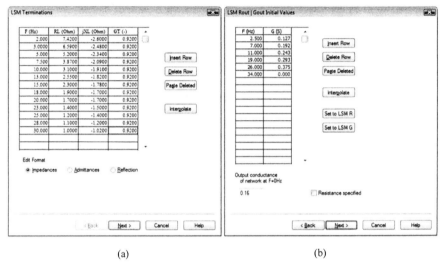

(a)   (b)

Figure 8.27   (a) The load terminations to be matched and (b) the initial values used for the output conductance in Example 8.7.

EXAMPLE 8.7   The gain-bandwidth constraints of a matching problem.

As an illustration of using the program LSM, the gain-bandwidth constraints imposed on a lowpass network by the input impedance of the transistor in Example 8.1 were established. The source resistance used was 6.25Ω. The goal was to achieve a flat response across the 2–30-MHz passband.

The data file "MRF406Yin.lsm" in directory "\artechLSM\Examples" was used to define the matching problem. The input impedance of the transistor is the load impedance in LSM. The termination data used is shown in Figure 8.27(a). The initial values specified for the output conductance of the network to be synthesized are shown in Figure 8.27(b). The output conductance at dc was set to the inverse of 6.25Ω, that is, 0.16 siemens.

Note the options to display or edit the termination data in impedance, reflection or admittance format in Figure 8.27(a), as well as the options to initialize the output resistance or conductance of the network by using the termination data specified (*Set to LSM R* and *Set to LSM G* buttons).

In order to determine the constraints, the gain specified [see Figure 8.27(a)] was decreased progressively until the ripple in the passband became very small. Note that the gain values can also be adjusted by copying (Control+C keys) and pasting a value (Control+V keys). This is useful when all the values are the same, as in this example. Also note that the gain must be bigger than zero and less or equal to one.

In addition to the data entered, the specifications shown in Figure 8.28 must also be made. Note the option to automatically adjust the last increment frequency in order to find the optimum value. The increment to be used and the number of times that this frequency should be incremented must be specified.

When only the gain-bandwidth constraints are of interest, it is not necessary to extract a network too. If the option to extract a network is selected, the number of elements in the network, as well as the number of transmission zeros at the origin (dc) must be specified. Because lowpass solutions are of interest here, the number of transmission zeros was set to zero.

Note that highpass solutions are not allowed when a single-matching problem is solved with LSM (that is, the number of transmission zeros cannot be equal to the number of elements). This follows because the output conductance or resistance was set to zero above the highest increment frequency. Single-matching highpass networks can, however, be designed indirectly by using the option to transform the specifications under an s to $1/s$ transformation [see the *Functions* menu in Figure 8.29(a)].

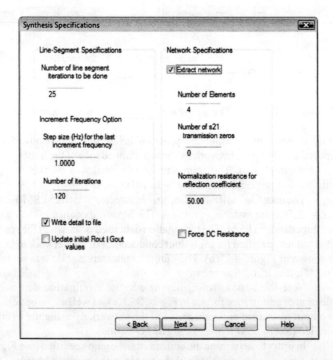

Figure 8.28   The best solution obtained (highest minimum insertion gain) with the program LSM to the matching problem solved in Example 8.7.

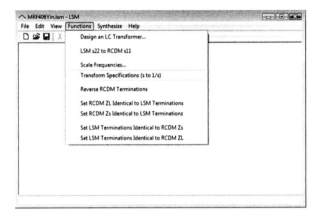

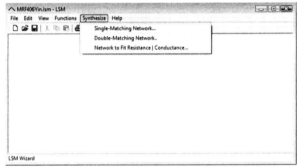

Figure 8.29   The functions menu and the synthesize menu provided in LSM.

Table 8.7   The Optimum Values of the Highest Increment Frequency and the Associated Normalized Least-Square Error and Maximum Deviation from the Prescribed Transducer Power Gain for the MRF406 as a Function of the Transducer Power Gain Specified

Transducer power gain specified	Optimum value of the highest increment frequency (MHz)	Maximum deviation from the prescribed transducer power gain (%)
1.00	36	14.2
0.98	36	12.0
0.96	36	9.5
0.94	37	7.1
0.92	38	4.9
0.90	38	3.2
0.88	56	1.9
0.86	79	1.3
0.84	119	1.1

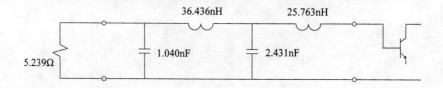

Figure 8.30   The best solution obtained (highest minimum insertion gain) with the program LSM the matching problem solved in Example 8.7.

The LSM wizard is launched automatically when a file is opened. The wizard can be closed by using the Cancel option, if necessary. Wizards for solving single-matching and double-matching problems were implemented [see Figure 8.29(b)] and the wizard of interest can be selected from the *Synthesize* menu [see Figure 8.29(b)].

The results obtained are summarized in Table 8.7. The optimum zero-resistance or zero-conductance frequency and the highest deviation from the gain targeted are listed in the table. Note that the gain ripple decreased as the gain was decreased.

If the criterion of minimum insertion loss across the passband is used, the best results will be obtained if the transducer power gain is chosen to be approximately 0.92. The insertion loss will then be less than 0.58 dB (G_{T_min} = 0.8746). The matching network associated with this gain level is shown in Figure 8.30.

The source resistance (5.239Ω) in Figure 8.30 is not 6.25Ω as required. When a lowpass network is synthesized, it should be possible to solve this problem by adding extra data points at the lower frequencies. The option to force the output resistance to the correct value (by impedance scaling) can also be selected, but the insertion loss will be degraded to 0.93 dB if this is done.

Note that the gain listed when the option to extract a network is chosen may differ from that when a network is not extracted [refer to Figure 8.31(b)]. The difference is introduced by the approximation step in the algorithm. When a network has been synthesized, the gain listed is that for the network, and not that associated with the optimized increment vector. The heading of the wizard page is changed accordingly.

In this example the gain associated with the actual network varies between 0.8888 and 0.9171, and the insertion loss is less than 0.51 dB, that is, when the gain targeted is 0.92. The output conductance of the network synthesized is also displayed in the table. When a network was not synthesized, the output conductance or resistance is that associated with the optimized increment vector.

When transmission zeros at the origin were specified, more than one network may have been synthesized. The different networks and the associated performance can be displayed by selecting the appropriate number from the Network List combo box [see Figure 8.31(a)].

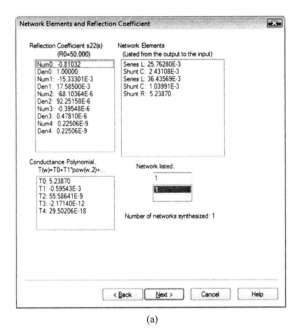

(a)

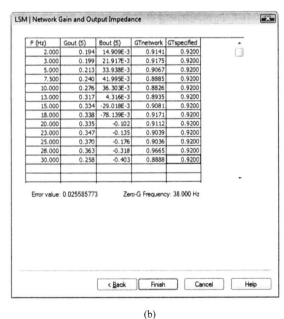

(b)

Figure 8.31 (a, b) The best solution obtained (highest minimum insertion gain) with the program LSM to the matching problem solved in Example 8.7.

8.4.2   The Reflection Coefficient Approach to Solving Double-Matching Problems

When both the load impedance and the source impedance are reactive, impedance-matching networks can be designed iteratively by using the algorithm developed by Yarman and Carlin [7].

In following this approach, the lossless matching network is modeled as a two-port network. When all the transmission zeros are at the origin or infinity, the scattering parameters (S-parameters) of the network are given by the following equations:

$$s_{11}(s) = h(s) / g(s) \tag{8.78}$$

$$s_{12}(s) = s_{21}(s) = \pm s^k / g(s) \tag{8.79}$$

$$s_{22}(s) = -(-1)^k [h(-s) / g(s)] \tag{8.80}$$

where k is an integer specifying the order of the zero of transmission at the origin, and $h(s)$ and $g(s)$ are polynomials.

Because the transfer and reflection parameters are related by the equation

$$s_{11}(s)s_{11}(-s) = 1 - s_{21}(s)s_{21}(-s) \tag{8.81}$$

it follows that

$$g(s)g(-s) = h(s)h(-s) + (-1)^k s^{2k} \tag{8.82}$$

It is clear from this equation, and the fact that the polynomial $g(s)$ must be positive-real, that $g(s)$ is a function of $h(s)$ and the order of the zero of transmission at the origin only. Because the S-parameters of the network are completely determined by $h(s)$, $g(s)$, and k, it follows that the network itself is defined when $h(s)$ and k are defined. The optimum network, therefore, can be determined by finding the optimum coefficients of the numerator polynomial $h(s)$ for a given value of k.

In order to optimize the coefficients of $h(s)$, an expression for calculating the transducer power gain at each relevant frequency is required. In terms of S-parameters, the gain is given by

$$G_T(\omega) = \frac{[1-|S_G(\omega)|^2] [1-|S_L|^2]|s_{21}(\omega)|^2}{|1-s_{11}(\omega)S_G(\omega)|^2 \left|1-\left[s22(\omega)+\dfrac{s_{21}^2(\omega)S_G(\omega)}{1-s_{11}(\omega)S_G(\omega)}\right]S_L(\omega)\right|^2} \tag{8.83}$$

$$= \frac{[1-|S_G(\omega)|^2] [1-|S_L|^2]|\omega^{2k}|}{|g(j\omega) - (-1)^k S_G S_L g(-j\omega) - S_G h(j\omega) + (-1)^k S_L h(-j\omega)|^2} \tag{8.84}$$

where

$$S_G(\omega) = \frac{Z_s(\omega) - R_0}{Z_s(\omega) + R_0} \qquad (8.85)$$

$$S_L(\omega) = \frac{Z_L(\omega) - R_0}{Z_L(\omega) + R_0} \qquad (8.86)$$

and R_0 is the normalizing resistance of the S-parameters.

The coefficients of $h(s)$ can be optimized by using a linear least-square routine. Because the gain is not a simple function of $h(s)$, the problem is more complex than before and the choice of the initial values can be critical.

Good results can be achieved if the results obtained by designing an impedance-matching network to match the more complex termination to a resistive source are used to determine the initial values.

The optimization can be done by using the double-matching wizard provided in LSM. As was the case with single-matching problems, the input data can be entered manually or can be imported. The available import options are provided under the *File* menu and are shown in Figure 8.32.

Note that the source termination and the load termination of the double-matching problem can also be imported by using the Touchstone .s1p import option. The termination imported is used as the load termination of the single-matching problem, but can be set to be the source or the load termination used by the double-matching wizard. This can be done by using the appropriate option in the *Functions* menu [see Figure 8.29(a) for the options provided]. It will be necessary to specify the two terminations at identical frequencies. Data should also only be supplied at the frequencies of interest and can be specified at up to 120 frequencies.

The specifications set in the previous version (RCDMfortran) can also be imported by LSM. Because the specifications were stored in ASCII text format, it is a simple matter to use one of these files as a template for a new problem too. The .s1p option may, however, be more convenient.

The RCDM.txt files and RCDMfano.txt files provided in the LSM examples folder (default: "c:\artechLSM\Examples) can be used as templates. Note that the text strings in the file must remain as they are. Blank lines will be ignored by LSM.

The input data for a double-matching problem consists of the following:

1. The source and load impedances to be matched and the transducer power gain required at each frequency (f_i, R_s, X_s, R_L, X_L, G_T).

2. The degree of the numerator polynomial $h(s)$ and the number of transmission zeros at the origin. The degree of the numerator is determined indirectly from the coefficients entered.

3. The initial values of the coefficients (h_0, h_1, ..., h_n).

4. The normalization resistance to be used to calculate the S-parameters of the matching network (usually 50Ω).

When initial values are assigned to the coefficients of $h(s)$, it is important to realize that there are some constraints on the values of the coefficients.

In the case of lowpass networks, the value of the transmission parameter $s_{21}(s)$ must be equal to 1 when $\omega = 0$ and the same input and output normalizing resistance (R_0) are used.

Because

$$s_{21}(s) = \pm \frac{s^k}{g(s)} \tag{8.87}$$

$$= \frac{\pm s^k}{g_n s^n + g_{n-1} s^{n-1} + ... + g_0} \tag{8.88}$$

and $k = 0$ for a lowpass network, it is clear that g_0 must equal 1 to ensure that $s_{21}(0)$ will be equal to 1.

This restriction on the value of g_0 imposes two constraints on the polynomial $h(s)$.

The first is that in determining initial values for $h(s)$, the numerator and denominator of the input reflection coefficient $s_{11}(s)$ must be scaled to ensure that g_0 will equal 1.

The second restriction follows from (8.82) by setting $s = 0$:

$$g_0^2 = h_0^2 + 1 \tag{8.89}$$

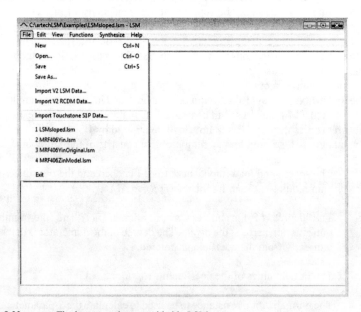

Figure 8.32   The import options provided in LSM.

Because $g_0 = 1$, it implies that h_0 must be equal to zero.

Following the reasoning above, it can be shown that for highpass networks h_n must be equal to zero. The numerator and denominator of the input reflection coefficient must also be scaled to ensure that $g_n = 1$ when initial values for the coefficients are determined.

When a bandpass problem is initialized, none of the coefficients needs to be equal to zero, but scaling is still required. It is obvious from (8.79) that if $h(s)$ and $g(s)$ are scaled by a factor a, the gain $|s_{21}|^2$ will change by a factor a^2. The correct scale factor for given polynomials $h_i(s)$ and $g_i(s)$ can be calculated by first calculating the input resistance (R_{in}, where $Z_{in} = R_{in} + jX_{in}$, and Z_{in} can be calculated from the reflection coefficient) of the network used for the initialization, terminated in the chosen normalization resistance, at any specific frequency. The value of a is then given by

$$a = \pm \frac{\omega^k}{|g_i(j\omega) - h_i(j\omega)|} \sqrt{\frac{R_0}{R_{in}(j\omega)}} \qquad (8.90)$$

This equation can be derived by starting with the expression $2R_{in} = Z_{in} + Z_{in}^{\,*}$, and using (8.78)–(8.89).

The constraints on the numerator and denominator coefficients of $\rho(s)$ are summarized in Table 8.8.

Table 8.8   The Constraints on the Numerator [$h(s)$] and the Denominator [$g(s)$] Coefficients of $\rho(s)$

Network	$h(s)$	$g(s)$
Lowpass	$h_0 = 0$	$g_0 = 1$
Highpass	$h_n = 0$	$g_n = 1$
Bandpass	scale factor a for $h(s)$	—

EXAMPLE 8.8   A double-matching example.

As a further example of using LSM, a highpass network will be designed to mismatch the source impedance in Table 8.9 to a 50-Ω load, as indicated in the table. The problem will first be transformed to lowpass form in order to solve it with LSM, after which the solution obtained will be used to initialize the double-matching routine.

The Single-Matching Problem

In order to design the required single-matching highpass network, the specifications in Table 8.9 were changed to those of the equivalent lowpass problem. This was done manually by using the transformation $s \to 1/s$. The new set of specifications is shown in Table 8.10. The frequencies in the table were obtained from the transformed frequencies [$\omega' = 1/\omega \to f' = 1/(4\pi^2 f)$] by using a scale factor of $4\pi^2 \times 10^9$.

Table 8.9   The Source Impedance, Load Impedance, and Transducer Power Gain Corresponding to the Problem Solved in Example 8.6

Frequency (MHz)	$R_s + jX_s$ (Ω)	$R_L + jX_L$ (Ω)	G_T
100	146.0 $-$ j114.0	79.1 $-$ j72.6	0.224
110	138.5 $-$ j112.5	73.6 $-$ j68.7	0.262
120	131.0 $-$ j111.0	68.0 $-$ j64.8	0.299
140	137.0 $-$ j103.0	63.2 $-$ j56.8	0.400
160	144.0 $-$ j88.0	59.6 $-$ j47.9	0.559
180	140.0 $-$ j88.0	57.5 $-$ j47.3	0.709
190	136.5 $-$ j92.0	55.0 $-$ j41.9	0.764
200	133.9 $-$ j96.0	53.5 $-$ j40.4	0.818

Because of the transformation, the impedances to be matched are the conjugates of those in Table 8.9.

This step could have been performed automatically by entering the data in Table 8.9 into LSM, and then using the *Functions* menu option to transform the specifications under an s to $1/s$ transformation. Frequency scaling can also be performed by using the option provided for this in the *Functions* menu [see Figure 8.29(a)].

Note that while double-precision calculations are done in LSM it is a good idea to keep the frequencies close to one. Ideally one could scaled the geometric mean frequency to be one, but this is not really necessary.

A minimum-impedance matching solution was synthesized to solve the problem as listed in Table 8.10. (Note that as a rule, both options should be tried.) The data entered was stored in the file "LSMsloped.lsm". This file can be found in the LSM examples folder. Note that while the impedance to be matched is the source impedance of the problem to be solved, it is taken to be the load impedance in LSM. Also note that an extra data point was added near dc in order to force the output resistance at dc towards 50Ω [refer to Figure 8.33(a)].

The synthesis specifications made are shown in Figure 8.34. The highest increment frequency (15 Hz) was set to be incremented in 5-Hz steps (range: 15–70 Hz). The number of elements to be used in the network was set to 4.

Table 8.10   The Source Impedance and Transducer Power Gain Corresponding to the Equivalent Lowpass Problem in Example 8.6

Frequency (Hz)	R_s (Ω)	X_s (Ω)	G_T
5.00	133.0	96.0	0.818
5.56	140.0	88.0	0.709
6.25	144.0	88.0	0.559
7.14	137.0	103.0	0.400
8.33	131.0	111.0	0.299
10.00	146.0	114.0	0.224

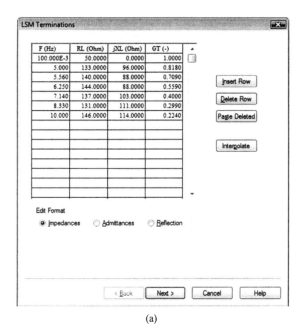

(a)

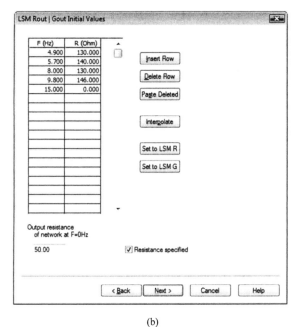

(b)

Figure 8.33   (a, b) The load terminations and the initial estimate for the output resistance of the single-matching network in Example 8.8.

The results obtained are shown in Figure 8.35. The least-square error is close to 5% and the optimum value for the zero resistance frequency was 30 Hz. The resistance function obtained [see Figure 8.35(a)] is

$$R(\omega) = 1/T(\omega)$$
$$= 1/(0.01984 - 3.9928 \times 10^{-5}\omega^2 + 4.94343 \times 10^{-8}\omega^4$$
$$- 1.3712 \times 10^{-11}\omega^6 + 1.1742 \times 10^{-15}\omega^8)$$

The poles of this function (see the file "LSMslopedoutput.txt") are

$$\omega = \pm 74.352 \pm j11.460$$
$$\pm 24.168 \pm j11.924$$

and the minimum-impedance function obtained is

$$Z(j\omega) =$$

$$\frac{207.176 \times 10^6 + j12.111 \times 10^6\omega - 97.352 \times 10^3\omega^2 - j2.082 \times 10^3\omega^3}{4.110 \times 10^6 + j151.621 \times 10^3\omega - 6.932 \times 10^3\omega^2 - j46.770\omega^3 + 1.000\omega^4}$$

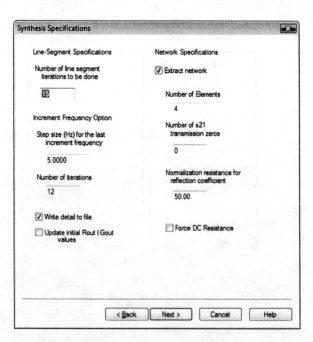

Figure 8.34   The synthesis specifications made for the single-matching problem in Example 8.8.

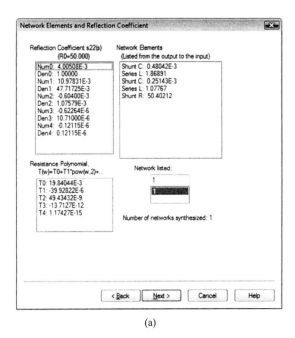

(a)

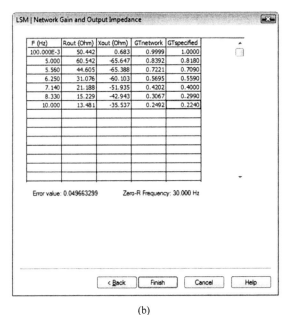

(b)

Figure 8.35   (a, b) The load terminations and the initial estimate for the output resistance of the single-matching network in Example 8.8.

As a function of s, this becomes

$$Z(s) = \frac{207.176 \times 10^6 + 12.111 \times 10^6 s + 97.352 \times 10^3 s^2 + 2.082 \times 10^3 s^3}{4.110 \times 10^6 + 151.621 \times 10^3 s + 6.932 \times 10^3 s^2 + j46.770 s^3 + 1.000 s^4}$$

(8.91)

The network obtained from program LSM is shown in Figure 8.36. It was synthesized in LSM by continued fractionation of the impedance function. The gain of the network is within 0.46 dB of that targeted.

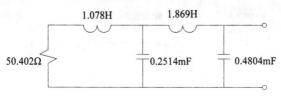

Figure 8.36   The network associated with (8.91).

The Double-Matching Problem

Initial values for the design of a double-matching network can be obtained by using the reflection coefficient for the network obtained from program LSM (refer to Figures 8.35 and 8.36). The reflection coefficient calculated in LSM is

$$s_{22}(s) = \frac{4.005 \times 10^{-3} + 10.978 \times 10^{-3} s - 0.604 \times 10^{-3} s^2 - 0.623 \times 10^{-6} s^3 - 0.121 \times 10^{-6} s^4}{1.0000 + 47.717 \times 10^{-3} s + 1.076 \times 10^{-3} s^2 + 10.710 \times 10^{-6} s^3 + 0.121 \times 10^{-6} s^4}$$

The numerator coefficients of $s_{22}(s)$ were used as initial values in the double-matching wizard. An option to copy the single-matching values to the double-matching data is provided in the *Functions* menu (*LSM s_{22} to RCDM s_{11}*) of LSM. This option was used in this example.

Note that single-matching network was synthesized from the load side towards the input side in LSM, while the reverse is done when a double-matching problem is solved. $s_{22}(s)$ in single-matching section is, therefore, $s_{11}(s)$ in double-matching section.

The double-matching problem is also defined in the data file "LSMsloped.lsm". The data entered is shown in Figure 8.37.

The matching network corresponding to the impedance function obtained is shown in Figure 8.38(a). The gain obtained is compared to the specifications in Figure 8.38(b). The maximum deviation from the specified gain response is 0.1 dB.

The final step in this example is to transform the results back to highpass form under the transformation $s \to 1/s$. Each inductor in Figure 8.39(a) must, therefore, be replaced with a capacitor with capacitance equal to the inverse of the inductance, and each capacitor must be replaced with an inductor with inductance equal to the inverse of the capacitance. Because a scale factor of $4\pi^2 \times 10^9$ was used to obtain the frequencies used in Table 8.10 from the transformed values, the network must also be frequency scaled with this factor. The final network is shown in Figure 8.39(b).

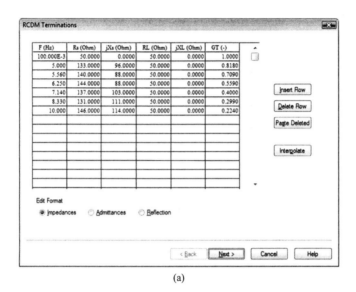

(a)

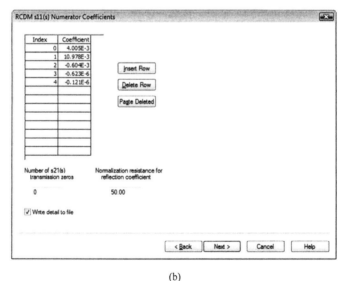

(b)

Figure 8.37   (a, b) The LSM specifications used for the double-matching problem solved in Example 8.8.

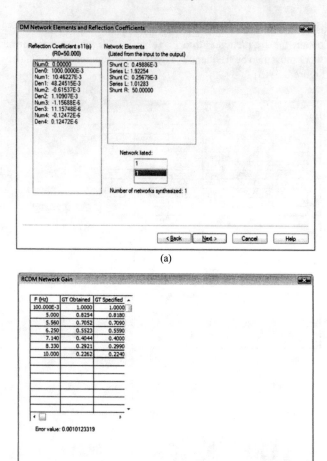

(a)

(b)

Figure 8.38   (a, b) The double-matching results obtained in Example 8.8.

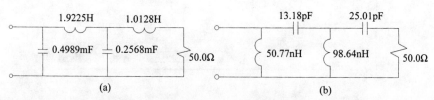

(a)   (b)

Figure 8.39   (a) The matching network designed by using the reflection coefficient technique and (b) the final matching after transformation and frequency scaling.

8.4.3 The Transformation-Q Approach to the Design of Impedance-Matching Networks

The narrowband impedance-matching technique described in Chapter 4 can be extended to increase the number of elements to an arbitrary number and to mismatch any complex load to any complex source by any specified amount at any single frequency [15].

In order to do this, it will be shown in Section 8.4.3.1 that the locus of input impedances for which the source impedance of a network will be mismatched to the load by a specified amount is a circle in the linear admittance plane or the impedance plane. The parameters of these circles will be derived here. The necessary extensions to the single-frequency technique will be made in Section 8.4.3.2.

The extended single-frequency matching technique forms an excellent basis for the iterative design of wideband impedance-matching networks. The main reason for this is that the range of each transformation Q is limited because high Q-factors will inevitably lead to a narrow bandwidth. Because of this fact, it is feasible to do a systematic search on the transformation Q-factors in search of the optimum combination, thus eliminating the need for a good initial solution. Furthermore, if the search is done thoroughly enough, the probability of finding the optimum solution will be very high.

With the systematic search completed, a number of the best sets of Q-factors can be optimized. This can be done with a least-square optimization routine, but better results in less time are obtainable by using a simple gradient optimization technique. The error optimized can be the mean-square value of the difference in the gain obtained and the specified gain, but a better alternative is to use the maximum relative deviation from the optimum as the error criterion. When this is done, the error value is determined by the worst performance in the frequency band of interest and not the average response.

The topologies of the networks synthesized by following this approach can easily be limited to lowpass form, highpass form, or to bandpass form with no shunt inductors or with no series capacitors. The last option is usually attractive in hybrid circuits at microwave frequencies.

The time required to solve a matching problem can be reduced significantly by constraining the gain at the frequency where the Q-factors are calculated to be higher than a specified minimum. The required run time is usually very short when networks with less than six elements are designed.

Major advantages of the transformation-Q technique over the techniques described previously are that many solutions instead of only one are obtained, that transformers are never required in the solutions, and that the probability of finding the optimum solution to a matching problem is very high when the search is done thoroughly enough. This approach is also very robust.

The transformation-Q technique can be extended easily to design more complicated networks, like distributed networks [15]. The basic information required to extend the technique to distributed networks is introduced in Sections 8.4.3.3 and 8.4.3.4. With this information any network synthesized to solve the matching problem at the transformation-Q frequency can be transformed easily to a distributed equivalent instead of a lumped equivalent. A commensurate equivalent (a distributed network in which all the lines have the same length) or a noncommensurate equivalent can be set up, as required.

When commensurate networks are synthesized, it is a good idea to allow different line lengths for the open-ended stubs, the shorted stubs and the main-line sections. This feature can be used to transform the stubs to equivalent lumped components, or to provide the short-circuit and/or open-circuit terminations required in class B, E, and F amplifiers.

Real-world components require pads. Any mixed lumped/distributed network designed without taking the pads into account is bound to require optimization. The transformation-Q synthesis procedures were extended in [15] to cater for these pads too. The pad capability also opens the door to synthesize matching networks in which the inductors are spiral inductors and the capacitors parallel-plate capacitors (MMIC applications).

8.4.3.1 Constraints on the Input Admittance of a Lossless Impedance-Matching Network If the Gain Is to Remain Constant at a Specific Frequency

The locus of input admittances for which the gain of a lossless impedance-matching network will remain constant can be derived by using the expression

$$|S_s|^2 = \left| \frac{Z_{in} - Z_s^*}{Z_{in} + Z_s} \right|^2 \tag{8.92}$$

$$= \left| \frac{Y_{in} - Y_s^*}{Y_{in} + Y_s} \right|^2 \tag{8.93}$$

where S_s is the input reflection parameter with the actual source impedance (Z_s) as normalizing impedance, and Z_{in} (Y_{in}) is the input impedance (admittance) of the matching network.

By substituting

$$|S_s|^2 = 1 - G_T \tag{8.94}$$

$$Y_{in} = G_{in} + jB_{in} \tag{8.95}$$

and

$$Y_s = G_s + jB_s \tag{8.96}$$

into (8.93), it follows that the locus of the input admittance for which the transducer power gain (G_T) will remain constant is a circle in the admittance plane. The center (G_0+jB_0) and radius (R_{Y0}) of this circle are:

$$G_0 + jB_0 = [2 / G_T - 1]G_s - jB_s \tag{8.97}$$

and

$$R_{Y0} = 2[1 / G_T^2 - 1 / G_T]^{1/2} G_s \tag{8.98}$$

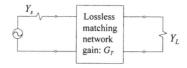

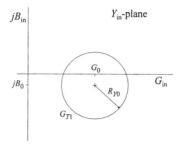

Figure 8.40   The locus of the input admittance for which the gain of a lossless matching network will remain constant at a specified frequency.

The gain of a lossless network will remain constant for all values of the input admittance that fall on the circumference of this circle. This is illustrated in Figure 8.40.

For all values of the input admittance that fall inside the constant gain circle, the gain will be higher than that on the circumference. The transducer power gain of a matching network, therefore, can be constrained to be higher than a specified minimum at any particular frequency by limiting its input admittance to the inside of the relevant constant transducer power gain circle.

Because (8.92) and (8.93) are of exactly the same form, the locus of input impedances for which the gain of a lossless network will remain constant is also a circle, and the parameters of this circle can also be obtained from (8.97) and (8.98) with $G_s + jB_s$ replaced with $R_s + jX_s$, and $G_0 + jB_0$ with $R_0 + jX_0$. The resulting equations are

$$R_0 + jX_0 = [2 / G_T - 1] R_s - jX_s \tag{8.99}$$

and

$$R_{z0} = 2[1 / G_T^2 - 1 / G_T]^{1/2} R_s \tag{8.100}$$

8.4.3.2   Extension of the Transformation-Q Impedance-Matching Technique

In an impedance-matching network, the resistance level is changed by each element except the last, which only serves to adjust the reactance or susceptance level. The change in the resistance of a four-element network with the first element a series element and no resonating sections is illustrated in Figure 8.41. The resistance is transformed in each transformation step by a factor of the form

$$D_n(\omega) = 1 + Q_n^2(\omega) \tag{8.101}$$

where

$$Q_n(\omega) = \frac{X_n(\omega) + X_{rn}(\omega)}{R_{rn}(\omega)} \tag{8.102}$$

or

$$Q_n(\omega) = \frac{B_n(\omega) + B_{rn}(\omega)}{G_{rn}(\omega)} \tag{8.103}$$

depending on whether the transformation under consideration is series-to-parallel or parallel-to-series, respectively (see Figure 8.42).

The factor $Q_n(\omega)$ is referred to as a transformation Q.

In (8.102), $X_n(\omega)$ is the reactance of the nth component, $X_{rn}(\omega)$ the effective reactance to the right of it, and $R_n(\omega)$ the effective resistance in series with it as illustrated in Figure 8.42(b).

Similarly, $B_n(\omega)$ is the susceptance of the nth component, $B_{rn}(\omega)$ is the effective susceptance to the right of it, and $G_{rn}(\omega)$ the effective conductance in parallel with it. This is illustrated in Figure 8.42(a).

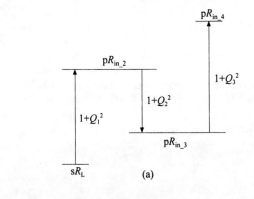

(a)

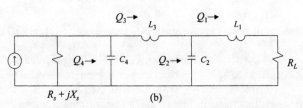

(b)

Figure 8.41   (a) Schematic illustration of the change in the resistance level of (b) a matching network.

A series-to-parallel transformation will always transform the associated resistance upward, while downward transformations are effected with shunt elements.

It follows from (8.102) and (8.103) that the sign of a transformation Q is positive when the effective reactance in series with the effective resistance is inductive, or when the effective susceptance in parallel with the effective conductance is capacitive.

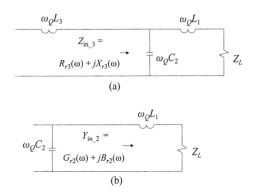

Figure 8.42   (a, b) Definition of the symbols in (8.102) and (8.103).

When the first element of an N-element network is a series element, the input resistance is given after $(N-1)$ transformations by the expression [15; refer to Chapter 4]

$$R_{in,N} = R_L \frac{1+Q_1^{\,2}}{1+Q_2^{\,2}} \frac{1+Q_3^{\,2}}{1+Q_4^{\,2}} \cdots [1+Q_{N-1}^2] \tag{8.104}$$

or

$$R_{in,N} = R_L \frac{1+Q_1^{\,2}}{1+Q_2^{\,2}} \cdots \frac{1+Q_{N-2}^2}{1+Q_{N-1}^2} \tag{8.105}$$

depending on whether the last matching element is a shunt element or a series element, respectively.

When the first element is a shunt element, the input conductance is given by the expression

$$G_{in,N} = \frac{G_L}{1+Q_1^{\,2}} \frac{1+Q_2^{\,2}}{1+Q_3^{\,2}} \cdots [1+Q_{N-1}^2] \tag{8.106}$$

or

$$G_{in,N} = \frac{G_L}{1+Q_1^{\,2}} \frac{1+Q_2^{\,2}}{1+Q_3^{\,2}} \cdots \frac{1+Q_{N-2}^2}{1+Q_{N-1}^2} \tag{8.107}$$

Because (8.104)–(8.105) are of the same form as (8.106)–(8.107), it is only necessary to consider the design of matching networks with a series element as the first element. Exactly the same procedure can then be followed to design networks in which the first element is a shunt element, after replacing all impedance specifications with the equivalent admittances.

In lowpass and highpass designs the number of Q-factors is equal to the number of elements in the network. In a bandpass network with N elements, but M resonating sections (series of parallel LC combinations), the number of Q-factors increases to $(N + M)$. This is illustrated for a three-element network in Figure 8.43.

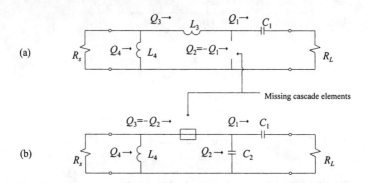

Figure 8.43   The influence of (a) series and (b) parallel resonating sections on the transformation Q-factors in a matching network.

When an element is absent, the associated Q is equal to the negative of the previous Q.

By using (8.104)–(8.107), it is very easy to calculate the input resistance of any impedance-matching network when the Q-factors are known.

In order to design a matching network to have a specified transducer power gain (G_T) at a particular frequency, it is only necessary to constrain the last two Q-factors to ensure that the input impedance (if the last element is a parallel element) will fall on the relevant gain circle as derived in the previous section.

When the last element is a series element (see Figure 8.44), the next to last Q should be constrained to ensure that the input resistance (R_{in}) will fall in the range

$$R_{in,min} \leq R_{in} \leq R_{in,max} \qquad (8.108)$$

The bounds on the next to last Q follow easily from the values of the previous Q-factors by using (8.105) in conjunction with (8.108). The resulting constraints are

$$Q_{N-1}^2 \geq \frac{R_L}{R_{in,max}} \frac{1+Q_1^2}{1+Q_2^2} \dots [1+Q_{N-2}^2] - 1 \qquad (8.109)$$

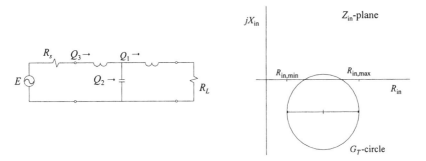

Figure 8.44 The constraints on the input resistance of a matching network where the last element is a series element when the transducer power gain should be higher than or equal to a specified minimum at a particular frequency.

and

$$Q_{N-1}^2 \leq \frac{R_L}{R_{in,min}} \frac{1+Q_1^2}{1+Q_2^2} \ldots [1+Q_{N-2}^2]-1 \qquad (8.110)$$

When the last element is a parallel element, the next to last Q should be constrained to ensure that the input conductance (G_{in}) will be within the constraints imposed by the constant gain circle on the admittance plane.

The resulting constraints are

$$Q_{N-1}^2 \leq \frac{1}{R_L G_{in,max}} \frac{1+Q_2^2}{1+Q_1^2} \ldots [1+Q_{N-2}^2]-1 \qquad (8.111)$$

and

$$Q_{N-1}^2 \leq \frac{1}{R_L G_{in,min}} \frac{1+Q_2^2}{1+Q_1^2} \ldots [1+Q_{N-2}^2]-1 \qquad (8.112)$$

The constraints on the last transformation Q can be derived from Figure 8.45. With the resistance (if the last element is a series element) or conductance (if the last element is a parallel element) in range, the reactance or susceptance must be such that the resulting impedance of admittance falls on the circumference of the gain circle.

When the last element is a series element, the reactance is constrained to

$$X_{in} = X_0 \pm R_{z0} \sin\left[\cos^{-1}\frac{R_{in}-R_0}{R_{z0}}\right] \qquad (8.113)$$

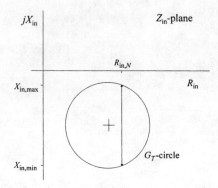

Figure 8.45   The constraints on the last transformation Q of a network if the gain is to be higher than a specified minimum.

The equivalent expression when the last element is a parallel element is

$$B_{in} = B_0 \pm R_{Y0} \sin\left[\cos^{-1} \frac{G_{in} - G_0}{R_{Y0}} \right] \qquad (8.114)$$

Because R_{in} or G_{in} is known, it is a simple matter to calculate the Q corresponding to this reactance or susceptance value.

From the viewpoint of matching at a single-frequency, there are no constraints on the first $N-2$ transformation Q. If the response is also of interest over a narrow passband, the Q-factors should, as a first-order approximation, be constrained to be smaller than twice the Q required for the circuit. Note that the change in Q from one transformation to the next is given by

$$\Delta Q = [Q_n + Q_{n-1}] \qquad (8.115)$$

The reason for the positive sign is that the sign of a Q of transformation changes under a series-to-parallel or parallel-to-series transformation. As an example of this, if the Q of a series combination is +3 (inductive), the Q of the parallel equivalent will be −3 (again inductive).

In summary, the following procedure can be followed to design an N-element matching network to match or mismatch a complex load by a specified amount to a complex source at a particular frequency and to have a specified (approximate) quality factor.

Design Procedure

Specifications: Load impedance Z_L, source impedance Z_s, transducer power gain G_T (at frequency f), and quality factor Q.

1. If the first element of the network is to be a shunt element, change the source and load impedances to $1/Z_s$ and $1/Z_L$, respectively, and assume that the first element is now a series element.

2. Choose any values for the first $N-2$ transformation Q-factors within the constraint that all the transformation Qs must be smaller than $2Q$, that is,

$$|Q_i| \le 2Q \tag{8.116}$$

3. If the last element is to be a parallel element, calculate the parameters of the constant gain circle on the admittance plane by using (8.97) and (8.98).
   If the last element is a series element, use (8.99) and (8.100) to calculate the parameters of the constant gain circle on the impedance plane.

4. Calculate the minimum and maximum allowable values for the input conductance or resistance by using the following equations, as applicable:

$$G_{in,max} = G_0 + R_{Y0} \tag{8.117}$$

$$G_{in,min} = G_0 - R_{Y0} \tag{8.118}$$

$$R_{in,max} = R_0 + R_{Z0} \tag{8.119}$$

$$R_{in,min} = R_0 - R_{Z0} \tag{8.120}$$

5. Determine the constraints on the next to last transformation Q by using (8.109) and (8.110) when the last element is a series element. Otherwise use (8.111) and (8.112).
   Choose a value for Q_{N-1} within these constraints and that imposed by (8.116).

6. When the last element is a series element, calculate the two possible reactance values corresponding to the last transformation Q ($X_{in,min}$ and $X_{in,max}$) by using (8.113). Otherwise, use (8.114) to calculate the allowable susceptance values ($B_{in,min}$ and $B_{in,max}$).
   Calculate $R_{in,N}$ or $G_{in,N} = 1/R_{in,N}$ by using (8.104) or (8.105), as applicable. Calculate the two possible values for the last transformation Q by using the following equations:

$$Q_{N,max} = \frac{X_{in,max}}{R_{in,N}} \tag{8.121}$$

$$Q_{N,min} = \frac{X_{in,min}}{R_{in,N}} \tag{8.122}$$

or

$$Q_{N,\max} = \frac{B_{\text{in},\max}}{G_{\text{in},N}} \qquad\qquad\qquad (8.123)$$

$$Q_{N,\min} = \frac{B_{\text{in},\min}}{G_{\text{in},N}} \qquad\qquad\qquad (8.124)$$

Choose either of the two possible Q values within the constraint imposed by (8.116) on each transformation Q.

7. Calculate the element values corresponding to the set of Q values. The reactance or susceptance of each component at the frequency where the Q values are calculated is given by an expression of the form

$$X_n = (Q_n + Q_{n-1})R_{rn} \qquad\qquad\qquad (8.125)$$

or

$$B_n = (Q_n + Q_{n-1})G_{rn} \qquad\qquad\qquad (8.126)$$

depending on whether it is a series or parallel element.

In these equations R_{rn} and G_{rn} are the effective series resistance and effective parallel conductance to the right of the component whose value is to be determined, respectively (refer to Figure 8.43, if necessary).

8. If the first element of the final network should be a shunt element, consider all inductors to be capacitors (i.e., 5 pH is 5 pF) and all capacitors to be inductors, and assign these values in sequence to the actual network.

As an illustration of this step, if element values of 3 pH (series element), 9 nF (shunt element), and 7 pF (series element) were obtained by following the procedure outlined above, the element values in the final circuit are 3 pF (shunt element), 9 nH (series element), and 7 pH (shunt element), respectively.

EXAMPLE 8.10   A transformation-Q example.

As an example of the application of the procedure outlined above, consider the design of a five-element matching network with the first element a series element, no resonating sections, and the following specifications:

$$Z_s = 20 - j20\Omega\ ;\ \ Z_L = 50 + j50\Omega$$

$$G_T = 0.89$$

$f_Q = 100\text{MHz}$

$Q = 5$

1. $Q_1 = 5 \quad (\Delta Q = 4)$

   $Q_2 = 5 \quad (\Delta Q = 10)$

   $Q_3 = -3 \quad (\Delta Q = 2)$

2. With no resonating sections and specifications as above, the last element of the network will be a series element, and therefore the constant gain circle on the input impedance plane is of interest. Application of (8.117) and (8.118) yields that the parameters of this circle are

$$R_0 + jX_0 = [2 / 0.89 - 1]20 + j20 = 24.94 + j20.00\Omega$$

and

$$R_{Z0} = 2\sqrt{1 / 0.89^2 - 1 / 0.89}\, 20 = 14.91$$

3. $R_{\text{in,min}} = 24.94 - 14.91 = 10.03\Omega$

   $R_{\text{in,max}} = 24.94 + 14.91 = 39.85\Omega$

4. Application of (8.109) and (8.110) yields that

$$|Q_4| \geq \left[\frac{50}{39.85} \frac{1+5^2}{1+5^2}(1+3^2) - 1 \right]^{1/2} = 3.398$$

$$|Q_4| \leq \left[\frac{50}{10.03} \frac{1+5^2}{1+5^2}(1+3^2) - 1 \right]^{1/2} = 6.989$$

$$Q_{n-1} = 5 \quad (\Delta Q = 2)$$

5. The two allowable values for the input reactance are found by using (8.113):

$$X_{\text{in}} = 20 \pm 14.91 \sin\left[\cos^{-1} \frac{19.23 - 24.94}{14.91} \right]$$

$$= 6.23\Omega;\ 33.77\Omega$$

In the equation above, R_{in} was found by applying (8.104):

$$R_{in} = 19.23\Omega \quad (10.03 \le 19.23 \le 39.85)$$

The last transformation Q is simply

$$Q_n = 6.23 / 19.23 = 0.32 \quad (\Delta Q = 5.32)$$

6. $X_1 = Q_1 R_1 - X_L = 5(50) - 50 = 200\Omega \quad (318.3\,\text{nH})$

$$Y_2 = [Q_2 + Q_1]G_{L2} = \frac{10(1)}{50(1+5^2)} = 7.69\,\text{mS} \quad (12.24\,\text{pF})$$

$$X_3 = [Q_3 + Q_2]R_{L3} = 2(50)\frac{1+5^2}{1+5^2} = 100\Omega \quad (1592\,\text{nH})$$

$$Y_4 = [Q_4 + Q_3]G_{L4} = 2\frac{1}{50}\frac{1+5^2}{1+5^2}\frac{1}{1+3^2} = 4.0\,\text{mS} \quad (6.37\,\text{pF})$$

$$X_5 = [Q_5 + Q_4]R_{L5} = 5.32(50)\frac{1+5^2}{1+5^2}\frac{1+3^2}{1+5^2} = 102.3\Omega \quad (162.8\,\text{nH})$$

The designed network is shown in Figure 8.46. The gain at 100 MHz is equal to 0.889 and the Q of the circuit is equal to 6.0.

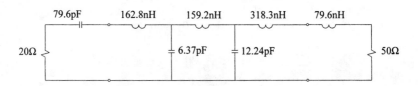

79.6pF   162.8nH   159.2nH   318.3nH   79.6nH

20Ω   6.37pF   12.24pF   50Ω

Figure 8.46   The network synthesized in Example 8.10.

8.4.3.3 Distributed Equivalents for Shunt Capacitors and Inductors

If the required inductance is low enough, a shunt inductor can be replaced to good approximation by a shorted, high-characteristic-impedance, transmission line, and vice versa. Similarly, a shunt capacitor can be replaced with an open-ended stub having low characteristic impedance if the required capacitance is small enough. Because the reactance or susceptance of a stub is dependent on the tangent of its electrical length, the accuracy with which these replacements can be made is dependent on the linearity of the tangent function. To give an indication of the frequency range over which this function can be considered linear, the value of $(\tan\theta - \theta)/\theta$ is summarized for several values of θ (radians) in Table 8.11. If a 10% deviation is acceptable, the maximum electrical length for an equivalent line is 30°.

In wideband cases, good results can be expected if shunt inductors are replaced exactly at the lowest frequency of interest, and shunt capacitors at the highest frequency of interest. This follows because the shunt susceptance is the highest at these frequencies and the influence of the component on the performance is then more pronounced.

The equations applying to replacing the lumped component exactly at a frequency f_x are

$$Z_{0L} \tan(\beta\ell) = X_{XL} \qquad \text{(inductive)} \tag{8.127}$$

and

$$Z_{0C} / \tan(\beta\ell) = X_{XC} \qquad \text{(capactive)} \tag{8.128}$$

where X_{XL} and X_{XC} are the reactances to be replaced at frequency f_x, and Z_{0L} (short-circuited stub) and Z_{0C} (open-ended stub) are the characteristic impedances of the stubs.

Table 8.11   The Value of $(\tan\theta - \theta)/\theta$ (in radians) as a Function of the Angle θ (in degrees)

θ (°)	$(\tan\theta - \theta)/\theta$ (%)	θ (°)	$(\tan\theta - \theta)/\theta$ (%)
5.0	0.3	35.0	14.6
7.5	0.6	37.5	17.2
10.0	1.0	40.0	20.2
12.5	1.6	42.5	23.5
15.0	2.3	45.0	27.3
17.5	3.2	47.5	31.6
20.0	4.3	50.0	36.0
22.5	5.5	52.5	42.2
25.0	6.9	55.0	48.8
27.5	8.5	57.5	56.4
30.0	10.3	60.0	65.4

8.4.3.4 A Distributed Equivalent for a Series Inductor

Series inductors in lumped designs are often replaced with high-characteristic-impedance transmission lines. When the series inductor forms part of a lowpass PI-section, significantly better results can be obtained by replacing the inductance and some of the capacitance with a series line. Similarly, shunt capacitors forming part of a lowpass T-section can also be replaced with series lines. These two possibilities are illustrated in Figure 8.47.

An exact transmission line equivalent for any symmetric lowpass T- or PI-section can be obtained at any particular frequency by equating the transmission matrix of the section to be replaced to that of a transmission line.

The transmission matrix of the T-section shown in Figure 8.47(a) is

$$\begin{bmatrix} 1-\omega^2 LC & j\omega L(2-\omega^2 LC)/(1-\omega^2 LC) \\ j\omega C & 1-\omega^2 LC \end{bmatrix} \tag{8.129}$$

By equating this to

$$\begin{bmatrix} \cos(\beta l) & jZ_0 \sin(\beta l) \\ jY_0 \sin(\beta l) & \cos(\beta l) \end{bmatrix} \tag{8.130}$$

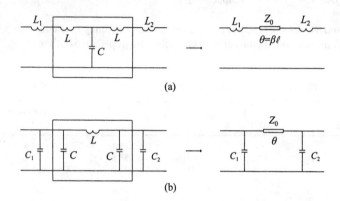

(a)

(b)

Figure 8.47   The partial replacement of (a) a lowpass T-section and (b) a lowpass PI-section with a series line.

it follows that a transmission line with the following parameters will be exactly equivalent to the T-section at the frequency of interest (ω):

$$L' = \frac{L}{1-\omega^2 LC}[2-\omega^2 LC] \tag{8.131}$$

$$C' = \frac{C}{1-\omega^2 LC} \tag{8.132}$$

$$Z_0 = \sqrt{\frac{L'}{C'}} \tag{8.133}$$

$$\beta l = \tan^{-1}(\omega\sqrt{L'C'}) \tag{8.134}$$

Excellent results can be expected when a T-section is replaced with a transmission line and the difference between the characteristic impedances and line lengths required for exact equivalents at the low and the high ends of the passband is negligible. Alternatively, the capacitance and inductance associated with a chosen line section at the lowest and at the highest frequency in the passband can be compared. The equations required for this purpose are

$$\omega L = Z_0 \frac{\sin(\beta l)}{1 + \cos(\beta l)} \tag{8.135}$$

$$\omega C = Y_0 \sin(\beta l) \tag{8.136}$$

where Y_0 is the inverse of Z_0.

The equations associated with the PI-section equivalent of Figure 8.47(b) are

$$L' = \frac{L}{1 - \omega^2 LC} \tag{8.137}$$

$$C' = \frac{C}{1 - \omega^2 LC}[2 - \omega^2 LC] \tag{8.138}$$

and

$$Z_0 = \sqrt{\frac{L'}{C'}} \tag{8.139}$$

$$\beta l = \tan^{-1}(\omega\sqrt{L'C'}) \tag{8.140}$$

The inverse relationships are

$$\omega L = Z_0 \sin(\beta l) \tag{8.141}$$

and

$$\omega C = Y_0 \frac{\sin(\beta l)}{1 + \cos(\beta l)} \tag{8.142}$$

It follows from the equations given above that the length of the equivalent line for a T- or PI-section is only a function of the normalized reactance $\omega L/Z_0$ and the normalized susceptance $\omega C/Y_0$, respectively. The following equations can be used to calculate the required normalized susceptance $\omega C/Y_0$ and the line length corresponding to a specified normalized value for the reactance of the inductor in a PI-section:

$$\frac{\omega C}{Y_0} = \frac{Z_0}{\omega L}\left[1 - \sqrt{1 - (\frac{\omega L}{Z_0})^2}\right]$$ (8.143)

and

$$\beta l = \tan^{-1}\frac{\omega L / Z_0}{\sqrt{1 - (\omega L / Z_0)^2}}$$ (8.144)

With ωC, ωL, and Y_0, Z_0 interchanged, the same set of equations applies to a T-section.

EXAMPLE 8.9   Replacing a lumped inductor with a line.

As an example of the application of the PI-section transformation, a transmission line equivalent for a series 2 nH inductor will be determined over the passband 2–8 GHz.

With $Z_0 = 150\Omega$, application of (8.143) and (8.144) yields that the required capacitance and the line length corresponding to an exact equivalent at 8 GHz are

$C = 0.051$ pF

and

$\beta l = 42.08°$

The PI-section equivalent for this line at 2 GHz ($\beta l = 42.08/4 = 10.52°$) can be found by using (8.106) and (8.107). The results are

$L = 2.18$ nH

and

$C = 0.049$ pF

which are close to the original values (within +9.0% and −7.3%, respectively).

Better results can sometimes (narrowband cases) be obtained by minimizing the error across the passband. This can be done by lowering the frequency at which the transformation is exact iteratively. By selecting this frequency as 5.8 GHz, the line length becomes 29.07° (at 5.8 GHz) and the difference in inductance becomes 3.9% at 2 GHz and −3.9% at 8 GHz. The difference in the parasitic capacitance reduces to −1.9% and 2.0%, respectively.

8.4.3.5 Optimization of the Transformation Qs of a Network

The transformation Q-factors corresponding to an initial solution for a matching problem can be optimized by using a linear least-square optimization routine. Although good results can be obtained by doing this, better results are obtainable through a slightly different approach.

The first improvement is to use the maximum relative deviation (MRD)

$$\text{MRD} = \text{MAX} \left| \frac{G_T(\omega)}{G_{T\_opt}(\omega)} - 1 \right| \tag{8.145}$$

as the error criterion instead of the mean-square error.

The main advantage of the MRD is that the maximum deviation from the optimum rather than the average deviation will be minimized. Because of this, the solution with the lowest insertion loss will be obtained when the ideal gain is set to unity.

The second improvement is to optimize the error by using the steepest-decent method. The results obtained in doing this were superior to those corresponding to the least-square method.

The gradient vector required for optimizing the Q values can be determined by calculating the change in MRD corresponding to a small increment in each Q.

The new set of Q values (Q_N) can be obtained from the previous set (Q_{N-1}) and the current MRD by using the equation

$$\overline{Q}_N = \overline{Q}_{N-1} - \frac{\alpha}{\sqrt{(\partial \text{MRD} / \partial Q_1)^2 + \dots + (\partial \text{MRD} / \partial Q_N)^2}} \begin{bmatrix} \partial \text{MRD} / \partial Q_1 \\ \cdot \\ \cdot \\ \cdot \\ \partial \text{MRD} / \partial Q_N \end{bmatrix} \tag{8.146}$$

where the optimum value of α can be determined iteratively by using the following method [17].

Start with a small value of α (α_1), and increase it during subsequent iterations (α_i) by using the expression

$$\alpha_i = \alpha_1 [1 + l + l^2 + \dots \ l^{i-1}] \quad i = 1,2,3, \dots \quad l = 1.5 \tag{8.147}$$

until the error value increases. This will result in the situation depicted in Figure 8.48. A quadratic curve can now be fitted to the last three coordinates, and the value of α (α_M), for which the error will be a minimum, can be estimated by using the expression

$$\alpha_M = \frac{1}{2} \frac{[\alpha_{n-1}^2 - \alpha_n^2]\text{MRD}_{n-2} + [\alpha_n^2 - \alpha_{n-2}^2]\text{MRD}_{n-1} + [\alpha_{n-2}^2 - \alpha_{n-1}^2]\text{MRD}_n}{[\alpha_{n-1} - \alpha_n]\text{MRD}_{n-2} + [\alpha_n - \alpha_{n-2}]\text{MRD}_{n-1} + [\alpha_{n-2} - \alpha_{n-1}]\text{MRD}_n}$$

$$\tag{8.148}$$

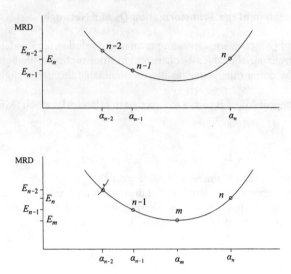

Figure 8.48   Estimation of the optimum scale factor in optimizing the MRD.

The actual value of the MRD at a_M can now be calculated. Depending on which of the four errors is now the largest, one of the four coordinates can be eliminated and the procedure can be repeated on the remaining three points.

The optimum value of a can be determined by continuing with this procedure until the improvement in the error value is negligible.

Excellent results were obtained by optimizing the Q values of transformation as outlined above.

8.4.3.6   An Algorithm for the Design of Impedance-Matching Networks by Using the Transformation Q-Factors of the Network

A procedure for designing a network to match a complex load to a complex source with a specified gain at a specified frequency was outlined in Section 8.4.3.2. By taking the transducer power gain to be the minimum expected gain at the frequency where the Q values are evaluated (usually the highest frequency in the passband or the frequency at which the gain required is a maximum), this narrowband technique forms the basis of an excellent approach to solving wideband impedance-matching problems.

It was shown that the first $(N-2)$ Q values in the single-frequency design can take arbitrary values and the constraints imposed on the last two Q values were derived. Since the range of possible transformation-Q values is limited in a wideband design, it is feasible to do a systematic search on these Q values in order to find solutions that yield good results over the whole passband. In this way, the dependence on a good initial solution is eliminated.

When the search is completed, a number of the best results obtained can be optimized as described in Section 8.4.3.3. If the search was done thoroughly enough, the optimum solution to any matching problem will be obtained. A further advantage is that the local minima corresponding to other initial solutions will also be obtained, and, consequently, a large choice between networks with different element values and topologies exist.

An idea of the required range of Q values can be obtained from the desired Q of the network when applicable, the maximum Q of the load and source impedances, and the analytically derived constraints on simple reactive loads as summarized in Table 8.5.

As a rule, a minimum value of -4.2 and a maximum value of 4.2 yield excellent results. When some of the Q values of solutions obtained exceed these values and the optimum solution is required, the bounds must be extended. This will seldom be necessary when a wideband network is designed. Increment values in the range from 0.4–0.6 are used.

The following algorithm can be used when this approach is followed.

Algorithm

1. Decide on the number of elements and the frequency at which the Q values are to be evaluated (f_Q).

   Estimate the range of possible Q values of transformation and specify the incremental value to be used.

   Estimate the minimum gain expected at f_Q.

   Specify the number of transformation-Q sets to be stored during the search (M).

2. Generate an allowable set of Q values by using the theory outlined in Section 8.4.3.2.

3. Synthesize the equivalent network and calculate the gain error (MRD). Compare the results with the previous results obtained and store the solution if it is better than the M best solutions previously stored.

4. Optimize the best results obtained in the search as described in Section 8.4.3.3.

EXAMPLE 8.11   A double-matching problem solved with the transformation-Q technique [15].

As an example of the results obtainable with the transformation-Q technique, consider the double-matching problem of Example 8.8.

With the gain set equal to the specified value of 0.818 at 200 MHz during the systematic search, and using minimum, increment, and maximum values of -4.4, 0.4, and 4.4, respectively, for the transformation-Q values, the maximum deviation from the specified response was found to be 0.05 dB (MRD = 1.23%) for the best four-element solution synthesized. This network is shown in Figure 8.49(a).

The Q values corresponding to this solution are 2.399, -1.797, 1.039, and -0.148, respectively.

The second best solution obtained is the network shown in Figure 8.49(b). The maximum deviation from the specified gain response is 0.06 dB (MRD = 1.43%), and the Q values are -0.777, 3.422, 2.360, and -0.130, respectively.

The maximum deviations for the other solutions shown in Figure 8.49 are 0.09 dB (MRD = 2.08%), 0.11 dB (MRD = 2.58%), 0.13 dB (MRD = 3.02%), and 0.15 dB (MRD = 3.41%), respectively.

The best three-element solution obtained under the constraint that the topology must be of highpass form is shown in Figure 8.50. The maximum deviation from the specified gain is 0.15 dB (MRD = 3.4%) and the Q values are -0.185, -0.570, and -0.931, respectively. This solution is basically the same as that obtained with the reflection parameter technique.

Having several solutions to choose from is an advantage both from the viewpoint of topology and sensitivity.

It is important to note that the solution with the best performance with the design values may not have the best worst-case performance too. It also does not necessarily follow that fewer elements will be better from a sensitivity viewpoint.

In this case, the MRD of the best four-element solution found is increased from 1.23% to 4.12% if the tolerance in all the lumped components is assumed to be 1%. A 1% change in the component values, therefore, leads to a 2.9% increase in the MRD in this case.

The MRD of the three-element solution shown increases from 3.4% to 5.5% with 1% tolerances in the component values.

The solutions shown so far are purely lumped. In practice, solder pads are also required for the lumped components. The transformation-Q approach can be extended to allow for this requirement too. A few of the best mixed lumped/distributed solutions obtained [15] are shown in Figure 8.51 with the pads used. The pads do not have a strong influence in this example, but will become a factor as the frequency is increased. The influence of the pads is also more severe when the dielectric constant of the substrate is high.

The MRD of the best solution is 1.09%. This value is increased to 4.32% when 1% tolerances are assumed for the lumped components.

The MRD values for the other three solutions shown in Figure 8.51 are 1.29% (4.29%), 3.0% (5.67%), and 3.18% (6.78%), respectively. Note that the worst-case performance of the second solution is fractionally better than that of the first solution (4.29% versus 4.32%).

The electrical parameters of the pads used for the shunt inductors are 57Ω and 0.38° (at 0.2 GHz), while those used for the shunt capacitor pads are 36.4Ω and 0.38°. The 57-Ω pads (0.23° or 0.36° long) were also used for the series inductors, while 71.2Ω pads (0.23° or 0.36° long) were used for the series capacitors.

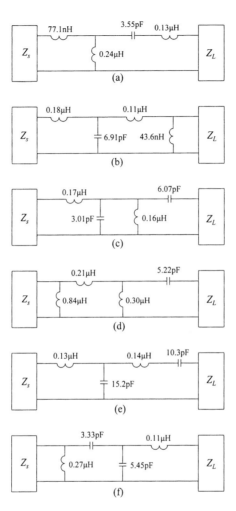

Figure 8.49 (a–f) Some of the solutions obtained with the transformation-Q technique for the matching problem of Example 8.8.

Figure 8.50 The best three-element solution (highpass topology) obtained with the transformation-Q technique for the matching problem of Example 8.8.

A low dielectric constant substrate was used ($\varepsilon_r = 2.99$; $h = 0.381$ mm). The width of these pads on the substrate used are 0.75 mm, 1.5 mm, 0.75 mm, and 0.5 mm, respectively.

Note that the length used for the series pads should be long enough to ensure sufficient separation between the shunt components. This is essential to prevent overlap in the artwork and coupling between the shunt components.

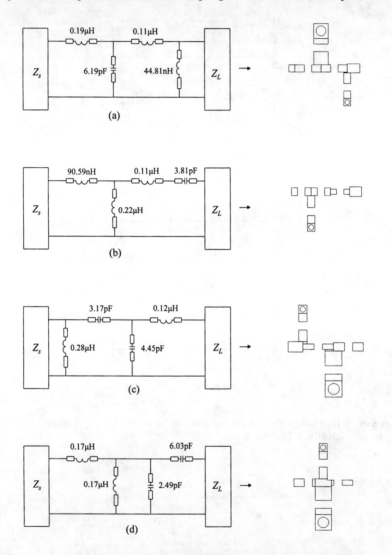

Figure 8.51   (a–d) The four best mixed lumped/distributed solutions obtained with the transformation-Q technique for the matching problem of Example 8.8.

EXAMPLE 8.12   Matching a 25-Ω source to a 100-Ω load (2–6 GHz).

A load resistance of 100Ω will be transformed to 25Ω over the passband 2–6 GHz in this example. The solutions were synthesized for a 10-mil microstrip substrate with $\varepsilon_r = 9.8$ by using [15].

The best solutions to a purely resistive matching problem are usually obtained with a commensurate distributed network. When the bandwidth required is large, the line length should be 90° long at the center frequency (arithmetic mean).

The commensurate solution shown in Figure 8.52(a) was synthesized by setting the line length equal to 90° at 4 GHz. This solution was obtained by setting the search range for the transformation Qs to the interval [-1.2, 1.2], with a step size of 0.1. Note that a 25-Ω pad was used to complete the input junction (the first element of the network synthesized was a shunt element).

The input VSWR of this solution is better than 1.06 over the passband ($|s_{11}| < -30.58$ dB; MRD = 0.08%, MRD$_{wc}$ = 0.35%). The size of this solution is 11.63 mm by 6.41 mm.

The noncommensurate solution shown in Figure 8.52(b) was obtained by setting the characteristic impedance of the main-line sections and the short-circuited stubs to 73Ω. The characteristic impedance used for the open-ended stubs was 33.5Ω. The search range for the transformation Qs was from -1.5 to 1.5, with steps of 0.25. The 100-Ω line on the output side was used to complete the junction associated with the open-ended stub.

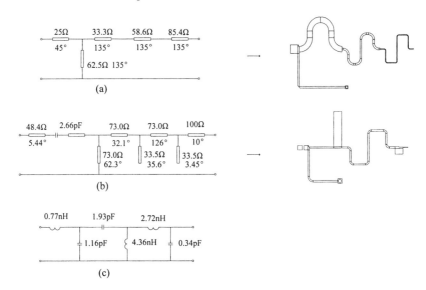

Figure 8.52   The best (a) commensurate (four elements), (b) noncommensurate (six elements), and (c) lumped (six elements) solutions obtained for matching 25Ω to 100Ω (2–6 GHz).

The input VSWR of this solution is better than 1.25 ($|s_{11}| < -19.16$ dB; MRD = 1.2%, MRD_{wc} = 1.78%). The outline size of this solution is 6.88 mm by 4.133 mm (a gap of 0.1 mm was used for the series capacitor).

The best six-element lumped-element solution obtained is shown in Figure 8.52(c). The input VSWR of this solution is better than 1.19 over the passband ($|s_{11}| < -21.36$ dB, MRD = 0.71%, MRD_{wc} = 1.24%).

8.5 THE DESIGN OF RLC IMPEDANCE-MATCHING NETWORKS

RLC impedance-matching networks are often used to compensate for the decrease in the gain of the transistors used with increasing frequencies. This can usually be done without reactive mismatching, which is an advantage of these networks over lossless networks.

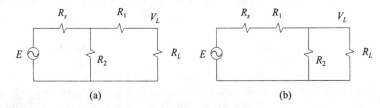

Figure 8.53   (a, b) Impedance matching with resistors.

RLC networks are usually designed by using a computer optimization program on a circuit with a suitable topology, after initial values have been assigned to its components.

The resistors in an RLC impedance-matching network have two functions: They provide the required attenuation at the lowest frequency in the passband, and they match the load impedance to the source impedance at this frequency. A minimum of one series and one parallel resistor are required in order to do this.

When only one series and one parallel resistor are used, initial values can be assigned to them by using the following set of equations:

$$A^2 = (E/V_L)^2 = \frac{4R_s}{G_T R_L} \tag{8.149}$$

$$G_1^2\left[1 + \frac{G_{in}}{G_s} - A\right] + G_1\left[2G_L\left(1 + \frac{G_{in}}{G_s} - A/2\right)\right] + G_L^2\left[1 + \frac{G_{in}}{G_s}\right] = 0 \tag{8.150}$$

$$G_2 = G_{in} - \frac{G_1 G_L}{G_1 - G_L} \tag{8.151}$$

where A is the inverse of the voltage gain (V_L/E), G_T is the required transducer power gain at the lowest frequency in the passband, $G_1 = 1/R_1$, $G_2 = 1/R_2$, and G_{in} is the required input admittance of the matching network at the lowest frequency (if a perfect match is required, $G_{in} = G_s$).

Equations (8.149)–(8.151) apply to Figure 8.53(a). The equations relevant to

Figure 8.53(b) can be obtained by replacing G_s with G_L and G_L with G_s in these equations.

In order to minimize the insertion loss at the higher frequencies in the passband, the resistors in an RLC network should be used in parallel with capacitors and in series with inductors, depending on whether they are used in a series or a parallel branch, respectively.

Apart from reducing the insertion loss, the capacitors and inductors used in the network also serve to match the load to the source at the higher frequencies.

The network shown in Figure 8.54 is a typical example of an RLC network. Note that the elements that are not combined with resistors are used as lowpass elements.

Initial values can be assigned to the lossless elements of the network chosen by considering the different elements to be part of independent L-, T-, and PI-sections. As an example of this, C_2 and L_2 in the network shown in Figure 8.54 form a lowpass L-section that should be designed to match the load to the source at the highest frequency in the passband. L_1 and C_1 should be designed to ensure that the insertion loss at the highest frequency will be as low as possible.

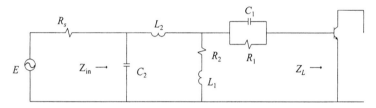

Figure 8.54   An example of an RLC impedance-matching network.

An alternative way of assigning initial values to the lossless components of an RLC network is to follow the iterative approaches outlined earlier for designing a lossless bandpass network that will match the source to the load at the intermediate and higher frequencies in the passband.

With initial values assigned to the lossless components and the resistors, an optimization program can be used to optimize the network.

EXAMPLE 8.13   Example of an RLC matching network.

The use of (8.149)–(8.151) will be illustrated by applying them to the following problem:

$R_L = 7.5\Omega$

$R_{in} = 6.25\Omega$

$R_s = 6.25\Omega$

$G_T = 0.19$

$$A = \frac{4R_s}{R_L G_T} = \frac{4(6.25)}{0.19(7.5)} = 4.19$$

$$G_1^2[2.00 - 4.19] + G_1\left[2\frac{2 - 4.19/2}{7.5}\right] + 1\frac{2}{7.5^2} = 0$$

$$G_1 = 0.1218; -0.1334\text{S}$$

$$G_2 = \frac{1}{6.25} - \frac{0.1218}{7.5[0.1218 + 1/75]} = 0.096\text{S}$$

The initial values of the resistors are therefore

$R_1 = 8.2\Omega$ and $R_2 = 10.4\Omega$

EXAMPLE 8.14   Matching networks for an HF power amplifier.

In this example a 5–20-MHz power amplifier will be designed with the Motorola MRF406 [20-W peak envelope power (PEP)] by using [1]. The operating power gain will first be leveled by using an series-shunt RLC network on the input side of the transistor. Mixed lumped/distributed matching networks will then be designed to maximize the output power and to provide a good input match over the passband. It will be assumed that the terminations were transformed to 12.5Ω with transmission-line transformers.

The load impedance required to maximize the output power and the associated input impedance and operating power gain are provided in the data sheet for the transistor. The estimated values are shown in Table 8.12.

It is convenient to convert the impedance and gain specifications in Table 8.12 to an equivalent set of unilateral S-parameters. The input reflection coefficient is chosen to correspond to the input impedance of the transistor when

Table 8.12   The Optimum Load Impedance of the MRF406, with the Measured Operating Power Gain and Input Impedance

Frequency (MHz)	Input impedance (Ω)	Load impedance (Ω)	Operating power gain (dB)
2.0	$7.5 - j2.6$	$8.314 - j4.263$	20.93
5.0	$5.2 - j2.4$	$6.212 - j4.914$	20.14
10.0	$3.1 - j1.9$	$4.971 - j4.476$	18.44
15.0	$2.3 - j1.75$	$4.471 - j4.028$	16.99
20.0	$1.7 - j1.7$	$4.272 - j3.536$	16.01
30.0	$1.0 - j1.0$	$3.484 - j2.445$	14.30

Table 8.13 The Equivalent Set of S-Parameters of the MRF406

Frequency (MHz)	s_{11} (−; °)		s_{21} (−; °)		s_{12} (−; °)		s_{22} (−; °)	
2.0	0.7398	186.1	5.224	327.6	0.0001	0.0	0.7167	190.0
3.0	0.7661	185.9	4.597	316.7	0.0001	0.0	0.7448	191.0
5.0	0.8120	185.6	3.707	299.1	0.0001	0.0	0.7809	191.4
7.0	0.8473	185.1	2.907	284.5	0.0001	0.0	0.8045	190.9
10.0	0.8834	184.4	2.238	264.7	0.0001	0.0	0.8205	190.3
15.0	0.9121	184.0	1.588	234.1	0.0001	0.0	0.8368	189.3
20.0	0.9343	183.9	1.210	206.0	0.0001	0.0	0.8433	188.1
30.0	0.9608	182.3	0.709	148.7	0.0001	0.0	0.8700	185.6

the optimum load is in place, while the output impedance is taken to be the conjugate of the optimum load impedance. The forward transmission parameter (s_{21}) is set to the value required to ensure that the maximum available gain (MAG) of this equivalent transistor is the same as the operating power gain of the transistor with the optimum load in place. The equivalent S-parameters for the MRF406 are shown in Table 8.13 in polar format (magnitude and angle).

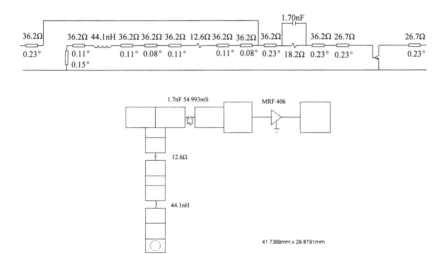

Figure 8.55 The RLC network used to level the operating power gain of the MRF406. (The electrical line lengths are specified at 20 MHz.)

With the equivalent S-parameters in place, the original targets can be realized by leveling the MAG of the equivalent transistor (lossy sections on the input side; no feedback allowed) and designing matching networks to minimize the input and the output VSWRs.

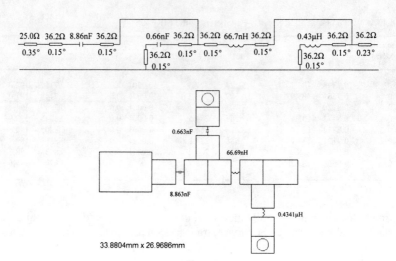

Figure 8.56   The mixed lumped/distributed input matching network synthesized for the MRF406. (The electrical line lengths are specified at 20 MHz.)

The RLC network designed to level the MAG of the equivalent transistor is shown in Figure 8.55. With this network in place, the MAG varies between 10.94 and 11.14 dB over the 5–20-MHz passband, and the input VSWR (relative to the 12.5-Ω source termination) varies between 1.61 and 3.06.

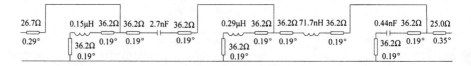

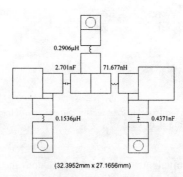

Figure 8.57   The mixed lumped/distributed load network synthesized for the MRF406. (The electrical line lengths are specified at 20 MHz.)

Leveling the MAG of the equivalent transistor corresponds to leveling the operating power gain of the actual transistor terminated in the optimum load impedance.

The mixed lumped/distributed input matching network is shown in Figure 8.56 (the electrical line lengths are specified at 20 MHz). The input VSWR varies between 1.24 and 1.46 over the passband ($|s_{11}| < -14.57$ dB).

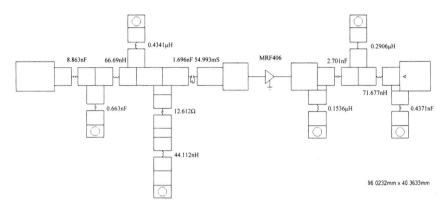

Figure 8.58   The artwork of the amplifier synthesized.

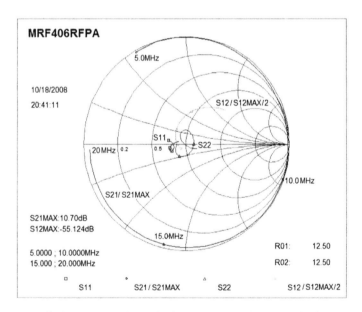

Figure 8.59   The final response associated with the equivalent transistor. (Note that the highest frequency on each trace is not marked; the Smith chart should be viewed as a polar plot only when the s_{21} trace is interpreted.)

Table 8.14   The *S*-Parameters of the Final Circuit (Equivalent Transistor)

Frequency	s_{11}		s_{12}		s_{21}		s_{22}	
(MHz)	(dB)	(°)	(dB)	(°)	(dB)	(°)	(dB)	(°)
2.0	−3.72	217.0	−110.56	10.0	−16.64	311.6	−0.01	133.9
5.0	−14.57	168.0	−76.96	237.0	10.91	120.0	−17.16	336.3
6.0	−16.67	174.8	−74.66	207.8	10.97	77.2	−20.15	133.2
7.0	−17.48	185.6	−72.59	189.3	10.84	47.1	−17.14	100.3
8.0	−17.23	194.4	−70.65	175.2	10.85	22.5	−18.16	72.5
9.0	−16.51	199.3	−68.87	163.0	10.91	0.9	−20.40	41.8
10.0	−15.76	200.8	−67.25	152.0	11.00	341.3	−22.41	3.9
11.0	−15.18	200.2	−65.75	142.0	11.00	323.0	−22.28	320.7
12.0	−14.78	198.2	−64.37	132.5	10.98	305.7	−20.64	288.6
13.0	−14.58	195.2	−63.10	123.4	10.95	289.3	−19.04	266.8
14.0	−14.58	191.7	−61.92	114.6	10.93	273.4	−17.91	250.5
15.0	−14.79	187.5	−60.81	106.0	10.91	257.9	−17.24	237.0
16.0	−15.12	182.9	−59.74	97.6	10.88	243.0	−17.03	224.8
17.0	−15.68	177.4	−58.71	89.2	10.87	228.2	−17.17	213.2
18.0	−16.52	170.9	−57.72	80.7	10.87	213.5	−17.64	201.2
19.0	−17.75	162.4	−56.78	72.1	10.88	198.8	−18.45	188.1
20.0	−19.49	150.4	−55.86	63.2	10.91	183.9	−19.58	172.4
25.0	−14.59	0.2	−51.95	12.2	10.80	103.7	−17.14	44.1
30.0	−4.19	306.9	−50.45	306.6	7.90	8.2	−8.24	352.0

The load network synthesized (five-element network) to match the output of the equivalent transistor (maximize the output power of the actual transistor) is shown in Figure 8.57 (the electrical line lengths are specified at 25 MHz). The output VSWR in the equivalent circuit varies between 1.16 and 1.32 over the passband ($|s_{11}| < −17.03$ dB).

The artwork of the amplifier designed is shown in Figure 8.58. (The gap spacings must be adjusted to accommodate the lumped components to be used.)

The final response associated with the equivalent transistor is shown in Figure 8.59 and tabulated in Table 8.14.

QUESTIONS AND PROBLEMS

8.1   The input impedance of a GaAs FET is tabulated in Table 8.15. Find an equivalent circuit for the input impedance by using LSM.

8.2   A lowpass network is to be used to match the input impedance of the transistor in Problem 8.1 to a 50-Ω source. Determine the minimum value of the insertion loss.

8.3   Design a network for matching the input impedance of the transistor in Problem 8.1 to a 50-Ω source. Do this with a lowpass as well as a bandpass network.

8.4   Determine how many different networks can be synthesized to have an input admittance of

$$Y_{IN} = \frac{2.125s^2 + 0.248s + 1.000}{1.992s^3 + 0.379s^2 + 1.531}$$

8.5   Determine the gain-bandwidth limitations of the power transistor with input impedance as given in Table 8.16. Design a network for matching the transistor to a 12.5-Ω source.

Table 8.15   The Input Impedance of a GaAs FET at Different Frequencies

Frequency (GHz)	Input impedance (Ω)
2	85.7-j17.0
3	132.6-j35.2
4	97.6-j73.2
5	62.8-j65.8
6	43.9-j39.8

8.6   Design a fourth-order 0.5-dB ripple Chebyshev interstage matching network with a positive slope of 6 dB / octave by following the Darlington approach. $R_s = 211\Omega$; $R_L = 50\Omega$.

The required slope can be obtained by adding a s^2 term to the numerator of the flat Chebyshev transducer power-gain function. When this is done the constant in the gain function must be adjusted to ensure that the maximum gain will be equal to one. This can be done iteratively.

Table 8.16   The Input Impedance of a Power Transistor at Different Frequencies in the Passband

Frequency (MHz)	Input impedance (Ω)
100	16.0-j9.0
150	15.5-j3.0
200	14.0-j0.0
250	11.0-j4.0
300	9.0-j9.0
350	7.5+j11.5
400	6.0+j15.0

8.7   A 50-Ω source must be matched to the impedance shown in Figure 8.60(a) over the frequency range 115–150 MHz. Design an impedance-matching network for this purpose. The insertion loss in the passband must be as low as possible.

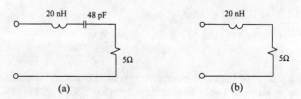

Figure 8.60   (a, b) The loads relevant to Problems 8.7 and 8.8.

8.8   Use Table 8.5 to determine the minimum insertion loss possible with the load shown in Figure 8.60(b), when a lossless three-element Chebyshev impedance-matching network is used. The cutoff frequency is to be 100 MHz.

8.9   Find the optimum values of the maximum gain and the passband ripple if the load and source shown in Figure 8.61 are to be matched with a four-element lowpass Chebyshev network. Assume $\omega_c = 1$ rad/s.

Figure 8.61   The load and source relevant to Problem 8.9.

8.10   Design a lowpass network to match the load shown in Figure 8.60(a) to a 15-Ω source over the frequency range 115–150 MHz.

8.11   Use Richards' transformation to design a commensurate distributed network to match the load shown in Figure 8.62 to a 25-Ω source over the passband 1,150–1,500 MHz. Remove any series inductive stubs in the design by using Kuroda's lowpass identities.

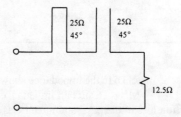

Figure 8.62   The load to be matched in Problem 8.11 (line lengths specified at the center frequency).

8.12 Design a lowpass network to match the reactive load and source shown in Figure 8.63 to each other over the frequency range 4–10 rad/s.

8.13 Design a lossless four-element network to mismatch a load of 50Ω to a 6.25-Ω source with $G_T = 0.79$ at 600 MHz.

Figure 8.63   The impedance to be matched in Problem 8.12.

8.14 Determine the poles and zeros of the RLC network shown in Figure 8.64. Plot the response for different values of L and C.

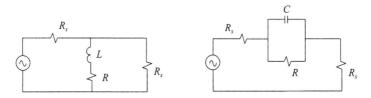

Figure 8.64   Simple RLC impedance matching networks.

8.15 The input impedance and operating power gain of an HF power transistor are given in Table 8.17 at several frequencies. Design an RLC network for matching the transistor to a 2.78-Ω source.

8.16 The matching network shown in Figure 8.65 was introduced in [18] for class-E amplifiers. C_e is used to augment C_i to increase the class-E output power obtainable. The two open-ended stubs filters the second and third harmonics from the output signal by presenting short circuits at these frequencies, while the series line transforms these short circuits to high impedances at Node A at the harmonic frequencies, and to the required Z_E (see Chapter 2) at the fundamental frequency. Given that Z_E at the fundamental frequency is known from the power requirements, derive equations for the characteristic impedances and line lengths of the three transmission lines [18]. Use the equivalent circuit shown in Figure 8.65(b) to do this.

8.17 The optimum load termination for a class-E amplifier is given as 8.5+*j*9.8Ω at 23.5 GHz [18]. Use (2.81) and (2.82) to calculate R_E and C_s in these equations, and adjust the required load at 23.5 GHz for the frequency range 22–25 GHz. Use the

Table 8.17 The Input Impedance and Operating Power Gain of the Transistor Relevant to Problem 8.15.

Frequency (MHz)	Input impedance (Ω)	Power gain (dB)
1.6	$4.8-j1.2$	27.8
2.5	$4.3-j1.6$	27.6
4.0	$3.8-j1.9$	27.4
5.0	$3.1-j1.9$	26.9
7.5	$2.4-j1.8$	26.1
10.0	$1.9-j1.4$	24.8
15.0	$1.6-j1.1$	23.1
20.0	$1.4-j0.7$	20.8
24.0	$1.3-j0.5$	19.5
28.0	$1.3-j0.4$	18.3

conjugates of these impedances as the source impedance in a real-frequency matching problem. Set up the load impedance by assuming it to consist of a 50-Ω load in parallel with the two open-ended stubs shown in Figure 8.65. Set the characteristic impedances initially to 75Ω (high). Design matching networks to provide class-E terminations close to those targeted at the fundamental frequencies (on the input side), and high impedances at (close to) the harmonic frequencies.

If a network ends in an open-ended stub on the output side, decrease the characteristic impedances of the open-ended stubs included in the load to absorb this stub.

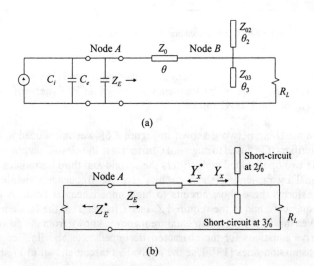

(a)

(b)

Figure 8.65 (a, b) The class-E matching network introduced in [18].

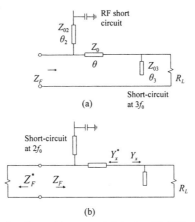

Figure 8.66   (a, b) A class-F matching network introduced in [18].

8.18   The matching network shown in Figure 8.66 was introduced in [18] for class-F amplifiers. The biasing is fed through Z_{02} and this line also provides low impedances at the even harmonics. Its length is initially set at $\lambda/4$ at the fundamental frequency. The open-ended stub at the load side short-circuits the third harmonic and its length is set initially at $\lambda/12$ at the fundamental frequency. The same initial length is used for the series lines in order to transform the short circuit at the third harmonic to the high third harmonic load impedance required for class-F operation. All of the lengths must be adjusted to allow for the transistor parasitics, because the open circuits and short circuits are required intrinsically across the current generator of the transistor. Assuming no transistor parasitics, show that Z_0 and Z_{02} are given by [18]

$$Z_0 = \frac{\sqrt{Z_F \left[R_L \left(1 + \tan^2 \theta \right) + Z_F \right]}}{\tan \theta} \tag{8.152}$$

$$Z_{03} = \frac{Z_0 \tan \theta_3 \left[Z_F^2 - Z_0^2 \tan^2 \theta \right]}{(Z_0^2 - Z_F^2) \tan \theta} \tag{8.153}$$

REFERENCES

[1]   Fano, R. M., "Theoretical Limitations on the Broadband Matching of Arbitrary Impedances," *Journal of the Franklin Inst.*, Vol. 249, January/February 1950, pp. 57–83, 139–154.

[2]   Youla, D. C., "A New Theory of Broad-Band Matching," *IEEE Trans. Circuits Syst.*, Vol. CT-11, March 1964, pp. 30–50.

[3]   Carlin, H. J., "A New Approach to Gain-Bandwidth Problems," *IEEE Trans. Circuits Syst.*, Vol. CAS-24, April 1977, pp. 170–175.

[4]   Lev, J., "Calculator Program Finds Fano Bandwidth," *Microwaves & RF*, September 1985.

[5]   Chen, W., and C. Satyanarayana, "General Theory of Broadband Matching," *IEE Proc. C* (GB), Vol. 129, No. 3, June 1982, pp. 96–102.

[6]   Carlin, H. J., and B. S. Yarman, "The Double Matching Problem: Analytic and Real Frequency Solutions," *IEEE Trans. Circuits Syst.*, Vol. CAS-30, No. 1, January 1983, pp. 15–27.

[7]   Yarman, B. S., and H. J. Carlin, "A Simplified 'Real-Frequency' Technique Applied to Broad-Band Multistage Microwave Amplifiers," *IEEE Trans. Microwave Theory Tech.*, Vol. MTT-30, No. 12, December 1982, pp. 2216–2222.

[8]   Carlin, H. J., and P. Amstutz, "On Optimum Broad-Band Matching," *IEEE Trans. Circuits Syst.*, Vol. CAS-28, No. 5, May 1981, pp. 401–405.

[9]   Bode H. W., *Network Analysis and Feedback Amplifier Design*, New York: Van Nostrand, 1945, pp. 205–207, 319.

[10]   Chen, W. K., *The Theory and Design of Broadband Matching Networks*, London: Pergamon Press, 1976.

[11]   Levy, R., "Explicit Formulas for Chebyshev Impedance-Matching Networks, Filters and Interstages," *Proc. IEEE*, Vol. 111, No. 6, June 1964.

[12]   Richards, P. I., "Resistor-Transmission-Line Circuits," *Proc. IRE*, February 1948, pp. 217–219.

[13]   Baher, H., *Synthesis of Electrical Networks*, New York: John Wiley & Sons, 1984.

[14]   Yarman, B. S., "Broad-Band Matching a Complex Generator to a Complex Load," Doctoral Dissertation, Cornell University, 1982, pp. 117–120.

[15]   *MultiMatch RF and Microwave Impedance-Matching, Amplifier and Oscillator Synthesis Software*, Stellenbosch, South Africa: Ampsa (PTY) Ltd; http://www.ampsa.com, 2008.

[16]   Carlin, H. J., and J. J. Komiak, "A New Method of Broadband Equalization Applied to Microwave Amplifiers," *IEEE Trans. Microwave Theory Tech.*, Vol. MTT-27, No. 2, February 1979, pp. 93–99.

[17]   Ha, T. T., *The Design of Microwave Solid State Amplifiers*, New York: John Wiley and Sons, 1981, pp. 178–179.

[18] Negra, R., F. M. Ghannouchi, and W. Bächtold, "Study and Design Optimization of Multiharmonic Transmission-Line Load Networks for Class-E and Class-F K-Band MMIC Power Amplifier," *IEEE Trans. Microwave Theory and Techniques*, Vol. 55, No. 6, June 2007.

SELECTED BIBLIOGRAPHY

Abrie, P. L. D., "Impedance Matching Networks and Bandwidth Limitations of Class B Power Amplifiers in the HF and VHF Ranges," Master's Thesis, University of Pretoria, November 1982.

Fletcher, R., and M. J. D. Powell, "A Rapidly Convergent Descent Method for Minimization," *Computer J.*, Vol. 6, 1963, pp. 163–168.

Ku, W. H., and W. C. Peterson, "Optimum Gain-Bandwidth Limitations of Transistor Amplifiers as Reactively Constrained Two-Port Networks," *IEEE Trans. Circuits Syst.*, Vol. CAS-22, June 1975, pp. 523–533.

Liu, L. C. T., and W. H. Ku, "Computer-Aided Synthesis of Lumped Lossy Matching Networks for Monolithic Microwave Integrated Circuits (MMICs)," *IEEE Trans. Microwave Theory Tech.*, Vol. MTT-32, No. 3, March 1984, pp. 282–290.

Matthaei, G. L., "Tables of Chebyshev Impedance Transforming Networks of Lowpass Filter Form," *Proc. IEEE*, August 1964.

Mellor, D. J., and J. G., Linvill, "Synthesis of Interstage Networks of Prescribed Gain Versus Frequency Slopes," *IEEE Trans. Microwave Theory Tech.*, Vol. MTT-23, No. 12, December 1975.

Pitzalis, O., and R. A. Gilson, "Tables of Impedance Matching Networks which Approximate Prescribed Attenuation Versus Frequency Slopes," *IEEE Trans. Microwave Theory Tech.*, Vol. MTT-19, No. 14, April 1971.

Schoeffler, J. D., "Impedance Transformation Using Lossless Networks," *IRE Trans. Circuit Theory*, CT-8, June 1961, pp. 131–137.

Van Valkenburg, M. E., *Introduction to Modern Network Synthesis*, New York: John Wiley and Sons, 1960.

CHAPTER 9

THE DESIGN OF RADIO-FREQUENCY AND MICROWAVE AMPLIFIERS AND OSCILLATORS

9.1   INTRODUCTION

Different classes of amplifiers and the associated load lines were considered in Chapter 2. It was also shown that, when load-pull data is not available, the load terminations required in linear amplifiers can be estimated by using the power parameters and boundary lines on the available I/V curve information. With the required terminations known, matching networks can be designed to transform the actual terminations to those required by using the design techniques introduced in Chapters 4–6 and 8. In narrowband power amplifiers the harmonic terminations can also be controlled to improve the power or the distortion performance. These issues were also examined in Chapter 2. Examples of class-A and class-B amplifiers were given at the end of Chapter 6, while class-E and class-F matching networks were considered at the end of Chapter 8.

In order to successfully design an amplifier it is necessary to ensure that it will not oscillate. Amplifier stability will be considered in detail in this chapter.

A minimum requirement to ensure stability is usually that the input and output resistance of each amplifier stage should be positive. It will be shown that the locus of load or source terminations for which the associated input or output conductance will be zero at any given frequency is a circle in the relevant plane, and the parameters of these stability circles will be derived. This will be done on the admittance plane, as well as on the Smith chart. Various stability factors will also be introduced, and calculation of the reflection gain and the loop gain of an amplifier stage will also be considered.

In addition to calculating the relevant stability factors, the loop gain should also be calculated for any amplifier stages to which feedback was applied. This also applies when transistors are combined to increase the output power or to lower the impedance levels (small-signal amplifiers) because of the possibility of odd-mode oscillations [1]. These precautions are also necessary when transistors are connected in parallel. Note that odd-mode oscillations are not influenced by the circuit terminations [1].

Inherent stability (stability for any passive load or source termination) is a desirable condition. It can frequently be effected by using frequency-dependent feedback and/or resistive loading without degrading the performance significantly. These networks can also serve to reduce gain slopes and the gain-bandwidth constraints associated with the impedances to be matched, as well as to force the optimum noise or power match closer to the optimum gain match [2]. Adding such networks to a transistor to make it more suitable for its intended purpose (preconditioning) is more often than not a good idea when wideband amplifiers are designed. Note that while gain slopes can also be removed by reactive mismatching, these networks are usually sensitive to component changes (high-Q networks).

When feedback or loading cannot be used, the matching networks should be designed to ensure conditional stability at all frequencies. This is frequently done when a high-power stage is designed. Note that the choice of topology can be critical to ensure stability outside the passband. Frequency selective resistive traps can sometimes also be used to prevent oscillations outside the passband.

It was shown in [3] that there is an upper limit on the available or operating power gain that will ensure positive resistance on both sides of the active device (single-sided matched condition). More general expressions, including an expression for the maximum double-sided mismatched gain were provided in [4]. The degree of stability (inside the passband) is also directly related to the available or operating power gain [5]. As one might suspect, the stability increases with decreasing gain. These issues will be considered in Section 9.3.

It is also often necessary to control the gain and/or the noise figure of an amplifier stage. It will be shown in Section 9.2 that the locus of terminations that will provide the same available or operating or transducer power gain is a circle on the admittance- or impedance-plane, and the Smith chart. It will also be shown that the contour of source terminations that will yield the same noise figure is also a circle.

When these circles are inside the unit circle on a Smith chart, the active gain or noise figure control problem can be transformed to an equivalent passive problem, that is, a passive matching problem with the same circle as the associated gain or noise figure problem exists. The impedance-matching techniques considered previously can be applied directly to these problems after the transformation. Allowance for the discrepancy in the error functions for the two cases (active and equivalent passive) can be made during the optimization phase of the networks, as is done in [2]. The equivalent passive problem will also be considered in Section 9.2.

Note that when the equivalent passive problem does not exist, the optimum point on the relevant contour should be selected to define the matching problem to be solved.

When a high dynamic range amplifier is required at RF frequencies, a lossless feedback amplifier should be considered. The performance obtainable with these amplifiers is excellent. It should be noted, however, that with careful design and the right choice of the transistor, similar performance can often be obtained with a cascade amplifier. Lossless feedback amplifiers will be considered in Section 9.10.

The design of reflection and balanced amplifiers will be also be considered in this chapter. For an excellent treatment on the design of distributed amplifiers refer to [6].

Many of the concepts and techniques used when amplifiers are designed also apply to oscillators. Oscillator design based on controlling the load line of the transistor and the loop gain will be considered in Section 9.13. For an in-depth treatment of oscillator design techniques refer to [7].

9.2   STABILITY

In order to design an oscillator or to prevent an amplifier from oscillating, it is necessary to know more about the conditions under which oscillations can occur. These conditions will be established in this chapter.

It will be shown that steady-state oscillations will occur when the input and output admittance of an active circuit is equal to zero. Oscillation is not possible at any frequency at which the input conductance is positive or, equivalently, the magnitude of the input and output reflection coefficients are smaller than unity. The boundary condition at which the input conductance is equal to zero or the magnitude of the reflection coefficient is equal to unity will be considered on the admittance plane and on the Smith chart, respectively. It will be shown that the stable and potentially unstable areas are separated by circles in both planes. The area inside the relevant circle is usually the area of potential instability.

When oscillations cannot occur with any passive termination, a transistor is said to be inherently stable.

The stability of a two-port can be considered by establishing the positions of the stability circles on the plane of interest or by calculating various stability factors defined for this purpose. Although essential, the stability factors by themselves do not always provide sufficient information on the stability of an amplifier. One of the reasons for this is that small changes in the component values can sometimes have a very pronounced effect on the stability factors. When transistors are combined in the circuit, odd-mode oscillation [1] may also be possible.

In order to get a clearer picture of the situation, the following can also be done:

1.   In addition to calculating the stability factors for the complete circuit, these factors should also be calculated for each transistor used in the circuit.

2.   The odd-mode [1] stability factors and loop gain associated with transistors connected in parallel or combined by using combination networks should be calculated (see Figure 9.1).

Note that (because of the symmetry) the circuit shown can be in oscillation with zero values for the input and output voltage. Also note, that the odd-mode oscillations in this circuit can be damped by connecting a suitable resistor between nodes A and B [1].

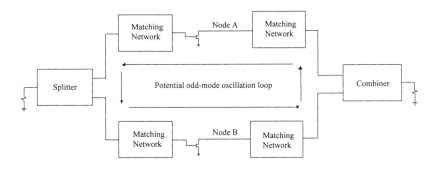

Figure 9.1   Odd-mode oscillation may be possible when circuits are combined or connected in parallel. The voltages at nodes A and B will be out of phase when the circuit is oscillating in an odd mode.

3.   The loop gain for each transistor stage to which external feedback is applied should be calculated and the associated gain and phase margins should be established. If more than one loop is used, the gain for each loop should be calculated.

Note that when transistors are connected in parallel, new loops are created. The (even-mode) loop gain around each of these loops should also be calculated.

It is important to realize that the (even-mode) loop-gain is dependent on the terminations used. The actual terminations of interest should be used (usually 50Ω).

4.   The reflection gain at the input and the output of each (modified) transistor can be calculated.

The reflection gain is also dependent on the terminations used.

5.   The series or shunt stabilizing resistance required on either or both sides of each transistor stage can be calculated (see Figure 9.2). In the well-behaved case, the series resistance required will decrease with frequency. Similarly, the shunt resistance will increase with frequency.

Note that the use of stabilizing components is not necessarily intended. The resistance required, however, provides a good idea of how severe the potential instability is.

Calculation of the stability circles, the stability factors, the stabilizing resistance, and the loop and reflection gain will be considered in this section.

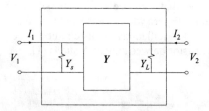

Figure 9.2   A two-port network augmented with its terminations.

9.2.1   Stability Circles on the Admittance Plane

Consider the two-port network in Figure 9.2. The terminal currents of the two-port augmented with its terminations can be calculated by using the equation

$$\begin{bmatrix} I_1 \\ I_2 \end{bmatrix} = \begin{bmatrix} y_{11} + Y_s & y_{12} \\ y_{21} & y_{22} + Y_L \end{bmatrix} \begin{bmatrix} V_1 \\ V_2 \end{bmatrix}$$

(9.1)

If the circuit is oscillating, the terminal voltages will not be equal to zero, even though the terminal currents are equal to zero.

In order for the voltages in (9.1) not to be equal to zero while the currents are equal to zero, the determinant of the extended Y-parameter matrix must be equal to zero. Were this not the case, the Y-parameter matrix would have had an inverse, and the only solution corresponding to zero values for I_1 and I_2 would have been zero values for both V_1 and V_2 as well.

A zero value for the determinant implies that

$$[y_{11} + Y_s][y_{22} + Y_L] - y_{12}y_{21} = 0 \tag{9.2}$$

leading to

$$y_{11} + Y_s - \frac{y_{12}y_{21}}{y_{22} + Y_L} = 0 \tag{9.3}$$

and

$$y_{22} + Y_L - \frac{y_{12}y_{21}}{y_{11} + Y_s} = 0 \tag{9.4}$$

Equations (9.3) and (9.4) are easily recognized as the equations for the input and output admittances of the augmented two-port, respectively. A necessary condition for any oscillation to occur, therefore, is that the input and output admittance of the augmented two-port must be equal to zero. This condition will apply at steady-state.

It is clear from (9.2) through (9.4) that if the input admittance of a network is equal to zero, the same will apply to its output admittance (as long as the product $y_{12} y_{21}$ is not equal to zero). It is also clear that whenever the resistive part of the input or output admittance is greater than zero at any particular frequency, oscillations cannot occur at that frequency. Considering only the input admittance for the moment, the locus of the load admittance for which the input conductance will be equal to zero can be derived easily by setting the real part of Y_{in} (Y_L) equal to zero:

$$\Re(Y_{in}) = G_s = g_{11} - \Re\left(\frac{y_{12}y_{21}}{y_{22} + Y_L}\right) = 0$$

where

$$g_{11} + jb_{11} = y_{11} \tag{9.5}$$

With

$$g_{22} + jb_{22} = y_{22} \tag{9.6}$$

$$y_{12}y_{21} = P + jQ \tag{9.7}$$

and

$$G_L + jB_L = Y_L \tag{9.8}$$

it follows that

$$G_s + g_{11} - \Re\left\{ \frac{(P + jQ)[(g_{22} + G_L) - j(b_{22} + B_L)]}{(g_{22} + G_L)^2 + (b_{22} + B_L)^2} \right\} = 0$$

This equation can be manipulated into the following form:

$$\left[G_L + g_{22} - \frac{P}{2(G_s + g_{11})} \right]^2 + \left[B_L + b_{22} - \frac{Q}{2(G_s + g_{11})} \right]^2 = \frac{|y_{12}y_{21}|^2}{4(G_s + g_{11})^2} \tag{9.9}$$

Equation (9.9) is the equation of a circle in the linear admittance plane. The center of the circle is given by

$$G_L + jB_L = \left[\frac{P}{2(G_s + g_{11})} - g_{22} \right] + j\left[\frac{Q}{2(G_s + g_{11})} - b_{22} \right] \tag{9.10}$$

while the radius is given by

$$R = \left| \frac{y_{12}y_{21}}{2(G_s + g_{11})} \right| \tag{9.11}$$

For all load admittances on the circumference of this circle, the input conductance will be equal to zero, and if the input susceptance is also equal to zero, the amplifier will oscillate. The input conductance is usually negative for all load admittances falling inside the stability circle. The exception occurs when $g_{11} < 0$.

If only passive loads are considered, the worst-case condition will clearly be when $G_s = 0$. Under this condition (9.10) and (9.11) simplify to

$$G_{Lsta} + jB_{Lsta} = \left[\frac{P}{2g_{11}} - g_{22} \right] + j\left[\frac{Q}{2g_{11}} - b_{22} \right] \tag{9.12}$$

and

$$R_{Lsta} = \left| \frac{y_{12}y_{21}}{2g_{11}} \right| \tag{9.13}$$

Whenever this stability circle lies to the left of the imaginary axis of the admittance plane, it will not be possible for an amplifier to oscillate at that particular frequency as long as its terminations are passive. The amplifier is then inherently stable.

Proceeding as above, the parameters of the stability circle in the source plane can be determined easily. The resulting equations are

$$G_s + jB_s = \left[\frac{P}{2(G_L + g_{22})} - g_{11}\right] + j\left[\frac{Q}{2(G_L + g_{22})} - b_{11}\right] \tag{9.14}$$

$$R = \left|\frac{y_{12}y_{21}}{2(G_L + g_{22})}\right| \tag{9.15}$$

The stable area will be outside this circle as long as $g_{22} \geq 0$. The worst-case condition is again associated with $G_L = 0$.

The Linville stability factor can be defined in terms of the parameters of the stability circle in the following way:

$$C = -\frac{R_{Lsta}}{G_{Lsta}} \tag{9.16}$$

$$= -\left|\frac{y_{12}y_{21}}{2g_{11}}\right| / \left(\frac{P}{2g_{11}} - g_{22}\right) \tag{9.17}$$

Whenever $0 \leq C < 1$, the stability circle will lie to the left of the imaginary axis of the admittance plane. If g_{11} is positive, the inside of the circle will represent the unstable area ($Y_{in} = y_{11}$ when $Y_L \to \infty$) and the device under consideration will be inherently stable.

The stability circle is plotted for different values of C in Figure 9.3.

The Linville stability factor is independent of the two-port terminations and is only a function of the parameters of the device used.

Another useful measure of stability is the Sterne stability factor. The Sterne stability factor takes the influence of the resistive loading by the load and source admittance into account:

$$K = \frac{g_{22} + G_L}{\dfrac{y_{12}y_{21}}{2(g_{11} + G_s)} + \dfrac{P}{2(g_{11} + G_s)}} \tag{9.18}$$

An amplifier will be stable at any frequency for which $K \geq 1$, that is, as long as the terminations used are in place.

Equation (9.18) is based on the fact that for inherent stability it is required that

$$R_{Lsta} + G_{Lsta} \leq 0 \tag{9.19}$$

that is, if $g_{11} \geq 0$.

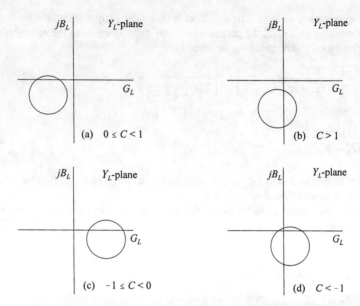

Figure 9.3   The relationship between the Linville stability factor and the position of a stability circle relative to the imaginary axis of the admittance plane. (The stable area is on the outside of each circle as long as $g_{11} > 0$.)

Whenever either or both of the terminations can change, it is advisable to set G_L and/or G_s in (9.18) equal to zero.

9.2.2   Stability Circles on the Smith Chart and Associated Stability Factors and Figures of Merit

The locus of load and source reflection coefficients for which the input or output conductance or resistance of an amplifier will be equal to zero are also circles on a Smith chart.

The equations for the stability circles on the Smith chart can be derived by using the expressions derived for $s_{11\omega}$ and s_{22a} in Chapter 1 [(1.86) and (1.87)]. For inherent stability it is required that

$$|s_{11\omega}| < 1 \tag{9.20}$$

and

$$|s_{22a}| < 1 \tag{9.21}$$

or any passive termination (i.e., for $|\Gamma_L| \leq 1$ and $|\Gamma_s| \leq 1$).
    It follows from these equations that [8]

$$|s_{11} - \Delta S_L| < |1 - s_{22} S_L| \tag{9.22}$$

and

$$|s_{22} - \Delta S_s| < |1 - s_{11} S_s| \tag{9.23}$$

where

$$\Delta = s_{11} s_{22} - s_{12} s_{21} \tag{9.24}$$

The parameters for the load stability circle (the values of S_L for which $|s_{11\omega}|=1$, that is, the load terminations for which the input impedance will be purely reactive) follow from (9.22) and are given by

$$C_L = \frac{(s_{22} - \Delta s_{11}^*)^*}{|s_{22}|^2 - |\Delta|^2} \tag{9.25}$$

$$R_L = \frac{|s_{12} s_{21}|}{\left||s_{22}|^2 - |\Delta|^2\right|} \tag{9.26}$$

where C_L is the center of the circle and R_L its radius.

Similarly, the parameters for the source stability circle (the values of S_s for which $|s_{22\omega}|=1$, that is, the source terminations for which the output impedance will be purely reactive) are given by

$$C_s = \frac{(s_{11} - \Delta s_{22}^*)^*}{|s_{11}|^2 - |\Delta|^2} \tag{9.27}$$

$$R_s = \frac{|s_{12} s_{21}|}{\left||s_{11}|^2 - |\Delta|^2\right|} \tag{9.28}$$

The stable area could be outside or inside the circle. The specific case can be established by observing the magnitudes of s_{11} and s_{22}, respectively. If, for example, $|s_{11}| < 1$, the input resistance is positive with a 50-Ω load; if the 50-Ω load falls outside the load stability circle, the area inside the circle is the unstable area.

It follows that the following conditions must be satisfied for a transistor to be inherently stable:

$$\left\| C_s \right| - \left| R_s \right\| > 1 \tag{9.29}$$

$$\left\| C_L \right| - \left| R_L \right\| > 1 \tag{9.30}$$

$$\left| s_{11} \right| \leq 1 \tag{9.31}$$

and

$$\left| s_{22} \right| \leq 1 \tag{9.32}$$

that is, the stability circles must lie outside the Smith chart and the input and output reflection coefficients associated with 50-Ω terminations must be passive.

An example of a Smith chart stability circle is provided in Figure 9.4.

Instead of formulating the inherent stability conditions in terms of stability circles, the Rollette stability factor (k) [9] can be used. Inherent stability is then established by the following conditions:

$$k = \frac{1 - \left| s_{11} \right|^2 - \left| s_{22} \right|^2 + \left| \Delta \right|^2}{2 \left| s_{12} \right| \left| s_{21} \right|} \geq 1 \tag{9.33}$$

$$\left| s_{12} s_{21} \right| < 1 - \left| s_{11} \right|^2 \tag{9.34}$$

$$\left| s_{12} s_{21} \right| < 1 - \left| s_{22} \right|^2 \tag{9.35}$$

These conditions can be derived by establishing the conditions under which $s_{11\omega}$ or s_{22a} will be passive for passive S_L or S_s, respectively [8]. This can be done by considering $\left| s_{11\omega} \right|$ or $\left| s_{22a} \right|$ when $\left| S_L \right| \leq 1$ or $\left| S_s \right| \leq 1$ (passive terminations), respectively. It was shown in [10], when $k \geq 1$, (9.34) implies (9.35), and vice versa. Only one of these conditions is therefore required for inherent stability.

A new criterion for evaluating the stability of a linear two-port was also introduced in [10]. This stability factor, μ, is defined as the minimum distance in the Γ_L-plane between the origin of the unit circle on the Smith chart and the unstable region (circle). The mathematical expression follows from this definition, and (9.25) and (9.26) after some manipulation [10]:

$$\mu = \frac{1 - \left| s_{11} \right|^2}{\left| s_{22} - s_{11}^* \Delta \right| + \left| s_{12} s_{21} \right|} \tag{9.36}$$

When $\mu > 1$, the two-port is inherently stable. Note that no other conditions are required, as was the case with k. Also note that because μ was defined as the minimum distance to the stability circle from the center of the Smith chart, it is also the largest load reflection coefficient for which the input resistance or conductance will be positive, that is, when $0 < \mu < 1$. When μ is negative the center of the Smith chart will be inside the unstable area. It also follows that the input resistance of the two-port will be positive for all load VSWRs smaller than

$$\text{VSWR}_{L\_max} = \frac{1+\mu}{1-\mu} \tag{9.37}$$

under the condition that $0 < \mu < 1$.

It is important to interpret this VSWR and the associated reflection coefficient modulus (μ) in terms of the normalization resistance used for the S-parameters in (9.36) (usually 50Ω). These parameters give no information about the actual mismatch of the transistor impedances to the transistor terminations in place, that is, when the transistor impedances are not close to the normalization resistance. (This is usually the case.) These numbers are, however, meaningful when a designed amplifier is considered and the normalization resistance is the same as the actual terminations.

A complementary stability factor, μ', was also defined for the source plane in [10]. The expression for μ' is

$$\mu' = \frac{1-|s_{22}|^2}{|s_{11} - s_{22}^*\Delta| + |s_{12}s_{21}|} \tag{9.38}$$

Similar to the case with μ, it also follows that the output resistance or conductance of the two-port will be positive for all source VSWRs smaller than

$$\text{VSWR}_{S\_max} = \frac{1+\mu'}{1-\mu'} \tag{9.39}$$

that is, when $0 < \mu' < 1$.

When the S-parameters used in (9.36) and (9.38) are normalized to the actual terminations of the circuit of interest instead of 50Ω, the load stability factor (LSF) and the source stability factor (SSF) defined in [2] are obtained. These factors indicate the mismatch that can be tolerated before stability may become an issue in terms of the actual terminations of interest.

A useful overall stability factor, K_s, was defined in [11] in terms of the S-parameters of the two-port and the preselected magnitudes for the load and source reflection coefficients, Γ_L and Γ_s:

$$K_s = \frac{1-|\Gamma_s s_{11}|^2 - |\Gamma_L s_{22}|^2 + |\Gamma_s \Gamma_L \Delta|^2}{2|\Gamma_s \Gamma_L| \cdot |s_{12}s_{21}|} \tag{9.40}$$

When $K_s > 1$, the two-port will be stable when the magnitudes of the reflection coefficients are smaller or equal to the values specified . Note that stability is guaranteed independent of the phase angles of Γ_L and Γ_s.

The parameters of the stability circles associated with a specified value for the modulus of the source reflection coefficient ($|\Gamma_s|$) were also derived in [11]. The derivation for the stability circle in the load plane is based on the condition

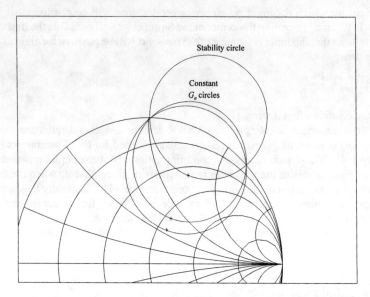

Figure 9.4   An example of a stability circle displayed with some constant gain circles (G_a circles) on a
Smith chart (source plane). Note that a stability circle can be considered as the gain circle
with infinite gain.

$$\left| \Gamma_s \, s_{11\omega} \right| = 1 \tag{9.41}$$

This reduces to the case previously considered when $|\Gamma_s|=1$. Note that (9.41) is independent
of the phase angle of Γ_s.

The center (C_{Lss}) and the radius (R_{Lss}) of the stability circle on the load plane are
given by [11]

$$C_{Lss} = \frac{s_{22}^* - \left| \Gamma_s \right|^2 s_{11} \Delta^*}{\left| s_{22} \right|^2 - \left| \Gamma_s \Delta \right|^2} \tag{9.42}$$

$$R_{Lss} = \frac{\left| \Gamma_s \right| \cdot \left| s_{12} s_{21} \right|}{\left| \left| s_{22} \right|^2 - \left| \Gamma_s \Delta \right|^2 \right|} \tag{9.43}$$

It is interesting to compare (9.42) and (9.43) to the equations associated with the unloaded
case [(9.25) and (9.26)]. When the modulus of the source reflection coefficient is equal to
one, the equations are the same, as expected.

When the interior of these circles is the stable area, the circles become smaller as the
modulus of the source reflection coefficient is reduced, and the centers of these circles lay
on a straight line between C_{Lss} and $1/s_{22}$ [11]. The interior of a circle will be the stable area
when

$$|s_{11}| < 1$$

and

$$|s_{22}|^2 - |\Gamma_s \Delta|^2 > 0$$

Because the equations are of the same form, it is a simple matter to transform the equations given above to apply to the source plane and a fixed value of $|\Gamma_s|$.

Amplifier stages are frequently designed by controlling the operating (G_ω) or available (G_a) power gain of a transistor. If the transistor is potentially unstable, the output resistance associated with a given value of G_ω (or output power level) and a conjugate match at the input (typical power amplifier case) is frequently negative. Similarly, the input resistance of a low-noise stage can be negative when G_a (or the noise figure) is controlled and the output of the transistor is conjugately matched. It was shown in [3], that these conditions can be avoided if the relevant gain is kept below a critical value and $0 < k < 1$. This value is the same for the operating and available power gains and can be used as a figure of merit for a transistor (maximum single-sided-matched stable gain). The maximum single-sided-matched stable gain is given by

$$G_{msmsg} = 2\,k\,\left|\frac{s_{21}}{s_{12}}\right| \tag{9.44}$$

that is, when $0 < k < 1$. Note that

$$G_{msmsg} > \left|\frac{s_{21}}{s_{12}}\right| = G_{MSG}$$

when $k > 0.5$.

A different figure of merit (the maximum controlled mismatch stable gain) that applies in also cases (no single-sided-match constraint) was presented in [4]:

$$G_{mcmsg} = G_{msg} \cdot \left[k + |\Gamma_{sm}||\Gamma_{Lm}| + \sqrt{(k + |\Gamma_{sm}||\Gamma_{Lm}|)^2 - (1 - |\Gamma_{sm}|^2)(1 - |\Gamma_{Lm}|^2)} \right] \tag{9.45}$$

that is, when $-1 < k < 1$. $|\Gamma_{sm}|$ and $|\Gamma_{Lm}|$ are the maximum moduli allowed for the source and load reflection coefficients, respectively. When arbitrary passive terminations are allowed, $|\Gamma_{sm}| = 1$ and $|\Gamma_{Lm}| = 1$, and the maximum arbitrary mismatched stable gain is given by

$$\tag{9.46}$$

$$G_{mamsg} = 2\,(k + 1)\,G_{msg}$$

The value of G_{mcmsg} associated with the conditions $|\Gamma_{sm}|=1$ and $|\Gamma_{Lm}|=0$ (arbitrary mismatch on the input side with a conjugate match in the output side), or $|\Gamma_{sm}|=0$ and $|\Gamma_{Lm}|=1$, corresponds with that given by (9.44) for G_{msmsg}.

9.2.3   The Reflection Gain Approach

Steady-state oscillations will occur in a circuit when [12]

$$|\Gamma_{rhs}\Gamma_{lhs}| = 1 \tag{9.47}$$

and

$$\text{ANGLE}[\Gamma_{rhs}] = -\text{ANGLE}[\Gamma_{lhs}] \tag{9.48}$$

where Γ_{rhs} is the reflection coefficient to the right of the point of interest, and Γ_{lhs} is the reflection coefficient to the left.

These conditions follow easily from Figure 9.5. If a_L is the signal incident on the load (RHS), it will be reflected as $b_L = \Gamma_{rhs} a_L$. Considering the conditions at steady-state, this reflected signal will be the incident signal on the LHS and, assuming no external signal to be present, will be reflected as $b_s = \Gamma_{lhs} \Gamma_{rhs} a_L$. This signal is in turn the incident signal on the load, which implies that $b_s = a_L$ (steady-state). This can only be the case for nonzero a_L if $\Gamma_{lhs} \Gamma_{rhs} = 1$. This condition is equivalent to the zero admittance oscillation condition at the common node.

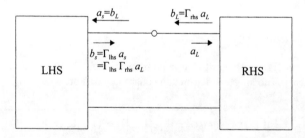

Figure 9.5   Illustration of calculating the reflection gain at a given point in a circuit.

This result can also be obtained from the expression for the transducer power gain of a one-port. By using the equation derived for the power available from a source and the constraint imposed by the load [(1.101) and (1.79)], it follows that

$$G_T = \frac{[1-|\Gamma_{rhs}|^2][1-|\Gamma_{lhs}|^2]}{|1-\Gamma_{lhs}\Gamma_{rhs}|^2} \tag{9.49}$$

Oscillation will occur when the gain approaches infinity, which will be the case when $\Gamma_{lhs}\Gamma_{rhs} = 1$, as was shown above.

At startup, the magnitude of $\Gamma_{lhs}\,\Gamma_{rhs}$ must be higher than unity.

Assuming the one side to be passive and the other to be active, (9.47) and (9.48) can be modified to

$$\left|\Gamma_{passive}\right| \geq \left|\frac{1}{\Gamma_{active}}\right| \tag{9.50}$$

and

$$\text{ANGLE}[\Gamma_{passive}] = \text{ANGLE}\,[1\,/\,\Gamma_{active}] \tag{9.51}$$

Note that (9.50) and (9.51) are steady-state conditions. If the magnitude of $\Gamma_{lhs}\,\Gamma_{rhs}$ is significantly larger than unity, start-up cannot be guaranteed, but the two-port will certainly be potentially unstable.

Conditions (9.50) and (9.51) can be detected easily if the inverted reflection coefficient of the active side is compared to the reflection coefficient of the other side (usually the resonator side) on a rectangular plot as a function of the frequency. The magnitude of the inverted reflection coefficient of the active side should be smaller than that of the passive side at the point where the phase traces cross (resonance point for the reflection coefficients).

Because no explicit frequency information is available when it is done, the common practice of considering only these quantities on a Smith chart is not recommended. The absence of explicit frequency information can be misleading and can lead to wrong conclusions.

9.2.4   The Loop Gain Approach

The loop gain of an oscillator or an amplifier stage can be calculated by using feedback theory (see Figure 9.6). The closed loop gain (A_{CL}) is given by

$$A_{CL} = \frac{A_{OL}}{1 + \beta A_{OL}} \tag{9.52}$$

where A_{OL} is the open-loop gain and $-\beta A_{OL}$ is the loop gain (that is, the gain around the loop). The loading effect of the feedback network on the relevant loops (at the relevant nodes) should be taken into account when the open-loop gain is calculated. The feed-forward effect associated with the sampling network should also be taken into account.

If the open-loop resistance in the input loop (series feedback case; open-loop conductance at the input node in the shunt feedback case) is positive, oscillations will start at any frequency at which the loop gain (gain around the loop) is greater or equal to 0 dB and the phase shift around the loop is $0°$ or a multiple of $360°$.

Oscillations are always possible if the sum of the open-loop resistance in the input loop and the resistance resulting from the feedback is negative (series feedback case).

Suitable reactance in series or parallel with this negative resistance (conductance) may inhibit oscillations and may even result in a Rollette factor bigger than unity.

It is important to realize that the reactance in the input loop will not necessarily be zero (at resonance) when the loop gain is in-phase. Resonance of the reactance (susceptance) in the relevant loop (at the relevant node) is only a necessary condition at steady-state (loop gain compressed to 0 dB).

It should also be realized that the gain will be different around different loops. This can be appreciated easily by considering a transistor with both current-series and voltage-shunt feedback loops (consider the case with significant voltage-shunt feedback and negligible current-series feedback).

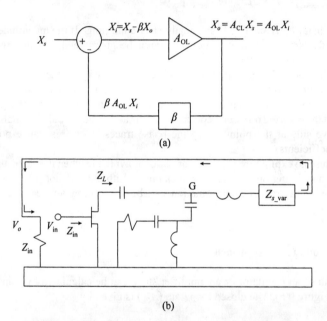

(a)

(b)

Figure 9.6   (a) Calculation of the loop gain of a feedback amplifier. (b) The circuit used to calculate the loop gain ($-\beta A$) of a series feedback oscillator [2]. (The actual ground of the circuit is at the point marked "G.")

The loop of interest is usually the loop associated with the gain compression. With a well-behaved load line, the main reason for gain compression will be compression of the transconductance caused by the voltage swing across the input junction (nonlinear transfer function).

The loop gain for an oscillator is shown in Figure 9.7. The startup frequency is listed with the loop gain, as well as the slope in the phase response at this frequency, below the plot. Negative resistance in the loop is indicated with the solid line drawn at the 0-dB gain level. Oscillation is not possible when the loop resistance is positive.

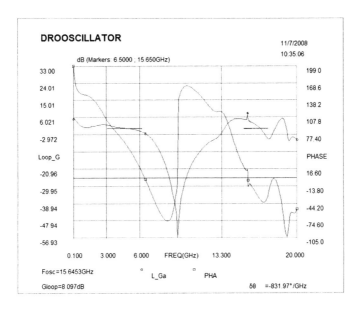

Figure 9.7   The small-signal loop gain calculated for a dielectric resonator oscillator [2].

Note that the phase response of this oscillator is well-behaved and that oscillations are not possible at the higher frequencies. (The loop phase again approaches zero at a higher frequency, but the loop gain is too low for oscillation when this happens.) Oscillations will start up at 15.6435 GHz with a loop gain of 8 dB. The slope in the phase response is $-832°/$GHz at this frequency.

The rate at which the phase is changing at the oscillation frequency is an indication of the loaded Q of the circuit. In the special case when a single-tuned response can be assumed and when the oscillation frequency is also the resonance frequency or a frequency close to it, the Q can be estimated by using the following equations for a parallel resonant circuit [8]:

$$Z_{in} = \frac{1}{G + j\omega C + 1/(j\omega L)}$$

$$= \frac{1}{G + j(\omega_0 + \Delta\omega)C + 1/(j(\omega_0 + \Delta\omega)L)}$$

$$= \frac{1}{G + j\omega_0 C + j\Delta\omega C + 1/(j(\omega_0(1 + \Delta\omega/\omega_0))L}$$

$$\approx \frac{1}{G + j\omega_0 C + j\Delta\omega C + (1 - \Delta\omega/\omega_0)/(j\omega_0 L)}$$

$$= \frac{R}{1 + j2Q(\Delta\omega/\omega_0)} \tag{9.53}$$

where the approximation applies at frequencies close to the resonant frequency.

This equation also explains why the phase of a resonant circuit is linear close to the resonant frequency. Note that at startup the oscillation frequency may not also be the resonance frequency. However, this will always be the case at steady-state.

It follows from (9.53) that the Q of an oscillator (single-tuned response) can be estimated as

$$Q = \frac{\pi}{360} (\Delta\theta / \Delta f) \, f_0 \tag{9.54}$$

where the phase slope $(\Delta\theta/\Delta f)$ is specified in degrees per gigahertz and the resonant frequency in gigahertz.

Note that the loaded Q will decrease as the transistor is driven into compression.

9.2.5 Stabilization of a Two-Port with Shunt or Series Resistance

Any transistor (two-port) can be stabilized by adding shunt or series resistance to it. It is sometimes necessary to add resistance on both sides of the transistor. This may be the case when the real part of y_{11}, y_{22}, z_{11}, or z_{22} is negative.

The shunt conductance required can be calculated by using the equations derived for the stability circles in terms of the Y-parameters (Section 9.2.1):

$$G_{22\text{sta}} = G_{L\text{sta}} + R_{L\text{sta}} \tag{9.55}$$

and

$$G_{11\text{sta}} = G_{s\text{sta}} + R_{s\text{sta}} \tag{9.56}$$

where $G_{11\text{sta}}$ is the shunt conductance required on the input side, $G_{22\text{sta}}$ the conductance required on the output side, $G_{L\text{sta}}$ the real part of the center of the stability circle on the load plane (admittance plane), $R_{L\text{sta}}$ the radius of this circle, $G_{s\text{sta}}$ the real part of the center on the source plane, and $R_{s\text{sta}}$ the radius of this circle.

The equations for the series resistance can be derived by deriving equations for the stability circles in the impedance plane first. The equations are identical in form to those for the shunt admittance. The only differences are that each Y-parameter should be replaced with the corresponding Z-parameter. The reason for this is clear if the expressions for the input/output admittance and impedance are compared:

$$Y_{\text{in}} = y_{11} - \frac{y_{12} \, y_{21}}{y_{22} + Y_L} \tag{9.57}$$

and

$$Z_{\text{in}} = z_{11} - \frac{z_{12} \, z_{21}}{z_{22} + Z_L} \tag{9.58}$$

Table 9.1 The Series and the Shunt Resistance Required to Stabilize a Transistor [2]

Frequency (GHz)	Source-side loading (Ω)		Load-side loading (Ω)	
	R_i	(R_o)	(R_i)	R_o
0.10	177.0		166.0	
0.50	118.0	200Ω // 1.07 pF	262.0	186Ω // 0.87 pF
1.00	58.4		145.0	
2.00	28.4		70.8	
3.00	17.6		44.2	
4.00	11.7		28.7	
5.00	7.95		18.8	
6.00	5.69		12.8	
7.00	4.11		8.84	
8.00	3.03		6.18	
9.00	2.52		4.87	
10.0	1.92		3.43	
11.0	0.82		1.32	

Frequency (GHz)	Source-side loading (Ω)		Load-side loading (Ω)	
	R_i	(R_o)	(R_i)	R_o
0.10	12.7k			64.5
0.50	1.46k	9.24Ω+4.89 pF		3.50	0.46Ω+0.31 nF
1.00	770.0			7.31
2.00	399.0			4.86
3.00	265.0			3.16
4.00	191.0			2.79
5.00	145.0			3.90
6.00	104.0			3.88
7.00	74.0			2.14
8.00	44.5		6.03k	0.10
9.00	11.8		279	1.09
10.0	0.8	111.0	66.6	1.38
11.0	0.48	7.38	6.88	0.51

The form of these two equations is identical and, because the input conductance and the input resistance are calculated by taking the real part of the two equations, respectively, the equations for the stability circles will also have the same form.

An example of the series and shunt resistance required to stabilize a Fujitsu FHX35LG transistor is provided in Table 9.1. If series loading is considered, the transistor can be stabilized by adding resistance in series with the input or the output side. A total of 177Ω is required on the input side to stabilize the transistor at 100 MHz, while only 0.82Ω is required at 11 GHz. A parallel combination of a 200-Ω resistor and a 1.07-pF capacitor in series with the input terminal will provide the required series resistance at 0.1 GHz and 11 GHz. Some adjustment in the capacitance may be required for inherent stability over the complete frequency range.

Note that the series resistance required is well-behaved and only a small amount of loading is required at the higher frequencies.

In contrast with the series case, stabilization by using shunt loading is simply not an option. The (shunt) resistance decreases with increasing frequency (greater loading is required at the higher frequencies), and loading is also required on the other side of the transistor in order to remove the negative conductance associated with y_{11} or y_{22}. The value required on the other side is listed under the headings (R_o) for the input side and (R_i) for the output side.

Note again that in general the intention is not necessarily to actually stabilize the transistor in this way, but rather to evaluate the degree of instability by getting an idea of the resistance required for inherent stability. Furthermore, even if the goal is inherent stability, better results can usually be obtained by using two modification sections instead of one.

9.3   TUNABILITY

When a designed amplifier is realized, it may be necessary to tune it to obtain the exact results predicted. When the influence of the reverse transfer gain of a transistor (s_{12}) is not negligible, its input impedance will be a function of the load termination, and, if the load changes, whether because of tuning, temperature drift, or a change in load, the input impedance will also change. The consequent dependence of the input match on the changes in the output circuit (and vice versa) can be useful (indirect control on the VSWRs), but is usually undesirable.

The tunability factor [13]

$$\delta = \left| \frac{\partial Y_{in} / Y_{in}}{\partial Y_L / Y_L} \right| \tag{9.59}$$

$$= \frac{|y_{12} y_{21} Y_L|}{|y_{22} + Y_L|^2 |Y_{in}|}$$

$$= \left| \frac{y_{12}}{y_{21}} \right| G_\omega \sec\theta_L / \sec\theta_{in} \tag{9.60}$$

where

$$\theta_{in} = \tan^{-1}[B_{in} / G_{in}] \tag{9.61}$$

and

$$\theta_L = \tan^{-1}[B_L / G_L] \tag{9.62}$$

is a measure of the relative dependence of the input match on changes in the output circuit.

It is obvious from (9.59) that the tunability factor is a strong function of the operating power gain. If the gain is decreased enough, the output circuit will usually have very little influence on the input circuit and vice versa.

A tunability factor of less than 0.3 is usually advisable [13]. When the two sides of a transistor are completely isolated ($s_{12} = 0$), the tunability factor will be equal to zero.

The relative change in the output admittance as a function of the relative change in the source admittance is given by

$$\delta' = \left| \frac{\partial Y_{out} / Y_{out}}{\partial Y_s / Y_s} \right| \tag{9.63}$$

$$= \frac{\left| y_{12} y_{21} Y_s \right|}{\left| y_{11} + Y_s \right|^2 \left| Y_{out} \right|} \tag{9.64}$$

The order of magnitude of the two tunability factors are usually the same.

An expression for the tunability expressed in terms of the reflection coefficients is given by [6]

$$\delta_{\Gamma} = \left| \frac{\partial \Gamma_{in} / \Gamma_{in}}{\partial \Gamma_L / \Gamma_L} \right|$$

$$= \frac{\left| s_{12} s_{21} \Gamma_L \right|}{\left| 1 - s_{22} \Gamma_L \right| \left| s_{11} - \Delta \Gamma_L \right|} \tag{9.65}$$

Because of tunability difficulties, the MAG or MSG of a transistor often cannot be realized. The maximum tunable gain (MTG) usually provides a more realistic idea of the gain obtainable with a transistor. The MTG can be established iteratively by decreasing the gain from its MSG or MAG value until the associated tunability factor is acceptable.

It should be noted that poor tunability is not always undesirable. When a low-noise stage is designed, the dependence of the input admittance on the load admittance can be used to improve the input VSWR associated with an optimum noise match by changing the load admittance appropriately. This effect can also be used to improve the output VSWR associated with an optimum power match.

9.4   CONTROLLING THE GAIN OF AN AMPLIFIER

The best way to control the gain of an amplifier is to control the operating power gain (G_o) and/or the available power gain (G_a) of each stage (refer to Figure 9.8). The maximum single-sided matched stable gain (G_{msmsg}) as defined in (9.44) will typically be used as an upper limit on the gain targeted.

If the noise figure is critical, the design should be started at the input side, and if the power performance is more important, the design should be started at the load side. The design can also be done from both sides and the two independent sections can then be linked up at some point [see Figure 9.8(c)]. This is a good option when the dynamic range required is high.

If the operating power gain or the available power gain of the last stage in the cascade is controlled, it is implicitly assumed that the other side of the (modified) transistor will be conjugately matched (in practice, a good match will suffice). The mismatch associated with the last matching network is incorporated in G_{Tnet3}, as shown in Figure 9.8.

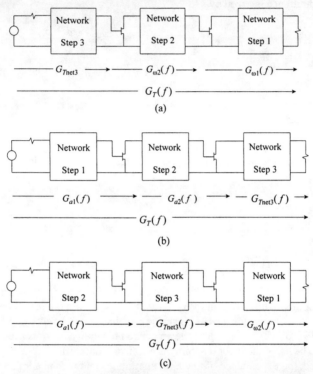

Figure 9.8   Calculation of the power gain of a multistage amplifier when the design is started (a) at the load, (b) at the source, and (c) from both sides (high dynamic range case).

The requirement of a conjugate match may be too restrictive when the last stage of an amplifier is designed, and in this case a better option could be to control the transducer power gain of this stage, with the matching networks already designed in place (refer to Figure 9.9). If this approach is followed to design a single-stage amplifier, the load termination can be designed for optimum power, after which the input network can be designed to control the transducer power gain of the amplifier. In this case, the source terminations associated with the best noise performance and acceptable matching can be selected on the different gain circles.

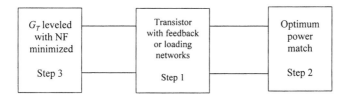

Figure 9.9   Illustration of a design procedure for a high dynamic range single-stage amplifier.

If the operating power gain, the available power gain, or the transducer power gain is controlled, the gain is controlled exactly and no approximations are made as is the case when the transistors used are assumed to be unilateral. It has been shown in [14] that the errors made by assuming a transistor to be unilateral are seldom negligible.

Before controlling the gain of a transistor with a lossless network, it is a good idea to first level its gain by using frequency-selective resistive networks (feedback or resistive loading sections; see Section 8.5 for some examples). The gain to be leveled is usually the MAG. Leveling the available power gain associated with the best noise figure (G_{an_opt}) is usually a better choice when a low-noise stage is designed. Similarly, leveling the operating power gain associated with the highest output power is usually a good idea.

Instead of using resistive networks, the gain can also be leveled by designing the impedance-matching networks (gain-control networks) to have positive gain slopes, but this route usually leads to sensitive designs and should be avoided if the goal is to design a first-time-right amplifier.

In order to control the operating or the available power gain of a stage it is necessary to establish what the source terminations or the load terminations should be to provide the required gain. The actual source or load impedance can then be transformed to that required by designing an impedance-matching network for this purpose.

It will be shown here that the contours of interest are circles on the admittance plane, the impedance plane, and the Smith chart. The centers and radii of the constant gain circles will be derived here.

When a transistor is inherently stable, the gain circles will be inside the Smith chart (passive terminations). It will be shown that in this case it is always possible to transform the active gain problem exactly to an equivalent passive impedance-matching problem. The equivalent passive problem can be solved by using standard impedance-matching techniques. As long as these problems are solved accurately, the solutions synthesized will also solve the original active problem.

If a wideband problem is solved by transforming the active problem to the equivalent passive problem, the deviation between the gain targeted and that obtained may not be insignificant. In this case an extra step should be introduced in which the solutions obtained from the equivalent passive problem are optimized for the best active performance [2].

At those frequencies at which the transistor is potentially unstable, the best point on each constant gain circle can be selected as targets. It should be noted that while the relevant gain will remain constant on the circumference of a constant gain circle, the other parameters of interest may change significantly. These parameters may include the noise figure, the power gain, the stability, the associated reflection coefficients (VSWRs), distortion, and various sensitivity factors.

9.4.1 Circles of Constant Mismatch for a Passive Problem

It was shown in Section 8.4.3.1 that the locus of load admittances for which the transducer power gain G_T of a passive source with internal admittance $Y_s = G_s + jB_s$ terminated in a passive load $Y_L = G_L + jB_L$ will remain constant is a circle in the linear admittance plane with center

$$G_0 + jB_0 = [2/G_T - 1]G_s - jB_s \qquad (9.66)$$

and radius

$$R_{Y0} = 2\,G_s\sqrt{1/G_T^2 - 1/G_T} \qquad (9.67)$$

Similarly, the locus of constant transducer power gain is also a circle on the Smith chart (see Figure 9.10). The parameters of this circle are given by

$$C_p = \frac{\Gamma_s^* G_T}{1 - |\Gamma_s|^2(1 - G_T)} \qquad (9.68)$$

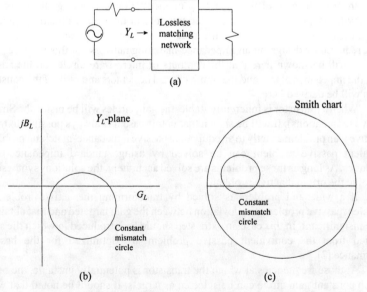

(a)

(b)   (c)

Figure 9.10   (a) The equivalent circuit relevant to the derivation of the constant mismatch circles and an example of these loci on (b) the admittance plane and (c) a Smith chart.

$$R_p = \frac{(1-|\Gamma_s|^2)\sqrt{1-G_T}}{1-|\Gamma_s|^2(1-G_T)} \tag{9.69}$$

where C_p is the center of the circle and R_p its radius.

Equations (9.68) and (9.69) can be derived from the expression for the transducer power gain of a one-port:

$$G_T = \frac{(1-|\Gamma_L|^2)(1-|\Gamma_s|^2)}{|1-\Gamma_s\Gamma_L|^2} \tag{9.70}$$

where Γ_L is the reflection coefficient of the load termination, and Γ_s is the reflection coefficient of the source termination. The derivation is repeated here for convenience.

It follows from (9.70) that

$$G_T|1-\Gamma_s\Gamma_L|^2 = (1-|\Gamma_L|^2)(1-|\Gamma_s|^2)$$

from which it follows that

$$|1-\Gamma_s\Gamma_L|^2 + \frac{1-|\Gamma_s|^2}{G_T}|\Gamma_L|^2 = \frac{1-|\Gamma_s|^2}{G_T}$$

that is,

$$|1-\Gamma_s\Gamma_L|^2 + \alpha|\Gamma_L|^2 = \alpha$$

which can be written as

$$(1-\Gamma_s\Gamma_L)(1-\Gamma_s^*\Gamma_L^*) + \alpha\Gamma_L\Gamma_L^* = \alpha$$

This expression can be written as

$$\left(\Gamma_L - \frac{\Gamma_s^*}{\alpha+|\Gamma_s|^2}\right)\left(\Gamma_L^* - \frac{\Gamma_s}{\alpha+|\Gamma_s|^2}\right) = \frac{\alpha-1}{\alpha+|\Gamma_s|^2} + \frac{|\Gamma_s|^2}{(\alpha+|\Gamma_s|^2)^2} \tag{9.71}$$

which is the equation for a circle on the Smith chart.

The center and radius of the circle can be obtained from this equation, and, after some simplification, the expressions given here are obtained.

The important point to grasp at this point is that while the problem of a conjugate match implies transforming a given load (source) termination into a specific input (output) impedance, the problem of getting a specified amount of mismatch is a circle problem. The load (source) termination can then be transformed to any point on the circumference of the relevant gain circle, and the gain of the passive network will be as specified.

9.4.2   Constant Operating Power Gain Circles

It will be shown here that the contours of constant operating power gain are circles on the admittance plane, as well as on the Smith chart. The equations for both cases will be derived here. The admittance plane case will be considered first.

The operating power gain of a transistor (see Figure 9.11) is given in terms of its Y-parameters by

$$
\begin{aligned}
G_\omega &= P_L / P_{in} \\
&= \left| \frac{y_{21}}{y_{22} + Y_L} \right|^2 \frac{G_L}{\mathrm{Re}\left(y_{11} - \dfrac{y_{12} y_{21}}{y_{22} + Y_L} \right)}
\end{aligned}
\tag{9.72}
$$

where P_L is the power dissipated in the load, and P_{in} is the power entering the input terminals of the amplifier.

With

$$ y_{12} y_{21} = P + jQ $$

as defined before, and

$$ y_{22} + Y_L = (G_L + g_{22}) + j(B_L + b_{22}) = G'_L + jB'_L $$

(9.72) becomes

$$
G_\omega = \frac{|y_{21}|^2 G_L}{g_{11} G_L'^2 + g_{11} B_L'^2 - P G'_L - Q B'_L}
\tag{9.73}
$$

By multiplying both sides of this equation with the denominator of the right-hand side and dividing them by $g_{11} G_\omega$, the following equation is obtained:

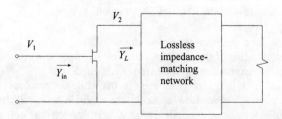

Figure 9.11   The circuit relevant to calculating the operating power gain of an amplifier stage.

$$G_L'^2 - \frac{PG_L'}{g_{11}} + B_L'^2 - \frac{QB_L'}{g_{11}} = \frac{|y_{21}|^2(G_L' - g_{22})}{g_{11}G_\omega} \tag{9.74}$$

This equation can be manipulated into the explicit form of the equation for a circle. The center of this circle is found to be

$$G_{L\omega} + jB_{L\omega} = \left[\left(\frac{P}{2g_{11}} - g_{22}\right) + \frac{\frac{|y_{21}|^2}{2g_{11}}}{G_\omega}\right] + j\left(\frac{Q}{2g_{11}} - b_{22}\right) \tag{9.75}$$

and its radius ($R_{L\omega}$) can be obtained from the equation

$$R_{L\omega}^2 = G_{L\omega}^2 - \left\{\left(\frac{P}{2g_{11}} - g_{22}\right)^2 - \left|\frac{y_{12}y_{21}}{2g_{11}}\right|^2\right\} \tag{9.76}$$

When the transistor is inherently stable [$0 \le C < 1$; $g_{11} > 0$; $g_{22} > 0$], the operating power gain circles lie entirely in the right-hand side of the admittance plane. When the transistor is potentially unstable, these circles cross over into the left-hand side of the plane, as is illustrated in Figure 9.12. Note that the gain circles cross the imaginary axis at the same two points, and that the gain circle corresponding to an infinite value for the gain is also the stability circle on the admittance plane, as derived in Section 9.2.

When a transistor is inherently stable, an expression for the maximum realizable power gain can be derived by calculating the operating power gain corresponding to the gain circle with radius equal to zero. With $R_{L\omega}$ set equal to zero in (9.76), the maximum realizable power gain is found to be

$$G_{\omega-\max} = \frac{|y_{21}|^2 / (2g_{11})}{G_{L\omega-\mathrm{opt}} - G_{Lsta}} \tag{9.77}$$

where $G_{L\omega\text{-opt}}$ is the real part of the load termination corresponding to the maximum realizable gain, and G_{Lsta} is defined by (9.12). Equation (9.77) can be simplified to

$$G_{\omega\_\max} = \left|\frac{y_{21}}{y_{12}}\right| \left|1/C - \sqrt{1/C^2 - 1}\right| \tag{9.78}$$

The load termination corresponding to the maximum realizable gain is given by

$$Y_{L\_\mathrm{opt}} = \sqrt{G_{Lsta}^2 - R_{Lsta}^2} + jB_{Lsta} \tag{9.79}$$

with B_{Lsta} as defined in (9.12).

When a transistor is potentially unstable, the maximum operating power gain obtainable is, theoretically, equal to infinity. The parameters of the constant operating power gain circles displayed on a Smith chart can be derived by using the expression for the operating power gain in terms of the load reflection coefficient (1.89):

$$G_\omega = \frac{|s_{21}|^2 [1 - |S_L|^2]}{|1 - s_{22} S_L|^2 - |s_{11}(1 - s_{22} S_L) + s_{12} s_{21} S_L|^2} \tag{9.80}$$

The center of each constant operating power gain circle is given by

$$C_\omega = \frac{g_\omega (s_{22}^* - \Delta^* s_{11})}{1 + g_\omega (|s_{22}|^2 - |\Delta|^2)} \tag{9.81}$$

and its radius by

$$R_\omega = \frac{(1 - 2k|s_{12} s_{21}|g_\omega + |s_{12} s_{21}|^2 g_\omega^2)^{1/2}}{|1 + g_\omega (|s_{22}|^2 - |\Delta|^2)|} \tag{9.82}$$

The normalized gain, g_ω, is given by

$$g_\omega = G_\omega / |s_{21}|^2 \tag{9.83}$$

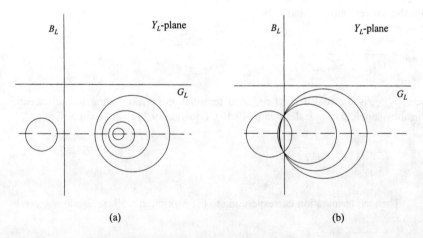

(a)   (b)

Figure 9.12   The position of the constant operating power gain circles relative to the imaginary axis of the admittance plane, illustrated for (a) an inherently stable transistor and (b) a transistor for which $C > 1$.

9.4.3 Constant Available Power Gain Circles

The contours of constant available power gain are also circles on the admittance plane or on the Smith chart. The equations for both cases will be presented here. The admittance plane case will be considered first.

The available power gain of an amplifier (see Figure 9.13) is given by

$$G_A = \frac{P_{av-O}}{P_{av-E}} = \left| \frac{y_{21}}{y_{11} + Y_s} \right|^2 \frac{\Re(Y_s)}{\Re(Y_o)} \tag{9.84}$$

where P_{av-O} is the maximum power available at the output of the amplifier and P_{av-E} is the maximum power available from the source.

Comparison of (9.84) and the expression for the operating power gain of an amplifier yields that if y_{11} is replaced with y_{22}, Y_s with Y_L, and Y_{out} with Y_{in}, the two expressions are identical. Because y_{21}, y_{11}, and y_{22} are constants, and the relationship between Y_s and Y_{out} is identical to that between Y_L and Y_{in}, it is possible to determine the locus of source admittances for which the available power gain of a transistor will be equal to a specified value by using the results obtained for the operating power gain. By following this approach, the center of a constant available power gain circle is found to be located at

$$G_{sa} + jB_{sa} = \left[\left(\frac{P}{2g_{22}} - g_{11} \right) + \frac{|y_{21}|^2}{2g_{22}} \right] + j \left(\frac{Q}{2g_{22}} - b_{11} \right) \tag{9.85}$$

and its radius (R_{sa}) can be obtained from the equation

$$R_{sa}^2 = G_{sa}^2 - \left\{ \left(\frac{P}{2g_{22}} - g_{11} \right)^2 - \left| \frac{y_{12}y_{21}}{2g_{22}} \right|^2 \right\} \tag{9.86}$$

When displayed on a Smith chart, the center of each constant available power gain circle is given by

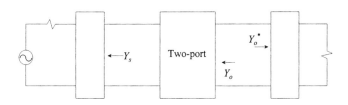

Figure 9.13   The equivalent circuit relevant to determining the available power gain of an amplifier.

$$C_a = \frac{g_a(s_{11}^* - \Delta^* s_{22})}{1 + g_a(|s_{11}|^2 - |\Delta|^2)} \tag{9.87}$$

and the radius by

$$R_a = \frac{(1 - 2k|s_{12}s_{21}|g_a + |s_{12}s_{21}|^2 g_a^2)^{1/2}}{|1 + g_a(|s_{11}|^2 - |\Delta|^2)|} \tag{9.88}$$

The normalized gain, g_a, is given by

$$g_a = G_a / |s_{21}|^2 \tag{9.89}$$

These equations are also identical in form to the operating power gain equations. This follows from the fact that the expressions for G_a and G_ω are also identical in form. An expression for G_a is shown below:

$$G_a = \frac{|s_{21}|^2 [1 - |S_s|^2]}{|1 - s_{11}S_s|^2 - |s_{22}(1 - s_{11}S_s) + s_{12}s_{21}S_s|^2} \tag{9.90}$$

9.4.4   Constant Transducer Power Gain Circles

The set of source admittances or reflection coefficients associated with a specified value of the transducer power gain is again a circle on the admittance plane or on the Smith chart. The same applies if the load admittance or reflection coefficient is considered.

    The derivation of the relevant equations for the admittance plane is based on the Y-parameter expression for the transducer power gain [refer to (1.11)]:

$$G_T = \frac{4|y_{21}|^2 G_L G_s}{|(y_{22} + Y_L)(y_{11} + Y_s) - y_{12}y_{21}|^2} \tag{9.91}$$

If the admittance Y_s is considered to be fixed, this equation can be used to find the constraints on Y_L to ensure that G_T will remain constant.

    It follows, after some manipulation, that the center of the circle is given by

$$G_{LT} + jB_{LT} = -Y_{out} + \left|\frac{y_{21}}{y_{11} + Y_s}\right|^2 \frac{2G_s}{G_T} \tag{9.92}$$

and the radius by

$$R_{LT}^2 = G_{LT}^2 - G_{out}^2 \tag{9.93}$$

where $Y_{out} = G_{out} + jB_{out}$ is the output admittance of the (modified) transistor terminated in the source admittance Y_s.

If the source admittance is taken to be the independent variable (fixed Y_L), the center of the circle is given by

$$G_{sT} + jB_{sT} = -Y_{in} + \left| \frac{y_{21}}{y_{22} + Y_L} \right|^2 \frac{2G_L}{G_T} \tag{9.94}$$

and the radius (R_{sT}) by

$$R_{sT}^2 = G_{sT}^2 - G_{in}^2 \tag{9.95}$$

where $Y_{in} = G_{in} + jB_{in}$ is the input admittance of the (modified) transistor terminated in the load admittance Y_L.

The parameters for the circles on the Smith chart can be derived by using the S-parameter expression for G_T [refer to (1.90)]:

$$G_T = \frac{|s_{21}|^2 (1 - |S_L|^2)(1 - |S_s|^2)}{|(1 - s_{11}S_s)(1 - s_{22}S_L) - s_{12}s_{21}S_sS_L|^2} \tag{9.96}$$

The center of the relevant circle on the load plane (S_s fixed) is given by

$$C_{LT} = \frac{1}{S_{out}\left(1 + \dfrac{|s_{21}|^2(1 - |S_s|^2)}{G_T|1 - s_{11}S_s|^2} \dfrac{1}{|S_{out}|^2}\right)} \tag{9.97}$$

and its radius by

$$R_{LT} = \frac{\sqrt{X_L\left(1 + X_L - \dfrac{1}{S_{out}S_{out}^*}\right)}}{1 + X_L} \tag{9.98}$$

where

$$X_L = \frac{|s_{21}|^2(1 - |S_s|^2)}{G_T|1 - s_{11}S_s|^2|S_{out}|^2} \tag{9.99}$$

The equations for the circles on the source plane (S_L fixed) are

$$C_{sT} = \cfrac{1}{S_{in}\left(1 + \cfrac{|s_{21}|^2(1-|S_L|^2)}{G_T|1-s_{22}S_L|^2}\cfrac{1}{|S_{in}|^2}\right)} \tag{9.100}$$

and

$$R_{sT} = \cfrac{\sqrt{X_s\left(1 + X_s - \cfrac{1}{S_{in}S_{in}^*}\right)}}{1 + X_s} \tag{9.101}$$

where

$$X_s = \frac{|s_{21}|^2(1-|S_L|^2)}{G_T|1-s_{22}S_L|^2|S_{in}|^2} \tag{9.102}$$

The transducer power gain circles can be used to control the transducer power gain and the noise figure of an amplifier when the load network has already been designed to optimize the power performance, and vice versa.

9.5   CONTROLLING THE NOISE FIGURE OF AN AMPLIFIER

It was shown in Section 2.2 that the noise figure of a (modified) transistor is determined by the source impedance presented at its input terminals by the circuit. It was also shown that the contours of constant noise figure are circles in the source plane.

It follows from

$$F = F_{min} + \frac{R_{nv}}{G_s}[(G_s - G_{s\_opt})^2 + (B_s - B_{s\_opt})^2] \tag{9.103}$$

[refer to (1.174a)] that the center of each constant noise figure circle is given by

$$G_F + jB_F = \left(G_{s\_opt} + \frac{F - F_{min}}{2R_{nv}}\right) + jB_{s\_opt} \tag{9.104}$$

while its radius (R_F) is given by

$$R_F^2 = G_F^2 - G_{s\_opt}^2 \tag{9.105}$$

An expression for the noise figure in terms of the reflection coefficient presented at the input terminals of the transistor can be derived by first modifying (9.103) to

$$F = F_{min} + \frac{R_{nv}}{G_s}\left|Y_s - Y_{s\_opt}\right|^2 \tag{9.106}$$

G_s in (9.106) can be replaced by using the following result:

$$\frac{G_s}{Y_0} = \frac{1-\left|\Gamma_s\right|^2}{\left|1+\Gamma_s\right|^2} \tag{9.107}$$

$\left|Y_s - Y_{s\_opt}\right|$ can be replaced by using the equality

$$\frac{Y_s - Y_{s\_opt}}{Y_0} = -\frac{2(\Gamma_s - \Gamma_{s\_opt})}{(1+\Gamma_s)(1+\Gamma_{s\_opt})} \tag{9.108}$$

Substitution in (9.106) yields the following expression for the noise figure:

$$F = F_{min} + \frac{4r_{nv}\left|\Gamma_s - \Gamma_{s\_opt}\right|^2}{(1-\left|\Gamma_s\right|^2)\left|1+\Gamma_{s\_opt}\right|^2} \tag{9.109}$$

where

$$r_{nv} = \frac{R_{nv}}{Z_0} \tag{9.110}$$

Straightforward manipulation of (9.109) yields that the center of each constant noise figure circle is given on the Smith chart by

$$C_F = \frac{\Gamma_{s\_opt}}{1+\alpha} \tag{9.111}$$

where

$$\alpha = \frac{F - F_{min}}{4r_{nv}}\left|1+\Gamma_{s\_opt}\right|^2 \tag{9.112}$$

The radius (R_F) of each circle is given by

$$R_F^2 = \frac{\alpha(1-\alpha-\Gamma_{s\_opt}\Gamma_{s\_opt}^*)}{(1+\alpha)^2} \tag{9.113}$$

If the available power gain associated with the optimum noise figure is too low or the associated reflection coefficients are unacceptable, the noise performance must be sacrificed to some extent.

The point with the highest gain on a constant noise figure circle is usually of interest. This point can be determined graphically by finding the noise figure circle that just touches the gain circle of interest. A better alternative is to tabulate the noise figure with the other parameters of interest at different positions around the constant gain circle (or vice versa).

Constant noise figure circles for an Avantek ATF35076 transistor, with series and shunt loading on the output side, are displayed graphically in Figure 9.14, as an example. The optimum noise figure reflection coefficients, as well as contours corresponding to 0.1-dB degradation in the noise figure, are displayed for a number in frequencies in the passband (3.5–4.5 GHz). Circles are displayed for the passband edges, as well as two intermediate frequencies. Note that the circles rotate counterclockwise with increasing frequency. Also note that the actual passband required (3.6–4.4 GHz) was extended in this example. This is generally a good idea.

The output reflection coefficients associated with the optimum source terminations (Z_{out} trace) and the conjugates of the input reflection coefficients associated with a conjugate match on the output side (Z_{in}^* trace) are also displayed. Because the Z_{out} and the Z_{in}^* traces are close to each other, the input VSWR associated with the optimum noise match will be good (around 2.0 in this case). This is the direct result of the series and shunt-loading networks used on the output of the transistor.

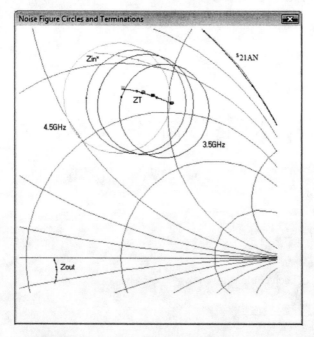

Figure 9.14   The optimum noise source terminations for a modified transistor (Z_T trace) and the noise circles associated with a degradation of 0.1 dB in the noise figure.

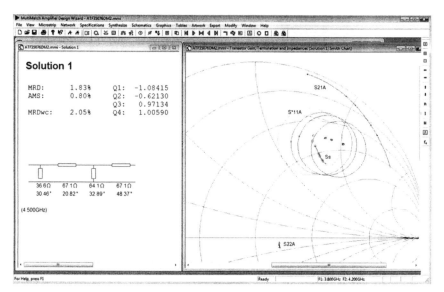

Figure 9.15 An example of a noise matching network synthesized for the circuit shown in Figure 9.14. The trace inside the circles is the reflection coefficients presented to the transistor at the insertion point by the network over the passband.

s_{21A} (normalized to be displayed inside the unit circle on the chart) is also displayed on the graph (the trace in the upper right-hand corner). The square of the magnitude of s_{21A} is also the available power gain of the modified transistor. Note that the gain is constant over the passband of interest. The is again the direct result of the resistive networks added to the transistor.

An example of a noncommensurate distributed network synthesized to minimize the noise figure is shown in Figure 9.15. The networks synthesized is insensitive to changes in the line lengths. The mini-max error associated with the noise figure response targeted will increase from 1.83% to only 2.05% with 1% changes in the line lengths. The source terminations present to the transistor at the insertion point of the matching network are indicated by the S_s trace. The input reflection with the load conjugately matched is indicated by the s_{11A}^* trace.

The Qs listed for the network are the transformation Qs discussed in Chapter 6. Note that the Qs are small.

Note that the shorted stub in the network can be lifted from ground and an rf ground can be provided by adding a large capacitor. This will provide a path for biasing the input of the transistor. The series line on the input side can also be split in two and a blocking capacitor can be inserted. Adjustment of the network may be required after these changes are made. The alternative is to synthesize mixed lumped/distributed networks.

The output reflection coefficients presented by the noise-matching network at the insertion point (the connection point of the two lines on the input side) are also shown in Figure 9.15. The graph shown should be compared with that in Figure 9.14.

9.6   CONTROLLING THE OUTPUT POWER OR THE EFFECTIVE OUTPUT POWER OF A TRANSISTOR

Load-pull equipment or a good large-signal model should be used to find the optimum terminations for a power amplifier when possible. The ideal terminations are usually a compromise between many factors. These include power, distortion, efficiency, stability, and the modulation scheme used.

When load-pull data is not available and a linear amplifier (class A, class B, class AB, class F, ...) is designed, the power performance (1-dB compression point) can be controlled by using the power parameter approach outlined in Chapter 2. An accurate small-signal model and the boundary lines to be used to constrain the load line on the I/V-plane (intrinsic) are required for this purpose.

The power parameter approach can be used to generate power contours for any transistor, with or without feedback and loading networks. When an amplifier stage is designed, the actual output power (P_{out}) is usually of interest, while the effective output power ($P_{out} - P_{in}$) is of interest when an oscillator is designed.

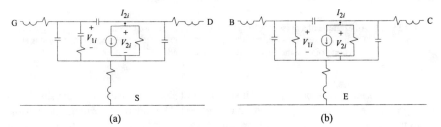

Figure 9.16   Typical small-signal models for (a) FETs and (b) bipolar transistors.

A small-signal model (see Figure 9.16) can be fitted to the measured S-parameters by optimizing initial values estimated for the components in the model to be used. Any information available on the physical transistor or its model (package parasitics, lines, and so forth) should be used to ensure that the model accurately represents the actual device.

The parameters used should be the parameters associated with the dc operating current and voltage at the power level of interest (the bias point usually shifts as the amplifier is driven harder).

The small-signal parameters at the correct operating point are usually adequate for the class A case. The class A parameters can also be used for the class B case if the transconductance is halved. This is required because of the relationship between the fundamental tone current and the actual current (half sinusoid). The same can be done for the class AB case when it is more class B than A. Note that it is a good idea to bias a class B stage slightly towards class A to set the small-signal gain close to the gain at medium power levels.

Large-signal S-parameters can also be used to create the required model when they are available.

If a model is fitted to a packaged transistor and no information is available on the package parasitics, the process is usually simplified by first fitting an intrinsic model only (no parasitics used) to the parameters at the lower end of the frequency range over which *S*-parameter data is available. The package parasitics can then be introduced in the second phase. When this is done, small values should be used as initial values for the unknown parasitics.

When a model is fitted, it is usually a good idea to optimize the fit to the *Y*-parameters of the device first. When a reasonable fit is obtained, the *S*-parameters can be targeted.

The error function used can also be an important factor in fitting a model. It is usually a good idea to start the process with a least-square error. During the final stages of the optimization process, the L_1 error (sum of the absolute values of the relative deviation from the target parameters at the different frequencies) is usually a good choice.

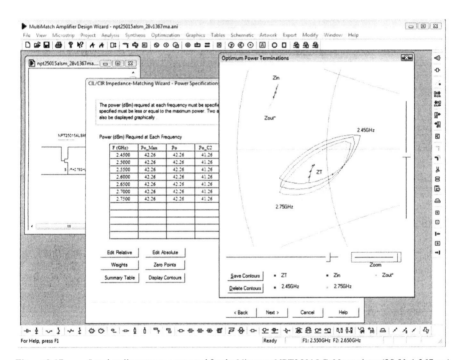

Figure 9.17   Load-pull contours generated for the Nitronex NPT25015 GaN transistor (28-V, 1,367-mA dc operating point).

Constant output power contours were generated for the Nitronex NPT25015 GaN transistor at a number of frequencies in the frequency range 2.45–2.75 GHz [2] and are displayed in Figure 9.17. The load terminations to be presented to the transistor for maximum linear power is showed in Figure 9.18.

The dc operating point was taken to be 28V, and 1,367 mA. The S-parameters required to fit the model used to calculate the power parameters were obtained by using a large-signal model available in Microwave Office. The model fitted is shown in Figure 9.19 and the I/V-curve boundary lines in Figure 9.20. The transistor S-parameters are displayed in Figure 9.21.

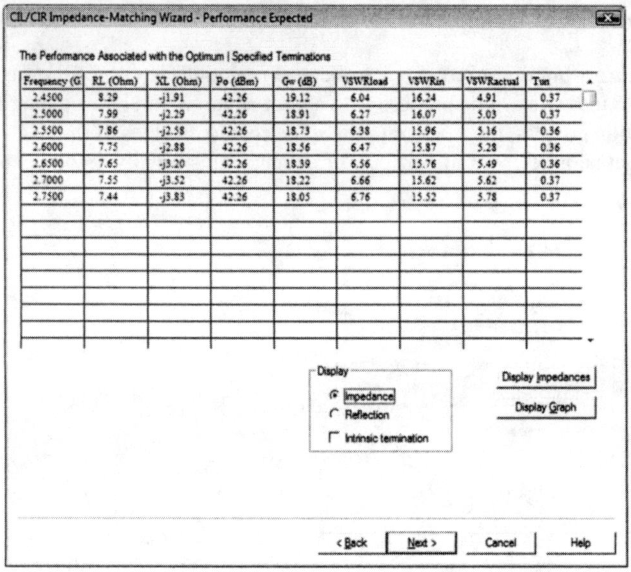

CIL/CIR Impedance-Matching Wizard - Performance Expected

The Performance Associated with the Optimum | Specified Terminations

Frequency (G)	RL (Ohm)	XL (Ohm)	Po (dBm)	Gw (dB)	VSWRload	VSWRin	VSWRactual	Tun
2.4500	8.29	-j1.91	42.26	19.12	6.04	16.24	4.91	0.37
2.5000	7.99	-j2.29	42.26	18.93	6.27	16.07	5.03	0.37
2.5500	7.86	-j2.58	42.26	18.73	6.38	15.96	5.16	0.36
2.6000	7.75	-j2.88	42.26	18.56	6.47	15.87	5.28	0.36
2.6500	7.65	-j3.20	42.26	18.39	6.56	15.76	5.49	0.36
2.7000	7.55	-j3.52	42.26	18.22	6.66	15.62	5.62	0.37
2.7500	7.44	-j3.83	42.26	18.05	6.76	15.52	5.78	0.37

Display
- Impedance
- Reflection
- Intrinsic termination

Display Impedances
Display Graph

< Back   Next >   Cancel   Help

Figure 9.18 The Nitronex NPT25015 terminations required for maximum linear power (28-V, 1,367-mA dc operating point).

The terminations required for maximum linear power (42 dBm) is around 8Ω. The expected operating power gain varies from 18.1 to 19.1 dB over the frequency range 2.45–2.75 GHz.

Note the VSWRs listed in Figure 9.18. The load VSWR listed is a measure of the degree of difficulty of the transforming the actual load ($Z_L = 50\Omega$ in this example) to that required ($Z_{LT} = 8.29 - j1.91\Omega$ at 2.45GHz). It is calculated by first calculating the mismatch between the conjugate of the required load termination (Z_{LT}^*) and the actual load (Z_L) by using (1.77) and then transforming the reflection coefficient to the equivalent VSWR by using

$$\text{VSWR} = \frac{1 + |S_L|}{1 - |S_L|} \tag{9.114}$$

The degree of difficulty of transforming the actual load termination to that required is, therefore, equivalent to a VSWR of around 6. The VSWR associated with the mismatch on the input side of the transistor is around 16.

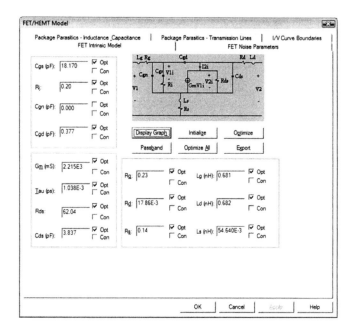

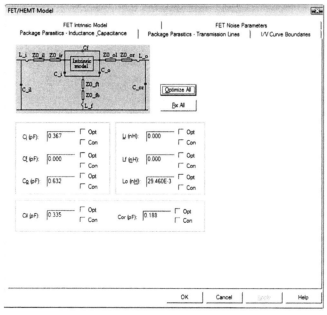

Figure 9.19 The small-signal model fitted to the Nitronex NPT25015 (28-V, 1,367-mA dc operating point).

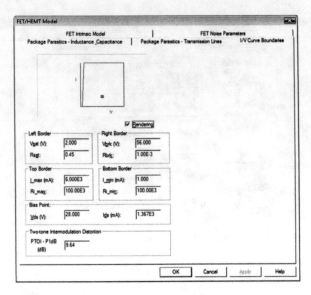

Figure 9.20   The Nitronex NPT25015 terminations required for maximum linear power (28-V, 1,367-mA dc operating point).

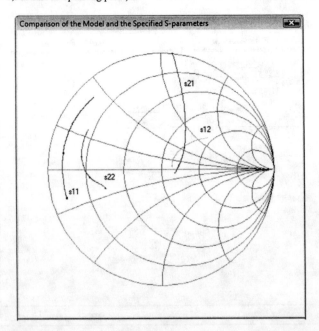

Figure 9.21   The small-signal *S*-parameters of the Nitronex NPT25015 GaN transistor (28-V, 1,367-mA dc operating point).

The value listed as VSWR$_{actual}$ is the actual VSWR (mismatch expressed as a VSWR) that can be expected on the output side of the transistor when the load matching network is in place and the input of the transistor is conjugately matched. Note that a VSWR of 5 to 6 is expected. This mismatch is also indicated graphically in Figure 9.17 (Z_T and Z_{out}^* traces).

An amplifier designed with the Nitronex NPT25015 transistor is shown in Figure 9.22. The expected performance is also shown. Note that resistive networks were used on the input side of the transistor to ensure inherent stability.

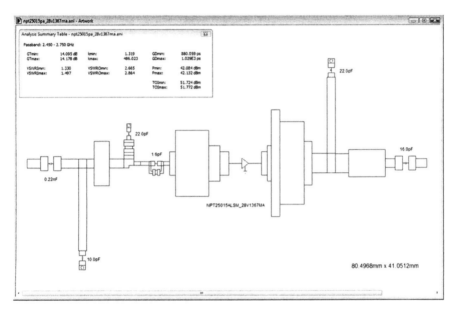

Figure 9.22   An amplifier designed with the Nitronex NPT25015 transistor (28-V, 1,367-mA dc operating point).

Note that when accurate nonlinear models are available for the transistors used, the results obtained with the power parameter approach can and should be refined with a nonlinear simulator. Also note that a nonlinear model may be accurate under limited conditions too (that is, for example at high power levels, but not a low levels).

It is also important to verify the performance of the matching networks synthesized by EM simulation at the higher frequencies, and also when the discontinuities are large. The networks in this example were simulated in the EM simulator provided by Sonnet Software. The results for the input network is shown in Figure 9.23. Note that, while it was not done in this example, it is a good idea to compare the *S*-parameters of each network simulated on the same plot.

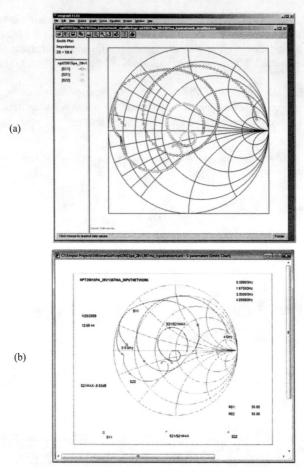

(a)

(b)

Figure 9.23   (a, b) Verification of the performance of the input network designed for the Nitronex NPT25015 amplifier in Sonnet Sofware's EM.

9.7   THE EQUIVALENT PASSIVE IMPEDANCE-MATCHING PROBLEM

It was shown in Section 9.4.1 that the locus of load impedances for which the transducer power gain of a voltage or current source terminated in a passive load will remain constant is a circle in the admittance plane or on the Smith chart. These constant transducer power gain circles always lie in the right-hand side of the admittance plane or inside the Smith chart. Similarly, it was shown that the constant operating, available, or transducer power gain contours for an active two-port are also circles, and, if the two-port is inherently stable, these circles will also be located inside the Smith chart or in the RHS of the admittance plane. The constant noise figure circles are always located inside the Smith chart.

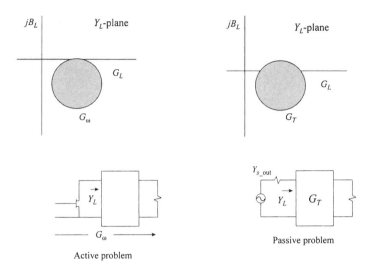

Figure 9.24   Illustration of the equivalence between a constant operating power gain circle and the circle corresponding to mismatching a one-port to a passive load.

By considering the active constant power gain or constant noise figure circles to be the gain circles of a passive source (one-port) terminated in a passive load, the problem of finding a network to transform a given load or source to fall on the circumferences of the relevant active circles can be transformed to that of matching a complex source to a complex load with a specified transducer power gain. This can be done at each of the frequencies of interest whenever the transistor used is inherently stable inside the passband.

The equivalence between a constant operating power gain circle and a passive constant transducer power gain circle is illustrated in Figure 9.24.

9.7.1   Constant Operating Power Gain Case

The equations relevant to finding the output admittance and transducer power gain equivalent to a given operating power gain circle can be derived by setting

$$G_{L\omega} + jB_{L\omega} = G_0 + jB_0 \tag{9.115}$$

$$R_{L\omega} = R_{Y0} \tag{9.116}$$

where $G_{L\omega}$, $B_{L\omega}$, and $R_{L\omega}$, and G_0, B_0, and R_{Y0} are the parameters of the constant operating power gain and the constant transducer power gain circles, respectively.

It follows from (8.96) and (8.97) that

$$G_s\left(\frac{2}{G_T} - 1\right) = G_0 \tag{9.117}$$

and

$$G_s\frac{2}{G_T}\sqrt{1-G_T} = R_{Y0} \tag{9.118}$$

G_s is eliminated if (9.117) is divided by (9.118):

$$\frac{\dfrac{2}{G_T} - 1}{\dfrac{2}{G_T}\sqrt{1-G_T}} = \frac{G_0}{R_{Y0}} = g_0 \tag{9.119}$$

After simplification of (9.119), it follows that

$$G_T^2 + 4(g_0^2 - 1)G_T - 4(g_0^2 - 1) = 0 \tag{9.120}$$

It follows from (9.120) that

$$G_T = \frac{-b \pm \sqrt{b^2 - 4(1)(-b)}}{2(1)}$$

$$= -\frac{b}{2}\left(1 \pm \sqrt{1 + \frac{4}{b}}\right) \tag{9.121}$$

$$= \frac{4(g_0^2 - 1)}{2}\left(1 \pm \sqrt{1 + \frac{4}{4(g_0^2 - 1)}}\right)$$

$$= -2(g_0^2 - 1) \pm 2(g_0^2 - 1)\sqrt{\frac{g_0^2}{g_0^2 - 1}}$$

$$= -2(g_0^2 - 1) \pm 2g_0\sqrt{g_0^2 - 1}$$

$$= 2 - 2g_0^2 + 2g_0\sqrt{g_0^2 - 1} \tag{9.122}$$

Because g_0 is bigger than one, only the positive sign in (9.122) will yield a value of G_T that is bigger than zero.

Equation (9.122) can also be written as

$$G_T = 1 - \left[g_0 - \sqrt{g_0^2 - 1} \right]^2 \tag{9.123}$$

from which it follows that

$$G_{s\_out} = R_{L\omega} G_{T\_out} / (2\sqrt{1 - G_{T\_out}}) \tag{9.124}$$

$$B_{s\_out} = -B_{L\omega} \tag{9.125}$$

$$G_{T\_out} = 1 - \left(\frac{G_{L\omega}}{R_{L\omega}} - \sqrt{\frac{G_{L\omega}^2}{R_{L\omega}^2} - 1} \right)^2 \tag{9.126}$$

While $G_{s\text{-out}}$ appears to be a function of $G_{T\text{-out}}$, it can be shown that (9.124) reduces to the output conductance associated with the highest value of the operating power gain and a conjugate match at the input $(Y_{s\text{-out}} = Y_{L\text{-opt}}^*)$.

If the Smith chart circles are considered, the equivalent passive problem is given by

$$\Gamma_{s\_out} = \Gamma_{L\_opt}^* \tag{9.127}$$

$$G_T = \frac{A_\omega}{2}(\sqrt{1 + 4 / A_\omega} - 1) \tag{9.128}$$

$$A_\omega = \frac{|C_\omega|^2}{R_\omega^2 |\Gamma_{s\_out}|^2} \left[1 - |\Gamma_{s\_out}|^2 \right]^2 \tag{9.129}$$

where Γ_{L_opt} is the load termination associated with the highest operating power gain.

9.7.2 Constant Available Power Gain Case

The equations necessary for transforming an available power gain circle to an equivalent load admittance (equivalent input admittance of the transistor) and transducer power gain are

$$G_{L\_in} = R_{sa} G_{T\_in} / (2\sqrt{1 - G_{T\_in}}) \tag{9.130}$$

$$B_{L\_in} = -B_{sa} \tag{9.131}$$

$$G_{T\_in} = 1 - \left(\frac{G_{sa}}{R_{sa}} - \sqrt{\left(\frac{G_{sa}}{R_{sa}}\right)^2 - 1} \right)^2 \tag{9.132}$$

If the Smith chart circles are considered, the equivalent passive problem is given by

$$\Gamma_{L\_in} = \Gamma_{s\_opt}^* \tag{9.133}$$

$$G_T = \frac{A_{av}}{2} (\sqrt{1 + 4/A_{av}} - 1) \tag{9.134}$$

$$A_{av} = \frac{|C_{av}|^2}{R_{av}^2 |\Gamma_{L\_in}|^2} \left[1 - |\Gamma_{L\_in}|^2 \right]^2 \tag{9.135}$$

where Γ_{s_opt} is the source termination associated with the highest available power gain.

9.7.3   Constant Noise Figure Case

The equations necessary for transforming a constant noise figure to an equivalent load admittance (equivalent input admittance of the transistor) and transducer power gain are

$$G_L = R_{YF} G_{Tn} / (2\sqrt{1 - G_{Tn}}) \tag{9.136}$$

$$B_L = -B_{sF} \tag{9.137}$$

$$G_{Tn} = 1 - \left(\frac{G_{sF}}{R_{sF}} - \sqrt{\left(\frac{G_{sF}}{R_{sF}}\right)^2 - 1} \right)^2 \tag{9.138}$$

and

$$\Gamma_{L\_in} = \Gamma_{s\_nopt}^* \tag{9.139}$$

$$G_T = \frac{A_F}{2} (\sqrt{1 + 4/A_F} - 1) \tag{9.140}$$

$$A_F = \frac{|C_F|^2}{R_F^2 |\Gamma_{L\_in}|^2} \left[1 - |\Gamma_{L\_in}|^2 \right]^2$$ (9.141)

where Γ_{sn_opt} is the source termination associated with the optimum noise figure.

9.8   RESISTIVE FEEDBACK AND LOADING

The main problem during amplifier synthesis is often not the impedance-matching networks to be designed, but rather the feedback and loading sections that should be added to the transistor before the matching is done [2, 15]. The resistive sections used usually strongly modify the transistor at the lower frequencies and have little influence at the higher frequencies where the gain is low and the noise figure is high.

Device-modification has the following advantages:

1.   The stability of the transistor can be improved. Inherent stability over the complete working frequency range of the transistor can often be obtained without degrading the potential performance significantly.

2.   The inherent gain slope of the transistor can be reduced or, ideally, removed over the passband of interest (frequency selective feedback and/or loading).

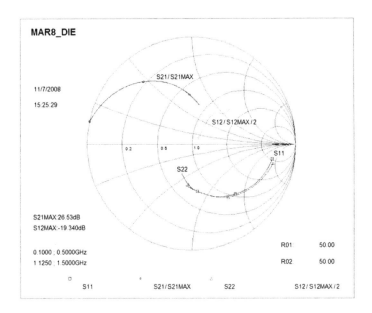

Figure 9.25   The *S*-parameters of the MAR8 die before modification.

3.   The gain-bandwidth constraints associated with the impedance-matching problems to be solved can be reduced.

4.   The optimum gain point can be forced to be closer to the optimum noise point. (When series feedback is used to do this, the feedback will usually be reactive.) This is usually essential if low noise figures with low VSWRs are required without using hybrid couplers or isolators.

5.   The optimum gain point can be forced to be closer to the optimum power point.

With reference to number 3, the difference between the actual impedance in place and that required to get the specified performance from the transistor can be expressed as a reflection coefficient or a VSWR, as explained in Section 9.6. Either of these parameters can be used as a first-order indication of the severity of the matching problem to be solved.

These points will be illustrated with three examples. In the first example a transistor will be modified for improved VSWRs, level gain, and inherent stability. In the second example a low-noise transistor will be modified to get the optimum gain match closer to the optimum noise match with simultaneous flat gain and inherent stability. In the third example a transistor will be modified to get the optimum power load closer to the optimum gain match, again with simultaneous flat gain and inherent stability.

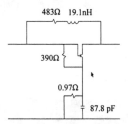

Figure 9.26   A lumped-element modification network for the transistor in Figure 9.25 [2].

EXAMPLE 9.1   Modifying a transistor for flat gain and low input and output VSWRs.

The S-parameters for a transistor (MAR8 die) are shown in Figure 9.25. The input and output VSWRs are poor and the gain is sloping downward over the passband (0.1–1.5 GHz). The transistor is also potentially unstable ($k \geq 0.53$).

The performance with the lumped-element modification network shown in Figure 9.26 is displayed in Figure 9.27. The 390-Ω resistor shown was used to remove the negative g_{11} of the transistor. The gain of the modified transistor is 17.96 ± 0.14 dB. The input VSWR is lower than 1.19 and the output VSWR is lower than 1.14. The modified transistor is just inherently stable.

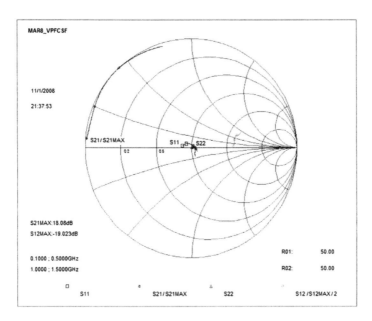

Figure 9.27   The *S*-parameters of the modified transistor.

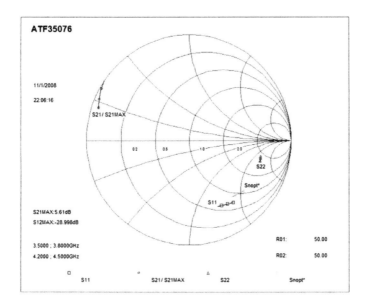

Figure 9.28   The *S*-parameters and the optimum noise impedance of the ATF35076 transistor before modification (passband 3.5–4.5 GHz).

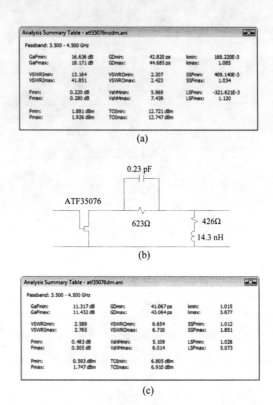

(a)

(b)

(c)

Figure 9.29   (a) The passband performance before modification [2], (b) the lumped-element modification network used [2], and (c) the passband performance after modification [2]. (The stability factors (k, SSF, LSF) are those over the complete frequency range.)

EXAMPLE 9.2   Modifying a transistor to get the optimum noise match closer to the optimum gain match.

The transistor to be modified (ATF35076) was used as the first stage in a low-noise amplifier. The goal was to level the available power gain associated with the optimum noise figure and to get the optimum noise match condition closer to that for optimum gain. Inherent stability was also required.

The passband performance before modification is listed in Figure 9.29(a). The values listed for the stability factors (k, SSF [2] and LSF [2]) are those over the complete frequency range for which data was available. Note that k is less than 1.085 and $G_a(Z_{sn\_opt})$ ("GaFmin") varies from 16.64 to 18.71 dB over the passband.

The source stability factor (SSF) and the load stability factor (LSF) [2] are generalized versions of μ' and μ [10], respectively.

The values listed for "VSWRImin" and "VSWRImax" are the smallest and largest VSWRs (relative to actual source termination; 50-Ω) that can be expected on the input side when the output of the transistor is conjugately matched. The output VSWRs are those associated with the optimum noise match, relative to actual load termination (50Ω in this example).

The *S*-parameters and the optimum noise impedance before modification are shown in Figure 9.28. It is clear from this figure that the input reflection with a 50-Ω load is severe and that the traces for the optimum noise match and the input reflection coefficient are far apart.

It is important to realize that the terminations for the transistor are taken to be 50Ω in Figure 9.28 and that the actual terminations will be different.

The performance associated with the lumped-element modification network displayed in Figure 9.29(b) is listed in Figure 9.29(c). Note that $G_a(Z_{sn\_opt})$ has been leveled (In this example the gain is actually level over a very wide band). The noise figure has been degraded slightly (better than 0.5 dB; previously better than 0.28 dB), and the modified transistor is inherently stable at all frequencies.

The *S*-parameters and the optimum noise impedance for the modified transistor are shown in Figure 9.30. Note that from the s_{11} and S_{nopt}^* traces (50-Ω load termination) the optimum noise match is now much closer to the optimum gain match.

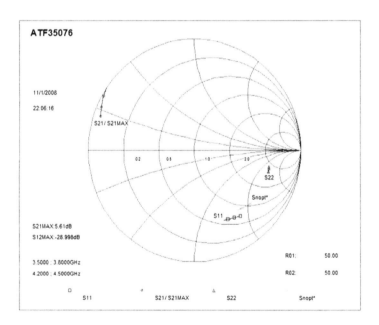

Figure 9.30 The *S*-parameters and the optimum noise impedance of the ATF35076 transistor after modification with lumped elements (passband 3.5–4.5 GHz).

Pads and connecting lines are required in a real modification network. The performance associated with a more realistic network is listed in (Figure 9.31).

Note that the gain is in this example was set to around 10 dB. The noise figure is around 0.55 dB.

Note that the "VswrI" values listed in Figure 9.31 define the range of the input VSWR values associated with the optimum noise match and a conjugate match on the output side of the transistor. The input VSWR will vary between 1.867 and 1.921 over the passband if the relevant matching problems can be solved perfectly. Similarly, the output VSWR (with the optimum noise matching network in place and before matching the output side) will vary between 1.2 and 1.3. The output VSWRs were calculated for a 50-Ω load. An output matching network is clearly not required in this example. Note that the drain biasing can be fed through the shunt resistive loading path if the dc ground is replaced with an RF ground (large capacitor). A blocking capacitor is still required.

The "VsNMax" value listed in Figure 9.31 is the maximum of the VSWR values calculated for the optimum noise impedances relative to the physical termination for the stage (50Ω in this case). These VSWRs serve as a measure of the degree of difficulty of the noise matching problem.

The S-parameters of the transistor after modification are shown in Figure 9.32.

It is important to realize that a distorted picture can be obtained if only the Smith chart results are interpreted. As mentioned above, the actual terminations of the modified transistor will not be 50Ω and the impedances associated with the actual terminations will be different. The performance associated with the actual termination should be evaluated and targeted during the optimization process.

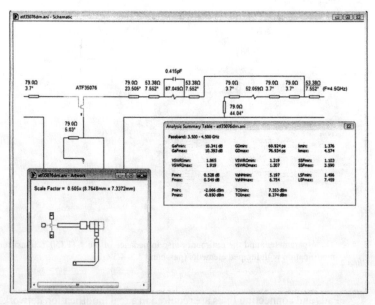

Figure 9.31   The topology of a more realistic modification network for the transistor in Example 9.2 with the associated performance (electrical line lengths specified at 4.5 GHz).

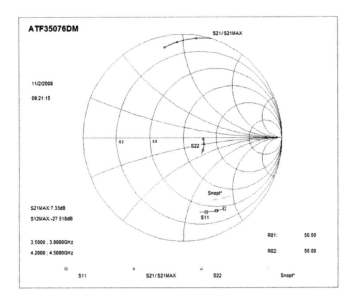

Figure 9.32   The *S*-parameters and optimum noise impedance associated with the distributed modification network [2].

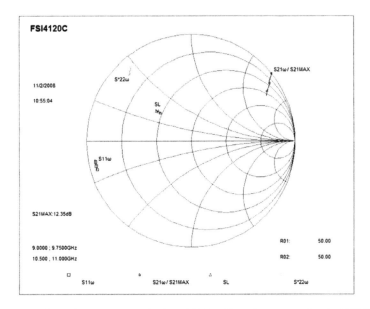

Figure 9.33   The optimum power terminations (S_L) for a Texas Instruments foundry FET [2] (V_{sat} = 0.55V; R_{sat} = 1.86Ω; R_{ds_max} = 100 kΩ = R_{ds_min}; dc operating point: 8V, 180 mA).

Table 9.2   The Estimated Optimum Power Termination of the Foundry FET with the Associated Output
Power and Small-Signal Operating Gain [2]

Frequency	Load termination	Output power	Power gain
(GHz)	(Ω)	(dBm)	(dB)
9.00	$24.0 + j15.2$	27.9	12.4
9.25	$23.6 + j15.3$	27.9	12.1
9.50	$23.3 + j15.3$	27.9	11.9
9.75	$22.6 + j15.4$	27.9	11.7
10.00	$22.3 + j15.4$	27.9	11.6
10.25	$21.9 + j15.4$	27.9	11.3
10.50	$21.4 + j15.4$	28.0	11.1
10.75	$21.0 + j15.4$	28.0	11.0
11.00	$20.8 + j15.4$	28.0	10.8

Note that while the output VSWR in this case is a measure of the mismatch
between the output impedance of the transistor (Z_{out}) and a 50-Ω load, it can also
be used as a measure of the difference between the actual load (50Ω in this case)
and the load required by the modified transistor (Z_{out}^* in this case). If interpreted in
this way, the VSWR becomes a measure of how difficult the associated matching
problem will be. With a predefined passband, this approach usually yields good
results. The alternative is to calculate the exact gain-bandwidth constraints
associated with the matching problem.

EXAMPLE 9.3   Modifying a power transistor to improve its stability and the
VSWRs associated with an optimum power match.

The optimum power termination and the associated small-signal operating power
gain for a Texas Instruments foundry FET (without modification) are shown in
Figure 9.33 and listed in Table 9.2. Note that the traces corresponding to the
optimum power match (S_L) and the optimum gain match ($s_{22\omega}^*$ trace) are far apart.
The operating power gain ($s_{21\omega}$ trace in the upper RHS corner) is also decreasing
with increasing frequency.

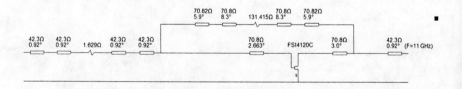

Figure 9.34   The modification circuit designed for the foundry FET [2].

Table 9.3   The Optimum Power Terminations of the Modified Foundry FET with the Associated
Output Power and Small-Signal Gain [2]

Frequency	Load termination	Output power	Power gain
(GHz)	(Ω)	(dBm)	(dB)
9.00	23.0 + j19.1	26.6	7.5
9.25	22.4 + j18.7	26.6	7.5
9.50	21.5 + j18.3	26.7	7.4
9.75	20.9 + j17.9	26.7	7.5
10.00	20.5 + j17.5	26.8	7.5
10.25	20.0 + j17.1	26.8	7.4
10.50	19.6 + j16.7	26.9	7.4
10.75	19.0 + j16.2	26.9	7.4
11.00	18.5 + j15.8	26.9	7.4

The transistor is potentially unstable.

The modification network used is shown in Figure 9.34. The electrical line lengths of the pads used are specified at 11 GHz. The optimum power termination and the gain after modification are shown in Figure 9.35. The numerical values are listed in Table 9.3.

The optimum power match (S_L trace in Figure 9.35) is now much closer to the optimum gain match ($s_{22\omega}^*$ trace). Note that the gain is now very flat (although it is on the low side). The maximum power obtainable has decreased by 1.4 dBm. The modified transistor is inherently stable ($1.379 \le k \le 2.389$).

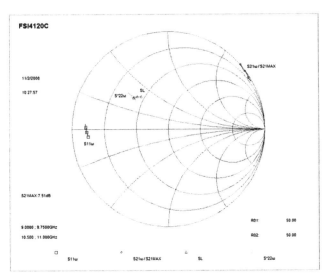

Figure 9.35   The optimum power termination and small-signal gain for a foundry FET (Texas Instruments FSI4120C) after modification [2].

9.9   DESIGNING CASCADE AMPLIFIERS

At this point the basic knowledge required to design single or multistage cascade type amplifiers are in place. A typical design cycle proceeds from the load side toward the source, or vice versa. A low-noise design is usually done by starting the design at the input side. When the output power is more important, the design is usually started at the load side.

When a multistage high dynamic range amplifier is synthesized, the design can be started at both sides and the two sections can then be linked up with an interstage matching network [refer to Figure 9.8(c)]. In the single-stage case the load network can first be designed for maximum output power after which the input network can be designed to level the gain with the noise figure as low as possible (this can be done by choosing the optimum noise figure points on the relevant constant gain circles).

The design of each stage consists of selecting a transistor for the stage, adding feedback and/or resistive loading sections to it (if required, and if possible), and synthesizing a lossless gain, noise figure, or power control network for it. If the associated matching problem is too difficult to be solved properly, the transistor should be modified more strongly or a different transistor should be used.

When the control network for each stage is designed, the performance around the relevant constant gain, noise figure, or output power circle should be evaluated. The options to match to a specific point on each circle (a point match) or to any arbitrary point on the circle (circle match) exist. If the performance is only acceptable in a narrow region on the circle circumference, a point-match should be enforced.

As an example of the options available around a typical circle, the performance of a transistor around a constant noise figure circle is displayed in Table 9.4. The following values are listed in the table as a function of the angle around the constant noise figure circle (Smith chart case):

(a)   The reflection coefficient at the point of interest (Γ_{L_mag}, Γ_{L_ang});

(b)   The available power gain (G_a);

(c)   The output power if the output side is conjugately matched;

(d)   The difference between the actual source termination (50Ω in this case) and the source termination required expressed as a VSWR;

(e)   The sensitivity of the noise figure to changes in the admittance presented at the input of the modified transistor (δ_n);

(f)   The sensitivity of the available power gain to changes in the source admittance (δ_a);

(g)   The sensitivity of the output match to changes in the source admittance (δ_{vs}).

Table 9.4   The Performance of a Modified Transistor Around a Constant Noise Figure Circle [2]

θ	G_a	Power	VSWR	δ_n	δ_g	δ_{vs}	Γ_{L_mag}	Γ_{L_ang}
(°)	(dB)	(dBm)	—	(%)	(%)	—	—	(°)
0.0	10.54	1.63	9.05	0.17	1.85	0.02	0.80	36.35
25.0	10.86	1.58	10.96	0.22	2.30	0.02	0.83	40.90
50.0	11.56	1.59	11.51	0.25	2.26	0.03	0.84	45.82
75.0	12.37	1.71	10.26	0.24	1.55	0.03	0.82	50.65
100.0	12.87	1.90	8.14	0.20	0.73	0.03	0.78	54.89
125.0	12.94	2.01	6.18	0.16	0.47	0.03	0.72	57.97
150.0	12.73	1.98	4.72	0.13	0.58	0.03	0.65	59.12
175.0	12.42	1.99	3.77	0.10	0.64	0.02	0.58	57.47
200.0	12.07	2.02	3.23	0.09	0.67	0.02	0.53	52.42
225.0	11.73	2.03	3.06	0.08	0.69	0.02	0.51	44.72
250.0	11.40	1.97	3.24	0.08	0.73	0.02	0.53	37.05
275.0	11.09	1.90	3.70	0.08	0.81	0.01	0.58	32.05
300.0	10.81	1.83	4.74	0.09	0.95	0.01	0.65	30.46
325.0	10.59	1.75	6.20	0.11	1.21	0.01	0.72	31.65
350.0	10.50	1.66	8.18	0.15	1.64	0.02	0.78	34.75

If matching to any point on a circle is acceptable (circle match), the equivalent passive problem can be defined for the circle as described in Section 9.7. Matching to a specific point may also be required. The highlighting in Table 9.4 is used to indicate the optimum point on the circle circumference for both cases.

The sensitivity factor is calculated by considering the change in the parameter of interest when the controlling admittance changes by 1%. Calculation of the operating power gain sensitivity factor (δ_ω) is demonstrated in Figure 9.36. The lowest and highest gain associated with the tolerance circle are G_{ω_min} and G_{ω_max}, respectively.

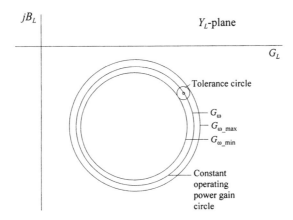

Figure 9.36   Calculation of the sensitivity factor associated with the operating power gain (δ_ω) [2].

The sensitivity factor (δ_ω) is calculated as the maximum of

$$\delta_{\omega 1} = \text{ABS}[(G_{\omega\_max} - G_\omega) / G_\omega] \tag{9.142}$$

and

$$\delta_{\omega 2} = \text{ABS}[(G_{\omega\_min} - G_\omega) / G_\omega] \tag{9.143}$$

High values for any of the sensitivity factors are undesirable. Note that the sensitivity factors calculated are indications of the sensitivity of the problem to be solved.

Different parameters like the output power and the operating power gain can also be displayed simultaneously on the same graph, but it is easier to make an appropriate selection from a table. As an example, constant operating gain contours and constant output power contours are displayed in Figure 9.37 for a Nitronex NPT25015 GaN transistor biased in class B mode. An approximate class B model based on the class A parameters provided in the Nitronex data sheet (bias point: 28V 200 mA) was used to generate the contours.

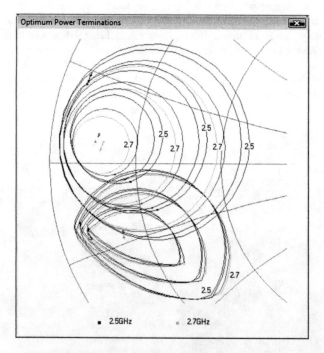

Figure 9.37   Constant operating power gain and constant output power contours for the Nitronex NPT25015 GaN transistor based on class B parameters estimated from the class A data provided in the data sheet.

CIL/CIR Impedance-Matching Wizard - Potential Performance on the Selected Contours

Stage Output Power Targeted: 44.00 dBm

Angle (°)	RL (Ohm)	XL (Ohm)	Gw (dB)	Tun	VSWRload	VSWRin	VSWRactual	Eff (%)
0.00	13.7704	-j10.75	10.21	59.73E-3	7.58	3.52	4.87	55.77
12.50	13.0942	-j8.97	10.57	61.11E-3	6.61	3.56	4.32	59.49
25.00	12.1833	-j7.67	10.84	62.72E-3	5.90	3.58	3.93	62.61
37.50	11.1953	-j6.77	11.03	64.41E-3	5.39	3.59	3.66	65.10
50.00	10.2779	-j6.17	11.16	65.97E-3	5.02	3.59	3.49	66.90
62.50	9.4172	-j5.79	11.23	67.49E-3	4.77	3.58	3.39	68.08
75.00	8.6312	-j5.56	11.26	68.94E-3	4.61	3.56	3.36	68.68
87.50	7.9029	-j5.41	11.25	70.37E-3	4.52	3.53	3.37	68.72
100.00	7.2021	-j5.36	11.19	71.94E-3	4.52	3.50	3.45	68.18
112.50	6.5138	-j5.38	11.08	73.73E-3	4.61	3.45	3.59	66.99
125.00	5.8174	-j5.47	10.91	75.95E-3	4.83	3.39	3.84	64.99
137.50	5.0910	-j5.63	10.63	78.91E-3	5.25	3.32	4.23	61.93
150.00	4.3219	-j5.89	10.20	83.13E-3	6.02	3.22	4.90	57.49
162.50	3.5063	-j6.29	9.51	89.32E-3	7.50	3.10	6.09	51.37
175.00	3.0892	-j7.03	8.80	94.89E-3	9.50	3.00	7.51	45.84
187.50	3.6740	-j7.98	8.85	92.70E-3	9.72	3.01	7.39	45.80

Frequency (GHz):

2.5000 GHz
2.5000 GHz
2.5500 GHz
2.6000 GHz
2.6500 GHz
2.7000 GHz

☑ Point Match

Selected angle: 87.50'

Display
- Impedance
- Admittance
- Reflection

Display
- Actual load
- Intrinsic load

Zoom In Zoom Out Reset

< Back   Next >   Cancel   Help

Figure 9.38   The estimated performance of a Nitronex NPT25015 GaN transistor around the 44 dBm power contour at 2.5 GHz (class B).

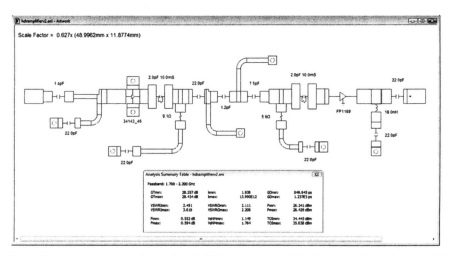

Figure 9.39   An example of a single-ended high-dynamic amplifier designed to be used with hybrid couplers.

An example of a data tabulated in [2] for a constant output power contour is given in Figure 9.38. The best point on the contour is highlighted. The default selection is controlled by using an error function, but the selection made can be changed by the user. Note the change in the gain and efficiency around the contour, as well the variation in the degree of difficulty (relative to the normalization resistance chosen) of the matching problems to be solved.

It is important to realize that the output power is highly dependent on the impedance presented to the transistor. If the load resistance is doubled, the power will be halved, while the gain will only change with 0.5 dB (passive mismatch). When a multistage power amplifier is designed it may be necessary to separate stages by using hybrid couplers or circulators in order to limit the potential change in the load termination of each power stage by the changes or tolerances in the other stages, or the external load terminations.

Note that the terminations can also change with temperature. The change in gain can usually be compensated by using temperature-dependent pads (at the cost of additional losses in the circuit).

An example of a high dynamic range two-stage amplifier designed with [2] is provided in Figure 9.39. The performance of the amplifier is also summarized in the figure. Because two of these amplifiers were combined by using hybrid couplers some margin was allowed on the input and output reflection coefficients of the single-ended amplifier. Note that the couplers can be designed first, and the single-ended amplifier can then be designed to allow for the actual source and load terminations presented to it by the couplers.

9.10   LOSSLESS FEEDBACK AMPLIFIERS

Lossless feedback networks implemented with transformers or directional couplers [6] can be used to remove the gain slope of a transistor instead of using resistive feedback or loading networks. When this is done, no (or very little) power is dissipated in the feedback networks; most of the power generated by the transistor ends up in the load. At the same time any distortion in the output voltage and/or the output current is reduced by the feedback.

Low input and output VSWRs can be obtained by choosing the correct turns ratios for the transformers. Most lossless feedback circuits are designed to minimize the VSWRs. It should be noted that the noise figure and the output power, and, therefore, the dynamic range, are not necessarily optimized when this is done.

A circuit using the principle of lossless feedback (Figure 9.40) was patented in November 1971 by D. Norton (Anzac Corporation) in the United States [16]. Two transformers are used and configured so that the load voltage is sampled by one and fed back as a voltage in series with the input of the amplifier, while the other one samples the load current and feeds it back as a current to the input of the amplifier (voltage-series and current-shunt feedback). The turns ratios of the transformers were chosen to create a directional coupler arrangement, the main purpose of which was to provide excellent VSWRs and to control the gain. Because of the directional coupler arrangement, any power entering the input port is directed to the input of the transistor and to the output port, and any power generated by the transistor is directed back to its own input as (negative) feedback and to the load as external power.

The main advantage of this circuit is that gain leveling at very low gain values can be obtained without a degradation in performance relative to that associated with the highest gain obtainable with this configuration. It has the disadvantage that any power incident on the output port (s_{12}) will be directed toward the input port and the output port of the transistor. The isolation of this amplifier, therefore, is usually poor, especially when an amplifier with low gain is designed.

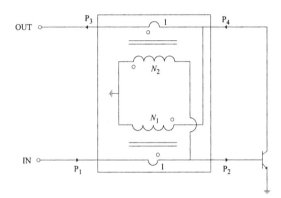

Figure 9.40   The lossless feedback amplifier patented by David E. Norton in 1971 [16].

The isolation problem was solved by Q-bit [17] by using two couplers instead of only one (Power Feedback Technology). In this arrangement (refer to Figure 9.41) the power incident on the output side is directed at the transistor and the termination of the input coupler, instead of the input port. The isolation, therefore, tends to be that of the transistor only plus the through losses of the two couplers (which should be small). However, some (most) of the power fed back is dissipated in the termination of the input coupler. This actually violates the principle of lossless feedback.

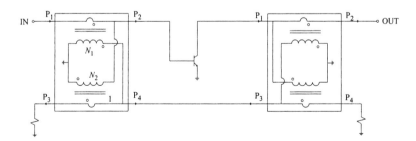

Figure 9.41   The lossless feedback circuit (Power Feedback Technology) patented by Q-bit [17].

The input power in this Q-bit circuit is directed at the input of the transistor and the termination of the input coupler, which degrades the noise figure (slightly) and dissipates some of the input power.

An alternative circuit (Figure 9.42) was introduced in [18]. The output voltage across the load and the output current are sampled with two of the windings of an impedance-matching transformer with three windings, and both the current and the voltage are fed back to the third winding, which is in series with the input terminal of the transistor (the winding that samples the current determines the current through the input winding, and the winding that samples the voltage determines the voltage across the input winding). The impedance associated with the third winding, therefore, is completely determined by the voltage and current sampled (alternatively, the input impedance required would determine the ratio between the voltage and the current feedback).

Ideally the input impedance of the transistor used should approximate a short-circuit in this arrangement, while its output impedance should look like an open circuit (a bipolar transistor used in the common-base configuration can usually be used to present such impedances).

It should be noted that while the output current of the transistor is actually sampled in this circuit, the transformer arrangement and the fact that the input current is very low compared to the output current ($i_b = i_c / \beta$) force the load current to be directly proportional to the transistor current.

With a correct choice of the turns ratio of the transformer, a two-way impedance match can be obtained easily with this arrangement.

This circuit is frequently used. It has the advantages that no power is dissipated in the terminations of directional couplers and that the impedances presented by the circuit to the transistor tend to approximate those required for optimum output power in a common-emitter or a common-base configuration at low currents and at the lower frequencies.

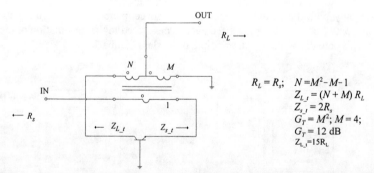

$R_L = R_s;$ $N = M^2 - M - 1$
$Z_{L\_t} = (N + M)\, R_L$
$Z_{s\_t} = 2R_s$
$G_T = M^2;\ M = 4;$
$G_T = 12$ dB
$Z_{L\_t} = 15R_L$

Figure 9.42   The impedance-matching transformer configuration patented by Norton and Podell [18].

The high load impedance at the collector ($15R_L$ when the terminations are equal and when the transformer is designed for an amplifier gain of 12 dB) tends to limit the bandwidth obtainable with this arrangement. The isolation is also not as good as that obtainable with the Q-bit circuit, but a two-stage design can be used to improve this (refer to Figure 9.43).

A higher gain cascade version of this type of amplifier is shown in Figure 9.43. The gain claimed for this amplifier is 19 dB over the bandwidth 70–570 MHz.

The single transistor version of this amplifier (see Figure 9.44) can also be used with the transistor in a common-emitter configuration instead of a common-base configuration (the correct configuration is obtained by simply rotating the emitter/winding combination to be the common branch). A transformer will be required on the input to transform the high input impedance downward as shown by Rohde in [19]. However, the input impedance will be a stronger function of the transistor parameters than was the case with the common-base configuration with its low (negligible) input impedance.

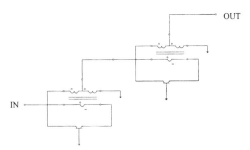

Figure 9.43   A cascade example of a high dynamic range amplifier using lossless feedback based on the impedance-matching transformer principle [6].

A better alternative would probably be to use the original configuration with an input transformer to provide an additional degree of freedom on the design parameters, if required.

An interesting variation on lossless feedback with transformers was also introduced by Rohde in [20]. In this variation the load current is sampled and fed back as a current, and the output voltage (actually the voltage across the transistor) is sampled and fed back as a voltage in series with the input (current-shunt and voltage-series feedback). The circuit is shown in Figure 9.45.

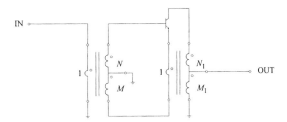

Figure 9.44   A different variation of the impedance-matching transformer lossless feedback amplifier [16].

If wideband performance is required, the best choice seems to be a modified version of the Norton coupler circuit (see Figure 9.46). The Norton coupler circuit was originally investigated for bipolar transistors only and was considered to be a good solution only if the transistor to be used had input and output impedances that were closely matched to the terminations presented to the coupler circuit. However, excellent results can be obtained by using FETs (capacitive input impedance; resistive output impedance) in this circuit. While the original coupler circuit used two identical transformers, it was found that better results could be obtained by increasing the turns ratio for the input transformer, that is, when a FET is used (an alternative is to increase the source impedance). This circuit is also not very sensitive to reduction of the coupling factor by leakage flux, and a simple shunt capacitor can be used to compensate for the effect as long as the coupling factor remains fairly good ($k > 0.9$). The isolation of the modified Norton coupler circuit also turns out to be much better than expected.

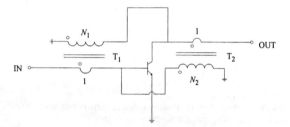

Figure 9.45   Another lossless feedback amplifier configuration introduced by Rohde [20].

It should be noted that the magnetizing inductance required in a modified Norton coupler amplifier is not only a function of the terminations and the lowest frequency at which acceptable performance is required, but is also a function of the transconductance and the input and output impedances of the transistor to be used. The transconductance also determines the isolation (reverse gain) of the amplifier. Lower transconductance values are associated with better isolation.

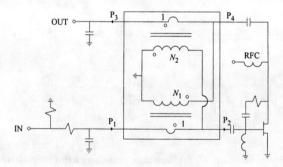

Figure 9.46   A modified version of the Norton coupler amplifier [2].

The performance obtainable with the modified Norton coupler circuit is impressive. Assuming a coupling factor of unity and no interwinding capacitance, an amplifier was designed over the passband 10 MHz to 1 GHz. The expected power gain was approximately 10.49 dB, the 1-dB compression point was close to 23 dBm, and the isolation was better than 19.7 dB. The expected efficiency was around 39%. The expected input VSWR was smaller than 1.81 over the whole band and less than 1.5 up to 625 MHz. The expected output VSWR was smaller than 1.5 over the whole band.

The performance of a manufactured prototype turned out to be close to that predicted except for the upper end of the passband, which was reduced to around 500 MHz. A modification to the circuit was also required to eliminate oscillations above 2 GHz.

The transistor was biased with an active biasing circuit. An active biasing circuit suitable for FETs is shown in Figure 9.47. V_C is set by the biasing resistors of the bipolar transistor and the junction voltage (V_{be}), while the drain current of the FET is almost equal to the current in R_d.

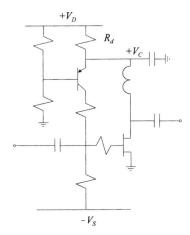

Figure 9.47   An active biasing circuit suitable for FETs.

9.11   REFLECTION AMPLIFIERS

At very high frequencies, Impatt, Gunn, and tunnel diodes are also used to provide amplification. These negative resistance single-port devices are usually used in combination with circulators and occasionally with 3-dB hybrid couplers. Only the circulator-type will be considered here.

The S-parameter matrix of an ideal circulator is given by

$$S = \begin{bmatrix} 0 & 0 & 1 \\ 1 & 0 & 0 \\ 0 & 1 & 0 \end{bmatrix} \tag{9.144}$$

This implies that

$$\begin{bmatrix} b_1 \\ b_2 \\ b_3 \end{bmatrix} = \begin{bmatrix} a_3 \\ a_1 \\ a_2 \end{bmatrix} \tag{9.145}$$

and, therefore, the energy incident at port 1 is always delivered to the load connected to port 2, the energy incident at port 2 to the load connected to port 3, and the energy incident at port 3 to the load connected to port 1. Consequently, the energy is propagated in a circular fashion around the circulator; hence the name circulator. These relationships are illustrated in Figure 9.48. The configuration of a circulator-type reflection amplifier is shown in Figure 9.49.

The transducer power gain of the amplifier is defined by

$$G_T = P_L / P_{av-E} \tag{9.146}$$

where P_{av-E} is the power available from the source. By using the relationships shown in Figure 9.49, it follows that

$$G_T = \frac{|b_3|^2}{|a_1|^2} = \frac{|a_2|^2}{|b_2|^2} = \frac{|b_{M1}|^2}{|a_{M1}|^2} = \frac{|b_{M2}|^2}{|a_{M2}|^2}$$

$$= \left| \frac{Z_{out} - Z_D^*}{Z_{out} + Z_D} \right|^2 \tag{9.147}$$

where $Z_D = -R_D + jX_D$ is the impedance of the negative resistance diode, and Z_D^* indicates its conjugate.

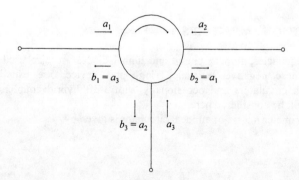

Figure 9.48   The relationships between the normalized incident and reflected components of an ideal circulator.

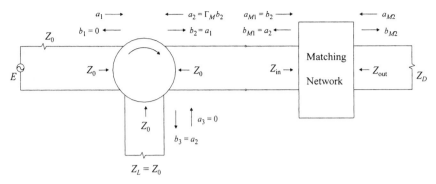

Figure 9.49   The configuration of a circulator-type reflection amplifier.

Equation (9.147) can be manipulated in the following way:

$$
\begin{aligned}
G_T &= \left| \frac{Z_{out} - (-R_D - jX_D)^*}{Z_{out} + (-R_D + jX_D)} \right|^2 = \left| \frac{Z_{out} + (R_D + jX_D)}{Z_{out} - (R_D + jX_D)^*} \right|^2 \\
&= 1 / \left| \frac{Z_{out} - (R_D + jX_D)^*}{Z_{out} + (R_D + jX_D)} \right|^2 \\
&= 1 / \left| \Gamma_{D+} \right|^2
\end{aligned}
$$

(9.148)

where Γ_{D+} is the reflection parameter of the network shown in Figure 9.50 with the source and load impedances shown as normalizing impedances. The problem of maximizing the gain of a circulator-type reflection amplifier, therefore, is equivalent to minimizing the mismatch between the source and the load shown in Figure 9.50.

When the amplifier is designed to have a specified gain versus frequency response, the gain of the equivalent matching network should be

$$
G_{TM} = 1 - 1 / G_T
$$

(9.149)

where G_T is the transducer power gain specification for the reflection amplifier.

EXAMPLE 9.7   Designing a matching network for a reflection amplifier.

As an example of designing the matching network of a reflection amplifier, a matching network will be designed for a Gunn diode (M/A-COM, MA-49110) with input impedance corresponding to the load impedance of the corresponding equivalent matching problem (as given in Table 9.5) and a gain of 10 dB across the passband 7–9 GHz.

With the required transducer power gain equal to 10 dB, the transducer power gain of the equivalent matching problem is found to be

$$G_{TM} = 1 - 1/G_T = 1 - 1/10.0 = 0.90$$

The specifications of the equivalent matching problem is shown in Table 9.5 and the designed matching network is shown in Figure 9.51. The maximum deviation from the specified gain response is 0.16 dB, and the transformation Q-factors corresponding to the solution are 1.183, 1.506, and 0.511, respectively.

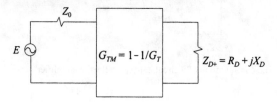

Figure 9.50   The matching problem to be solved when the amplifier in Figure 9.49 is designed.

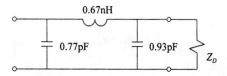

Figure 9.51   The designed matching network for the reflection amplifier of Example 9.7.

9.12   BALANCED AMPLIFIERS

In a balanced amplifier, the input signal is split into two or more amplifiers, and the output signals of these amplifiers are combined to a single load, with isolation between the individual amplifier ports in both cases. The most commonly used configuration is shown in Figure 9.52.

The S-parameter matrix of a 3-dB, 90° hybrid divider is given by [21], with the ports numbered as in Figure 9.52. For the divider, the energy incident at port 1, therefore, is delivered to the loads connected to ports 2 and 3 with a 90° phase shift between the two components. The energy incident at ports 2 and 3 in the combiner is routed to port 1, again with a 90° phase shift between the two components.

$$S_{Hd} = 0.707 \begin{bmatrix} 0 & j & 1 \\ j & 0 & 0 \\ 1 & 0 & 0 \end{bmatrix} \tag{9.150}$$

and that for a 3-dB, 90° hybrid combiner by

Table 9.5   The Specifications for the Output Matching Network of the Reflection Amplifier

Frequency (GHz)	Source impedance (Ω)	Load impedance (Ω)	Transducer power gain
7.0	50.0 + j0.00	10.0 + j3.0	0.900
7.5	50.0 + j0.00	12.0 + j7.0	0.900
8.0	50.0 + j0.00	15.0 + j10.0	0.900
8.5	50.0 + j0.00	19.0 + j13.0	0.900
9.0	50.0 + j0.00	25.0 + j15.0	0.900

$$S_{Hd} = 0.707 \begin{bmatrix} 0 & 0 & j \\ 0 & 0 & 1 \\ j & 1 & 0 \end{bmatrix} \tag{9.151}$$

The S-parameter matrix of the amplifier is given in terms of the S-parameters of the two individual amplifiers by [21]

$$S_T = 0.5 \begin{bmatrix} s_{11,1} - s_{11,2} & j(s_{12,1} + s_{12,2}) \\ j(s_{21,1} + s_{21,2}) & -s_{22,1} + s_{22,2} \end{bmatrix} \tag{9.152}$$

It is clear from this equation that if amplifiers 1 and 2 are identical, the input and output reflection parameters of the balanced amplifier will be equal to zero, even when the reflection parameters of the individual amplifiers are not equal to zero. As long as the individual amplifiers are almost identical, the input and output VSWRs of a balanced amplifier will therefore be very low, independent of the VSWRs of the individual amplifiers.

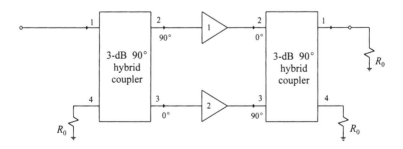

Figure 9.52   The most commonly used balanced amplifier configuration.

The transducer power gain of the balanced amplifier is given by

$$G_T = 0.25 \left| s_{21,1} + s_{21,2} \right|^2 \tag{9.153}$$

When the individual amplifiers are identical, this reduces to

$$G_T = \left| s_{21,1} \right|^2 \tag{9.154}$$

which is identical to the gain of a single amplifier.

Although the gain of the balanced amplifier is therefore identical to that of each individual amplifier in the ideal case, the output power is twice that obtainable by using only a single-ended stage.

Should one of the amplifiers comprising the balanced amplifier fail, the gain will be reduced to one-fourth of its original value. This can be proved easily by setting $s_{21,1}$ in (9.153) equal to zero. In some applications this advantage can be an important factor when deciding whether a balanced or single-ended amplifier should be used.

9.13   OSCILLATOR DESIGN

Oscillators can be designed by controlling the reflection coefficient (negative resistance) or the loop gain of the transistor [2, 6, 8, 12]. The better alternative usually is to control the loop gain. Independent of how the design was done, both conditions should be checked to ensure that spurious oscillations will not occur.

The two basic oscillator configurations are shown in Figure 9.53. Voltage-shunt feedback is used in Figure 9.53(a), while current-series feedback is used in Figure 9.53(b).

In order to control the output power of an oscillator, the load termination presented to the transistor should be controlled too. The load termination can be controlled easily by first modifying the basic configurations to those shown in Figure 9.54 [22]. In the case of the series feedback, the original ground connection was floated and a virtual ground was introduced. No physical change is required in the shunt feedback circuit.

In the series feedback case, any transmission lines used should first be converted to lumped T- or PI-section equivalents before the ground connection is changed.

Any extension lines should be kept as short as possible. The extra phase shift around the loop will reduce the frequency range over which oscillation is possible and will also increase the start-up time.

A simplified flow diagram of the oscillator design process is shown in Figure 9.55.

In order to control the output power, power contours can be generated for the transistor by using the power parameter approach described in Chapter 2 or by using a nonlinear simulator. A suitable load line can then be selected, after which a feedback network can be designed to provide this load termination to the transistor with the loop gain required. Ideally, the load line at steady-state should be controlled (which imply using a large-signal model or large-signal S-parameters), but excellent results can usually also be obtained by using the small-signal parameters.

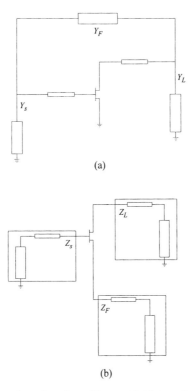

(a)

(b)

Figure 9.53 The basic configurations for oscillators with (a) shunt feedback and (b) series feedback.

The output power of an oscillator will increase initially as it is driven harder into compression, after which it will decrease. The transistor will be driven harder into compression as the loop gain or the negative resistance in the input loop (series feedback case) is increased.

The gain compression associated with the maximum effective output power can be estimated by assuming the power saturation characteristic to be governed by an exponential law function [23]. Under the assumptions made, this point is only a function of the small-signal gain associated with the load termination chosen. The relevant equations are derived in Section 9.13.1.

If the compression required is relatively low (a few decibels), the compression at steady-state will be approximately the same as the loop gain at start-up. In this case the loop gain at start-up can be used directly to force the transistor to its peak power point.

Substantial compression is frequently required to extract the maximum output power from an oscillator. It is important to realize that in these cases the load termination presented to the transistor and the oscillator frequency will change as the transistor is driven into compression. In order for this change and the change in the oscillation frequency to be small, the conditions listed in Section 9.13.1 must apply.

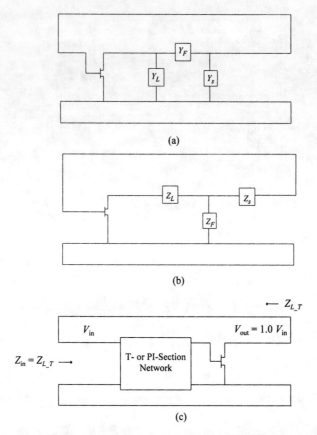

Figure 9.54   (a, b) The two oscillator topologies shown in Figure 9.53 modified for the purpose of calculating the transistor load-termination and the loop gain [2]. (c) The voltage and impedance at steady-state.

If these conditions do not apply, a better approach would be to make use of the fact that, with a well-behaved load line, the main nonlinear effect in the transistor would be the compression of the transconductance (G_m). The transconductance in the small-signal model can therefore be reduced until the large-signal operating gain is compressed as required. The feedback network can then be designed with the associated set of S-parameters instead of the small-signal parameters. In this case the steady-state load line is controlled instead of the load line at start-up.

When the goal is low phase-noise and not power, the steady-state compression should be kept low. If this is done, the conversion efficiency (mixing effects) will be low, with a corresponding effect on the upconversion of the flicker noise. A well-behaved load line for the transistor is still desirable as it will prevent running into nonlinear effects associated with a poor choice of the load line.

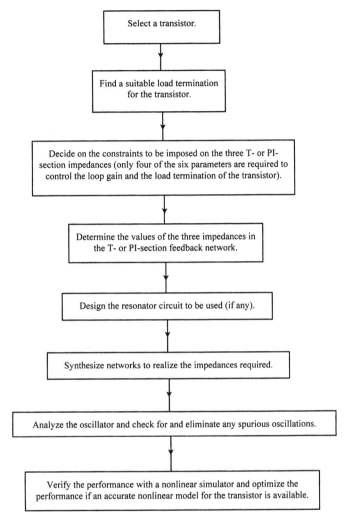

Figure 9.55   A simplified flow diagram of the oscillator design process outlined [2].

If low phase-noise is required, extra care should be taken to maximize the loaded Q (or equivalently the slope in the phase of the loop gain response) of the oscillator. This will reflect on the choice of the resonator to be used, as well as the load line chosen (higher parallel or lower series resistance will be associated with higher Qs). In simple cases the loaded Q at start-up can be estimated from the loop gain response by using (9.54). Instead of trying to estimate the loaded Q, a better option would be to control the slope in the loop phase directly.

Table 9.6   An Example of a Table of the T-section Impedances Required at a Specific Frequency (3.5 GHz) as a Function of the Loop Gain [2]

Loop gain	R_L	X_L	L_F, C_F	X_F	L_s, C_s	X_s
(dB)	(Ω)	(Ω)	(nH, pF)	(Ω)	(nH, pF)	(Ω)
-0.0927	49.620	1.808	23.300 pF	-1.952	4.651 nH	102.356
0.9073	49.573	2.029	20.765 pF	-2.190	4.665 nH	102.594
1.9073	49.521	2.277	18.508 pF	-2.457	4.677 nH	102.861
2.9073	49.463	2.555	16.495 pF	-2.757	4.691 nH	103.161
3.9073	49.397	2.866	14.701 pF	-3.093	4.706 nH	103.494
4.9073	49.324	3.216	13.103 pF	-3.471	4.723 nH	103.875
5.9073	49.241	3.608	11.678 pF	-3.894	4.743 nH	104.298
6.9073	49.149	4.049	10.408 pF	-4.369	4.764 nH	104.773
7.9073	49.015	4.543	9.276 pF	-4.902	4.789 nH	105.306
8.9073	48.928	5.097	8.267 pF	-5.500	4.816 nH	105.905
9.9073	48.798	5.719	7.368 pF	-6.172	4.846 nH	106.576
10.9073	48.651	6.417	6.567 pF	-6.925	4.881 nH	107.329
11.9073	48.486	7.200	5.853 pF	-7.770	4.919 nH	108.174
12.9073	48.301	8.078	5.216 pF	-8.718	4.962 nH	109.122

Note: The selected loop gain is equal to the estimated compression required to maximize the output power.

The feedback network (refer to Figure 9.53) must be designed to provide the required load line and loop gain at start-up or at steady-state (or an approximation of steady-state). Two of the three impedances (series case) or admittances (shunt feedback case) are usually assumed to be purely reactive (i.e., at least during the initial stages of the design), while the output power is extracted from the third impedance or admittance.

Because the load line is known, the input impedance of the transistor is also known, and it follows that the terminations for the T- or PI-section feedback are known. With the terminations and the gain of the transistor known, equations can be derived for the components that will provide the required loop gain, as well as the required load termination. This is done in Section 9.13.2 [2].

An example of a table of the Z_L, Z_F, and Z_s values required (series feedback case) at 3.5 GHz to realize different values of the loop gain and a specified load termination is given in Table 9.6. In this case Z_F and Z_s were chosen to be purely reactive. The highlighted loop gain is equal to the estimated compression required to maximize the output power.

Table 9.7 gives the Z_L, Z_F, and Z_s values generated for this oscillator from 3.5 to 4.5 GHz after selecting the loop gain estimated for peak power. The required terminations are displayed on a Smith chart in Figure 9.56. Table 9.6 shows the T-section impedances required at a specific frequency as a function of the loop gain.

Note that the trace for at least one of the sets of impedances to be used must rotate counterclockwise around the Smith chart in order to ensure frequency stability (i.e., the oscillator must lock at the frequency of interest and not drift around in frequency). Such impedances will be referred to as of the varactor type.

The equivalent statement in terms of the loop phase versus frequency response (displayed on a rectangular plot) is that the phase trace must pass through zero without any jitter and must not cross the zero-degree line again before the loop gain is too low for oscillation.

With the T- or PI-section impedances known over the frequency range of interest, networks must be synthesized to approximate each of the impedances over the frequency range of interest. One would generally select a combination that would result in one fixed-valued component, a varactor, or a resonator circuit and a complex impedance (to be realized with an impedance-matching network).

When a voltage-controlled oscillator (VCO) is designed, better results can usually be obtained with two varactors and one impedance-matching network.

The impedance associated with the load termination is often taken to be the actual load (50Ω), but this is clearly not optimum. In general, an impedance-matching network is required to realize the impedance required.

The reactances required can be realized with capacitors, inductors, transmission lines, varactor diodes, or resonators, depending on the requirements. The design of high-Q resonator networks is considered in Section 9.13.3, while that of varactor networks is considered in Section 9.13.5.

If a resonator is used, the resonator impedance must be transformed to present the impedance required at the relevant position. This can often be done by simply using a transmission line with the correct characteristic impedance and length. This is illustrated in Section 9.13.4.

One would generally use series-tuned varactor networks in a series-feedback oscillator and parallel-tuned networks in a shunt feedback oscillator. The particular choice would depend on the component values and the behavior outside the oscillation band. When a series tuned network is used in a shunt feedback oscillator, and vice versa, losses in the varactor network could have a serious stabilizing effect on the circuit. If such a choice was made, be sure to check the effect of such losses on the performance of the circuit.

Table 9.7   An Example of a Table of the T-Section Impedances Required to Provide the Specified Load Termination and the Specified Loop Gain over the Oscillation Band (VCO with Two Varactors) [2]

Frequency	R_L	X_L	L_f, C_f	X_F	L_s, C_s	X_s
(GHz)	(Ω)	(Ω)	(nH, pF)	(Ω)	(nH, pF)	(Ω)
3.50	49.149	4.049	10.408 pF	−4.369	4.764 nH	104.773
3.60	49.114	4.085	9.979 pF	−4.430	4.452 nH	100.707
3.70	49.078	4.126	9.565 pF	−4.497	4.177 nH	97.097
3.80	49.041	4.165	9.178 pf	−4.564	3.916 nH	93.492
3.90	49.004	4.212	8.797 pF	−4.639	3.684 nH	90.275
4.00	48.965	4.257	8.438 pF	−4.715	3.464 nH	87.051
4.10	48.923	4.327	8.053 pF	−4.820	3.273 nH	84.305
4.20	48.878	4.403	7.679 pF	−4.935	3.095 nH	81.683
4.30	48.832	4.475	7.335 pF	−5.046	2.925 nH	79.038
4.40	48.786	4.552	7.002 pF	−5.166	2.772 nH	76.645
4.50	48.738	4.632	6.685 pf	−5.290	2.630 nH	74.352

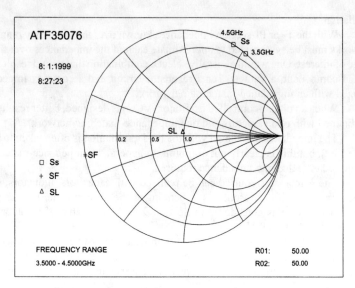

Figure 9.56   The T-section impedances in Table 9.7 displayed on a Smith chart [2]. Note that at least one of the sets of impedances should rotate counterclockwise around the Smith chart to ensure frequency stability (Z_F in this case).

Care should be taken when deciding on the impedance to be approximated with a fixed capacitor or inductor. Ideally the choice made should result in a topology that cannot sustain oscillations at very low or very high frequencies.

When suitable networks have been fitted to the target impedances, the oscillator should be analyzed to confirm its performance and to check for any spurious oscillations. Because loops may be present, the analysis should be done fairly densely. Both the loop gain and the reflection gain performance should be checked.

If an accurate nonlinear model for the transistor used is available, the oscillator performance should be verified and optimized with a nonlinear simulator.

An example of a dielectric resonator oscillator (DRO) designed as described here is shown in Figure 9.57. The oscillator was designed to oscillate at 15.65 GHz with the output power higher than 10 dBm (Bias point: 2V, 20 mA). The performance was realized with slight adjustments in the supply voltage and the puck position.

Note that because a nonlinear model for the transistor used was not available, a nonlinear simulator was not used.

The loop gain performance of the oscillator is shown in Figure 9.58. Oscillations seem to be possible around 6 GHz too. However, a modification was made to the basic oscillator circuit (a gap capacitor was inserted between the transistor and the resonator circuit) to delay the change in the loop phase in this area, and the gain margin in this case is actually quite large. Interestingly enough, the circuit does oscillate around 6 GHz if the change introduced is not made. While the spurious oscillation is undesirable, the fact that it can be predicted with such accuracy and can be eliminated with relative ease serves as a validation for the loop gain approach.

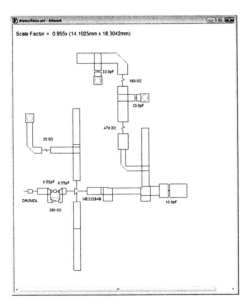

Figure 9.57   The artwork of a DRO oscillator [2] (Courtesy of Plessey Avionics, Retreat, South Africa).

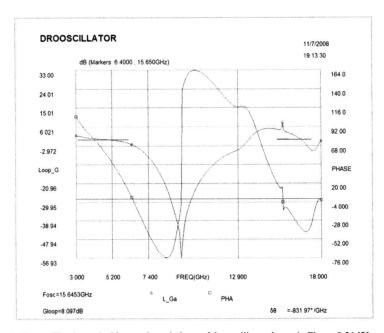

Figure 9.58   The theoretical loop gain and phase of the oscillator shown in Figure 9.56 [2].

The spurious oscillation can also be eliminated by using a different (more expensive) transistor.

9.13.1   Estimation of the Compression Associated with the Maximum Effective Output Power

If the power gain of a transistor is considered as a function of the drive level, it is clear that the gain is equal to the small-signal operating power gain (G_{os}) when the input power is low and the output power will approach the saturation limit when the input power is high (see Figure 9.59). Assuming the transition to be exponential, the output power could be described by the following equation [23]:

$$P_{out} = P_{sat}[1 - e^{-G_{os}P_{in}/P_{sat}}]$$

(9.155)

The maximum effective output power ($P_{out} - P_{in}$) is delivered by the transistor when

$$\frac{\partial(P_{out} - P_{in})}{\partial P_{in}} = 0$$

that is, when

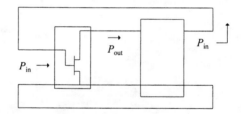

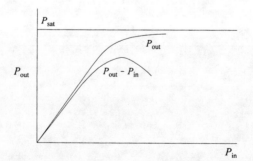

Figure 9.59   Typical saturation characteristics for a transistor.

$$\frac{\partial P_{out}}{\partial P_{in}} = 1$$

Applying this to the equation above yields

$$P_{in} = P_{sat}[\ln(G_{\omega s}) / G_{\omega s}]$$ (9.156)

and

$$P_{out\_max} = P_{sat}(1 - 1 / G_{\omega s})$$ (9.157)

from which it follows that

$$P_{osc\_max} = P_{out\_max} - P_{in}$$

$$= P_{sat}[1 - 1 / G_{\omega s} - \ln(G_{\omega s}) / G_{\omega s}]$$ (9.158)

The corresponding value of the large-signal operating power gain ($G_{\omega l}$) at this maximum effective output power point is given by

$$G_{\omega l} = P_{out\_max} / P_{in}$$

$$= (G_{\omega s} - 1) / \ln(G_{\omega s})$$ (9.159)

The ratio of the small-signal and the large-signal operating power gain is therefore

$$G_{\omega s} / G_{\omega l} = [G_{\omega s} / (G_{\omega s} - 1)]\ln(G_{\omega s})$$ (9.160)

If an oscillator is synthesized with a set of small-signal S-parameters, a first-order approximation for the loop gain that will result in the maximum possible output power is the square root of this ratio, that is,

$$G_{loop\_opt} = \sqrt{\frac{G_{\omega s}}{G_{\omega l}}} = \sqrt{\ln(G_{\omega s})[G_{\omega s} / (G_{\omega s} - 1)]}$$ (9.161)

This approximation would apply to the degree that the following conditions [2] apply.

Series Feedback Case

$$|Z_{o\_loop}| = |Z_s + Z_F + z_{11}| >> |Z_F|$$ (9.162)

and

$$\left| Z_F / G_{\text{loop\_opt}} \right| \gg \left| z_{12} \right| \tag{9.163}$$

Shunt Feedback Case

$$\left| Y_{o\_\text{loop}} \right| = \left| Y_s + Y_F + y_{11} \right| \gg \left| Y_F \right| \tag{9.164}$$

and

$$\left| Y_F / G_{\text{loop\_opt}} \right| \gg \left| y_{12} \right| \tag{9.165}$$

9.13.2   Derivation of the Equations for the T- and PI-Section Feedback Components Required

With the transmission parameters of the transistor in Figure 9.60 assumed to be

$$\begin{bmatrix} A_t & B_t \\ C_t & D_t \end{bmatrix}$$

it follows that [1]

$$Z_{\text{in}} = \frac{A_t Z_{LL} + B_t}{C_t Z_{LL} + D_t} \tag{9.166}$$

and

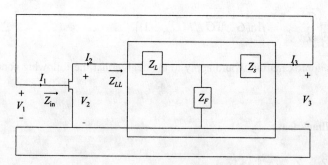

Figure 9.60   The T-section network in an oscillator with series feedback.

$$A_I = \frac{z_{21t}}{z_{22t} + Z_{LL}} \tag{9.167}$$

where Z_{LL} is the specified load termination for the transistor. Z_{LL} is given in terms of Z_{in} and the T-section impedances by

$$Z_{LL} = Z_L + \cfrac{1}{Y_F + \cfrac{1}{Z_s + Z_{in}}}$$

$$= Z_L + \frac{Z_s + Z_{in}}{1 + Y_F(Z_s + Z_{in})} \tag{9.168}$$

while the loop gain (G_{loop}) is given by

$$G_{loop} I_1 = I_3 = \frac{Z_F}{Z_F + (Z_s + Z_{in})} I_2$$

$$= \frac{A_I I_1}{1 + Y_F(Z_s + Z_{in})} \tag{9.169}$$

from which it follows that

$$Z_F = -\frac{Z_s + Z_{in}}{\cfrac{A_I}{G_{loop}} + 1} \tag{9.170}$$

This expression can be written as

$$Z_F = -\frac{Z_s}{\cfrac{A_I}{G_{loop}} + 1} - \frac{Z_{in}}{\cfrac{A_I}{G_{loop}} + 1} \tag{9.171}$$

Equation (9.170) can be used to simplify (9.168):

$$Z_{LL} = Z_L + \frac{Z_s + Z_{in}}{1 - \left(\cfrac{A_I}{G_{loop}} + 1 \right)}$$

$$= Z_L - \frac{G_{\text{loop}}}{A_I}(Z_s + Z_{\text{in}}) \tag{9.172}$$

which can be rearranged to give an expression for Z_L in terms of Z_s:

$$Z_L = \frac{G_{\text{loop}}}{A_I} Z_s + \left(Z_{LL} + \frac{G_{\text{loop}}}{A_I} Z_{\text{in}} \right) \tag{9.173}$$

Equations (9.173) and (9.171) can be combined to eliminate Z_s and give an expression for Z_L in terms of Z_F:

$$Z_F = -\frac{1}{\dfrac{A_I}{G_{\text{loop}}} + 1} \frac{Z_L - \left(Z_{LL} + \dfrac{G_{\text{loop}}}{A_I} Z_{\text{in}} \right)}{\dfrac{G_{\text{loop}}}{A_I}} - \frac{Z_{\text{in}}}{\dfrac{A_I}{G_{\text{loop}}} + 1} \tag{9.174}$$

which can be simplified to

$$Z_L = -\left[1 + \frac{G_{\text{loop}}}{A_I} \right] Z_F + Z_{LL} \tag{9.175}$$

Equations (9.171), (9.173), and (9.175) can be used to solve for the required components once the relevant constraints on Z_L, Z_s, or Z_F have been established. The real parts of two of these impedances are usually assumed to be zero or to be fixed, after which these equations can be solved for the remaining values.

The derivation of the equations for the PI-section components proceeds similarly to that for the T-section components.

With the transmission parameters of the transistor in Figure 9.61 assumed to be

$$\begin{bmatrix} A_t & B_t \\ C_t & D_t \end{bmatrix}$$

it follows that

$$Y_{\text{in}} = \frac{C_t Z_{LL} + D_t}{A_t Z_{LL} + B_t} \tag{9.176}$$

and

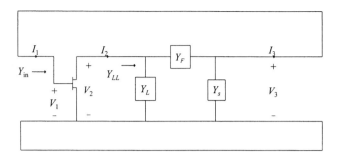

Figure 9.61   The PI-section network in an oscillator with parallel feedback.

$$A_v = \frac{1}{A_t + B_t Y_{LL}}$$ (9.177)

where Y_{LL} is the specified load termination for the transistor. Y_{LL} is given in terms of Y_{in}, and the PI-section impedances by

$$Y_{LL} = Y_L + \frac{1}{Z_F + \dfrac{1}{Y_s + Y_{in}}}$$

$$= Y_L + \frac{Y_s + Y_{in}}{Z_F (Y_s + Y_{in}) + 1}$$ (9.178)

while the loop gain (G_{loop}) is given by

$$G_{loop} V_1 = V_3 = \frac{\dfrac{V_2}{Y_s + Y_{in}}}{\dfrac{1}{Y_s + Y_{in}} + Z_F} = \frac{A_v V_1}{1 + Z_F (Y_s + Y_{in})}$$ (9.179)

from which it follows that

$$Y_F = \frac{Y_s + Y_{in}}{\dfrac{A_v}{G_{loop}} - 1}$$ (9.180)

Equation (9.180) can be written as

$$Y_F = \frac{1}{\dfrac{A_v}{G_{\text{loop}}} - 1} Y_s + \frac{Y_{\text{in}}}{\dfrac{A_v}{G_{\text{loop}}} - 1} \tag{9.181}$$

which provides a simple expression for Y_F in terms of Y_s.

Equation (9.180) can also be used to simplify (9.178):

$$Y_{LL} = Y_L + \frac{Y_s + Y_{\text{in}}}{Z_F (Y_s + Y_{\text{in}}) + 1}$$

$$= Y_L + \frac{Y_s + Y_{\text{in}}}{\left(\dfrac{A_v}{G_{\text{loop}}} - 1\right) + 1}$$

$$= Y_L + \frac{G_{\text{loop}}}{A_v} Y_s + \frac{G_{\text{loop}}}{A_v} Y_{\text{in}} \tag{9.182}$$

which can be rearranged to provide an expression for Y_L in terms of Y_s:

$$Y_L = -\frac{G_{\text{loop}}}{A_v} Y_s + \left(Y_{LL} - \frac{G_{\text{loop}}}{A_v} Y_{\text{in}} \right) \tag{9.183}$$

Equations (9.181) and (9.183) can be combined to eliminate Y_s, which will give an expression for Y_L in terms of Y_F too:

$$Y_L = \left(\frac{G_{\text{loop}}}{A_v} - 1 \right) Y_F + Y_{LL} \tag{9.184}$$

Equations (9.181), (9.183), and (9.184) can be used to solve for the required components once the relevant constraints on Y_L, Y_s, or Y_F have been established. The real parts of two of these impedances are usually assumed to be zero or to be fixed, after which these equations can be solved for the remaining values.

9.13.3 High-Q Resonator Circuits

High-Q resonator circuits can be realized by using dielectric resonators, cavity resonators or a magnetically biased yittrium iron garnet (YIG) sphere.

The YIG resonator is a high Q, ferrrite sphere of yittrium iron garnet [$Y_2Fe_2(FeO_4)_3$] that can be tuned over a wide band by varying the biasing dc field. In a YIG-tuned oscillator, a YIG sphere is normally used to control the inductance of a coil in the resonant circuit. Because YIG is a ferrimagnetic material, its effective permeability can be controlled with an external dc magnetic field, thus controlling the oscillator frequency. YIG oscillators can be made to tune over more than a decade of bandwidth, while varactor-tuned oscillators are limited to a tuning range of about an octave [8].

Cavity resonators are usually realized with low-loss coaxial line or waveguide. The simplest coaxial cavity is a quarter wavelength (λ/4) shorted stub. The signal is coupled into the cavity with a shorted loop or an open probe. The resonant frequency is usually adjusted with a tuning screw near the open end. It can be shown [8] that the impedance near the resonance frequency and the Q of the resonator are given by

$$Z_{in} \approx \frac{1}{\dfrac{\alpha l}{Z_0} + j\pi \dfrac{\Delta\omega}{2\omega_0} Y_0} \tag{9.185}$$

$$Q \approx \frac{\pi}{4\alpha l} = \frac{\beta}{2\alpha} \tag{9.186}$$

Equations (9.185) and (9.186) are derived by starting with the expression

$$Z_{in} = Z_0 \tanh[(\alpha + j\beta)l] \tag{9.187}$$

Open-circuited λ/2 resonators are often used on microstrip. The input impedance and Q in this case are given by [8]

$$Z_{in} \cong \frac{1}{\dfrac{\alpha l}{Z_0} + j\pi \dfrac{\Delta\omega}{\omega_0} Y_0} \tag{9.188}$$

$$Q \approx \frac{\pi}{2\alpha l} = \frac{\beta}{2\alpha} \tag{9.189}$$

The width and length of the smallest rectangular waveguide cavity is λ/2 (TE[101] mode). The rectangular cavity is a waveguide version of a short-circuited λ/2 transmission-line resonator [8].

Because of the small size and low cost, dielectric resonators are frequently used at microwave frequencies. The high dielectric constant of the resonator puck ensures that most of the fields are contained within the dielectric, but there is some fringing from the sides and ends of the puck. The fringing fields provide a convenient means of coupling to a microstrip line. The spacing between the puck and the microstrip conductor determines the amount of coupling.

Only dielectric losses are present in the puck, and Qs of several thousand can be realized. Metallic shielding is required to minimize radiation losses. The Q can be increased by with a dielectric spacer under the puck.

The resonant frequency of the puck can be adjusted by using an adjustable metal plate above it.

The lowest order resonant mode for a dielectric puck is the $TE_{01\delta}$ mode. This mode couples easily to a microstrip line. The resonant frequency for a puck can be estimated by solving the following transcendental equation iteratively [8]:

$$\tan\frac{\beta L}{2} = \frac{\alpha}{\beta} \tag{9.190}$$

where

$$\alpha = \sqrt{\left(\frac{2.405}{a}\right)^2 - k_0^2} \tag{9.191}$$

$$\beta = \sqrt{\varepsilon_r k_0^2 - \left(\frac{2.405}{a}\right)^2} \tag{9.192}$$

$$k_0 = \frac{2\pi f}{c} \tag{9.193}$$

In these equations f is the required resonant frequency, ε_r is the dielectric constant of the puck material, L the height of the puck, and a its radius.

The resonant frequency must lie in the interval $[f_1, f_2]$, where [8]

$$f_1 = \frac{2.405c}{2\pi\sqrt{\varepsilon_r}a} \tag{9.194}$$

and

$$f_2 = \frac{2.405c}{2\pi a} \tag{9.195}$$

Equations (9.194) and (9.195) are necessary conditions to ensure that the roots in (9.191) and (9.192) can be taken.

The unloaded Q of a dielectric puck can be estimated as [8]

$$Q = 1/\tan\delta \tag{9.196}$$

Two commonly used DRO configurations are shown in Figure 9.62. The equivalent circuits associated with the coupled sections are shown in Figure 9.63.

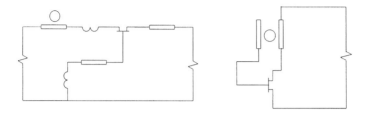

Figure 9.62   Two commonly used dielectric resonator oscillator configurations.

9.13.4   Transforming the Impedance Presented by a Resonator Network to That Required in the T- or PI-Section Feedback Network

When a resonator is used, the impedance presented by the resonator must usually be transformed to present the target impedance in the T- or PI-section feedback network. This can often be done by simply using a transmission line with the correct characteristic impedance and length.

The principle is illustrated in Figures 9.64 and 9.65. The trace on the right-hand side of Figure 9.65 (trace A; the smaller arc) is the measured impedance of a dielectric resonator puck coupled to a microstrip line (15.6–15.7 GHz). The impedance at the oscillation frequency (15.65 GHz) is in the area of a second marker shown on this trace (point a). The target impedance in the T-section feedback network designed is in the area of the cursor displayed on the left-hand side of this figure (trace B, point b). The measured impedance was transformed to this point by cascading a transmission line with suitable characteristic impedance and length to the resonator circuit as shown in Figure 9.64.

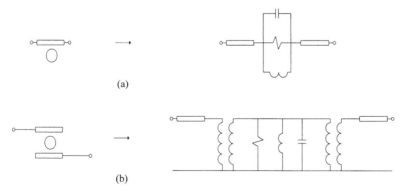

Figure 9.63   The equivalent circuits associated with (a) a dielectric puck coupled to a microstrip line and (b) a puck used to couple two microstrip lines.

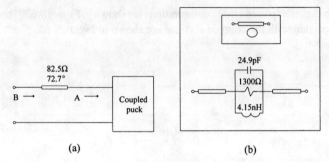

(a)	(b)

Figure 9.64   (a) The transmission line used to transform the impedance presented by a dielectric resonator puck coupled to a microstrip line as shown in Figure 9.63 (electrical line length specified at 15.65 GHz). (b) The equivalent circuit fitted to resonator circuit.

Note that the line losses should be taken into account when the resonator impedance is transformer. These losses were ignored in Figure 9.65.

The equivalent circuit fitted to the resonator circuit is shown in Figure 9.64(b). If wideband measurements are not available, an equivalent circuit can be used to check for spurious oscillations.

The size of the modified resonator loop is a function of the characteristic impedance chosen.

Note that if the characteristic impedance used was 50-Ω, the original loop will simply be rotated around the origin as the line length is increased. This follows because the electrical length of the line added is basically constant over the narrow frequency range considered.

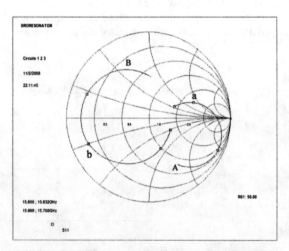

Figure 9.65   An example of transforming the impedance presented by a resonator circuit to the target impedance in the T-section feedback network at the oscillation frequency [2].

9.13.5   Designing Varactor Circuits to Realize the Varactor-Type Reactance Required

The varactor-type reactances (impedance rotating counterclockwise on the Smith chart) required in the T- or PI-section feedback network must be realized with varactor networks. The design of the varactor network usually consists of finding a varactor with a tuning range bigger than the minimum required and calculating the loading capacitance or inductance required.

The loading inductance or capacitance required is usually realized with lumped elements.

The effect of the parasitic inductance of the varactor diode should be included when the tuning circuit is designed.

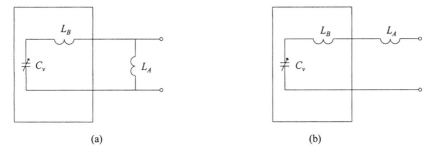

(a)   (b)

Figure 9.66   Loading a varactor with (a) a parallel inductor and (b) a series inductor to obtain the series reactance required at the passband edges.

The basic varactor networks are shown in Figure 9.66.
With a maximum achievable varactor capacitance ratio of

$$\rho_{v\_max} = C_{v\_max} / C_{v\_min} \qquad (9.197)$$

and specified limiting values of the series reactance required at the passband edges ($X(\omega_H)$ and $X(\omega_L)$), the loading capacitance or inductance required can be calculated.
Considering the series loading circuit [Figure 9.66(b)], it follows that

$$X(\omega_H) = X_A(\omega_H) + \omega_H L_B - \frac{1}{\omega_H C_{v\_min}} \qquad (9.198)$$

and

$$X(\omega_L) = X_A(\omega_L) + \omega_L L_B - \frac{1}{\omega_L C_{v\_max}} \qquad (9.199)$$

where L_B is the package inductance of the varactor.

These equations can be manipulated to give expressions for the minimum and maximum values of the varactor capacitance. Substitution of these expressions in (9.197) yields an expression that can be solved to obtain the value of X_A.

It follows from this equation that

$$L_A = \frac{\alpha X(\omega_L) - X(\omega_H)}{\alpha \omega_L - \omega_H} - L_B \tag{9.200}$$

with

$$\alpha = \frac{\omega_L C_{v\_max}}{\omega_H C_{v\_min}} = \frac{\omega_L}{\omega_H} \rho_{v\_max} \tag{9.201}$$

If L_A is found to be negative, a series capacitor is required instead. The value of the capacitor is given by

$$C_A = \frac{\dfrac{1}{\omega_H} - \dfrac{\alpha}{\omega_L}}{\omega_H L_B - \alpha \omega_L L_B - X(\omega_H) + \alpha X(\omega_L)} \tag{9.202}$$

The value of C_A or L_A required in the parallel case can be found by starting with the equations

$$-B(\omega_H) = \frac{1}{\omega_H L_A} + \frac{1}{\omega_H L_B - \dfrac{1}{\omega_H C_{v\_min}}} \tag{9.203}$$

and

$$-B(\omega_L) = \frac{1}{\omega_L L_A} + \frac{1}{\omega_L L_B - \dfrac{1}{\omega_L C_{v\_max}}} \tag{9.204}$$

These equations can be manipulated to give expressions for the minimum and maximum values of the varactor capacitance. Substitution of these expressions in (9.197) yields a quadratic expression that can be solved to obtain the value of the loading component (L_A or C_A).

If an inductor is required, its value is given by [22]

$$L_A = \frac{1}{P \pm \sqrt{P^2 - Q}} \geq 0 \qquad (9.205)$$

where

$$P = \frac{1}{2}\left\{\frac{\omega_L}{X(\omega_L)} + \frac{\omega_H}{X(\omega_H)} - \frac{1}{L_B}\right\} \qquad (9.206)$$

and

$$Q = \frac{\omega_L \omega_H}{X(\omega_L)X(\omega_H)} - \frac{1}{L_B}\frac{(\omega_H / X(\omega_L)) - \omega_L \rho_{v\_max} / X(\omega_H))}{(\omega_H / \omega_L) - \omega_L \rho_{v\_max} / \omega_H} \qquad (9.207)$$

9.13.6 Considerations Applying to Oscillators with Low Phase Noise

The phase noise of an oscillator can be minimized by doing the following:

1. Select a transistor with a low noise figure and low flicker noise;

2. Bias the transistor correctly;

3. Maximize the loaded Q or the rate at which the loop phase passes through zero;

4. Keep the conversion gain of the oscillator low.

The conversion gain can be kept low by keeping the output power well below saturation (limiting the loop gain) and/or by linearizing the transconductance with resistive feedback (a small series resistor can be used in the source or the emitter). The conversion gain will also be low if the loop gain at start-up is kept low.

If the amplitude of the oscillation tends to be unstable because of these measures, a linear automatic gain control (AGC) loop can be designed to stabilize it. A pin diode circuit is probably the best option.

It is also possible to reduce the level of the upconverted flicker noise by designing suitable low-frequency circuitry for the oscillator [24]. The aim of such circuitry is to reduce the level of the flicker noise across the nonlinear junctions.

When a fixed-frequency oscillator is designed and large-signal information is available, AM-to-PM conversion can be minimized if a network is designed to ensure rectangular crossing of the impedance versus frequency and impedance versus amplitude traces displayed on a Smith chart.

QUESTIONS AND PROBLEMS

9.1   Calculate the parameters of the input and output stability circles on the admittance-plane for the Nitronex NPT25015 GaN transistor biased at 28V, 200 mA. The S-parameter data is provided in Table 9.8. Also calculate the Rollette and Linville stability factors, as well as μ and μ'.

Calculate the upper limits of the input and output VSWRs (relative to 50-Ω normalization) for which oscillations will not be possible at the different frequencies.

Table 9.8   The S-parameters (28V, 200 mA) of the Nitronex NPT25015 GaN Transistor as Provided in the Nitronex Datasheet

Frequency (GHz)	s_{11} (magnitude; °)		s_{12} (magnitude; °)		s_{21} (magnitude; °)		s_{22} (magnitude; °)	
0.8	0.88	−176	0.017	16.4	6.01	72.6	0.57	−172
1.0	0.88	180	0.016	8.8	4.85	65.1	0.57	−174
1.2	0.89	176	0.017	14.3	4.02	59.1	0.58	−176
1.4	0.89	173	0.017	14.6	3.42	52.7	0.59	−177
1.6	0.89	170	0.019	24.0	2.98	46.9	0.60	−178
1.8	0.89	167	0.018	27.4	2.63	41.0	0.62	180
2.0	0.89	164	0.017	28.1	2.35	35.5	0.64	178
2.2	0.90	161	0.021	33.2	2.11	29.9	0.67	176
2.4	0.90	158	0.021	33.4	1.92	23.8	0.69	174
2.6	0.90	155	0.023	36.0	1.73	18.2	0.71	172
2.8	0.90	152	0.025	37.5	1.58	12.9	0.73	169
3.0	0.90	150	0.026	35.2	1.45	8.01	0.74	167
3.2	0.91	147	0.028	36.3	1.33	3.33	0.75	165
3.4	0.91	145	0.030	35.5	1.25	−1.09	0.76	162
3.6	0.91	143	0.032	35.6	1.18	−5.46	0.76	160
3.8	0.90	140	0.035	34.6	1.12	−9.83	0.77	158
4.0	0.90	138	0.038	32.3	1.09	−14.2	0.77	156

9.2   Show that, when it is inherently stable, the maximum available power gain of a transistor is given by (9.72) and the optimum source admittance by the conjugate of the load admittance as given by (9.79).

$$G_{a,max} = \frac{|y_{21}|^2}{2g_{22}} \Bigg/ \left[G_{sa,opt} - \left(\frac{P}{2g_{22}} - g_{11} \right) \right] \tag{9.208}$$

$$Y_{sa,opt} = G_{sa,opt} + jB_{sa,opt}$$

$$= \left| \frac{y_{21}y_{12}}{2g_{22}} \right| \sqrt{1/C^2 - 1} + j\left(\frac{Q}{2g_{22}} - g_{11} \right) \tag{9.209}$$

where

$$P + jQ = y_{12}y_{21} \tag{9.210}$$

The equations for the maximum operating power gain are similar (g_{11} and g_{22} should be exchanged).

9.3   Explore various ways to make the transistor in Problem 9.1 inherently stable. Calculate the maximum operating power gain, as well as the maximum available power gain obtainable from the transistor after stabilization. Also calculate the maximum single-sided matched stable gain of the transistor before and after stabilization. Assume a passband of 2.5–2.7 GHz for the gain calculations.

Find the parameters of the operating and available power gain circles on the admittance plane with gain equal to the maximum single-side matched stable gain, before and after stabilization.

9.4   Find the maximum tunable operating gain and available power gain for the transistor in Problem 9.3, both before and after stabilization.

9.5   Find the source impedance and transducer power gain, which is equivalent to the operating power-gain circle of Problem 9.5 (after stabilization), and the load admittance and transducer power gain which is equivalent to the available power gain circle (after stabilization).

9.6   Fit a model to the *S*-parameters provided in Table 9.8.

Assume the load line to be constrained by $V_{sat}=2V$, $R_{sat}=0.45\Omega$, $I_{ds,max} = 6A$, $V_{ds,brk} = 56V$ and dc operating point 28V, 200 mA. Calculate the maximum linear output power (maximum output power just before current and/or voltage clipping) of the transistor. Assume the model to remain the same if the dc current is increased to 250 mA. Calculate the maximum linear output power for this case.

9.7   Design a conditionally stable class A power amplifier (dc operating point 28V, 200 mA) with the transistor used in Problem 9.1 for the passband used in Problem 9.3 (2.5–2.7 GHz). The output power must be as high as possible and the amplifier must exhibit no negative resistance with the load termination and/or source termination in place.

9.8   An linear amplifier is required for a multisignal application. The average output power is 1.5W (31.76 dBm) and the peak level is 10.3 dB above that (see the datasheet for the Nitronex NPT25015 GaN transistor). Assume that some clipping of the signal can be tolerated and that the maximum power level can be reduced to 6 dB above 1.5W (37.76 dBm). Also assume that the class A parameters provided at 28V 200 mA for the transistor in Problem 9.1 apply at 250 mA too. Assuming class A operation for the average signal and class B operation for the peaks, design an amplifier to provide the required class B power (38.76 dBm), with maximum class A gain for the average signals. Ensure that the class A power obtainable is above 31.76 dBm.

Note from [25] that larger input signals are associated with increased levels of harmonic distortion and gain compression.

9.8   (a) Calculate the thermal noise available per unit bandwidth from a resistor at a temperature of 290°K. (b) Calculate the noise power available at the output terminals of a transistor with a 0.5-dB noise figure and available power gain of 15 dB if the bandwidth (ideal filter response) is equal to 1 MHz. (c) Repeat the last calculation for the case where the gain response is of the same form as that of a single-tuned resonant circuit.

9.9   (a) Calculate the parameters of the 2.0 dB constant noise-figure circle of a transistor on the admittance plane, if $F_{min} = 1.2$ dB, $R_n = 100\Omega$, and $G_{sn,opt} + jB_{sn,opt}$ $= (20.0 + j10.0)$ mS. (b) Calculate the noise figure of the transistor if its source admittance is equal to $(30.0 + j0.0)$ mS.

9.10   Use NEC's NE3210S01 HJ FET biased at 2V 10 mA to design a single-stage 4–6 GHz low-noise amplifier. The .s2p data file provided by the manufacturer is available in the LSM Examples folder. The datasheet can be downloaded from the Internet.

9.11   Odd-mode oscillations (refer to Figure 9.1 and [1]) are possible when two or more transistors are connected in parallel. Examine the stability for the even and the odd modes when two of the transistors in Problem 9.10 are connected in parallel. Do this by using the stability factors defined, as well as by calculating the loop gain around the different loops.

9.12   Design a reflection amplifier with a gain of 12 dB by using a Gunn diode with the specifications given in Table 9.9 over the passband 5–7 GHz.

Table 9.9   The Impedance of the Gunn Diode Relevant to Problem 9.12 (M/A-COM MA 49139)

Frequency (GHz)	Diode Impedance (Ω)
5	$-10.0 - j13.0$
6	$-9.0 - j2.5$
7	$-9.0 + j5.0$

9.13   Calculate the (single-tuned) loaded Q at start-up for the oscillator in Figure 9.58 by using (9.54). Calculate the gain and phase margins around 6.4 GHz (spurious oscillation).

9.14   Fit a model to the S-parameters of NEC's NE33284A transistor (datasheet available on the Internet), biased at 2V, 10 mA. Use the I/V-curves provided in the datasheet to decide on the four boundary lines to be used for the load line. Find the load termination for maximum linear output power at 15 GHz. Estimate the compression required to maximize the output power of a 15 GHz oscillator when the load termination presented to the transistor is the optimum power line. Design series and parallel feedback oscillators based on the load termination and the loop gain calculated.

REFERENCES

[1] Freitag, R. G., "A Unified Analysis of MMIC Power Amplifier Stability," *1992 IEEE MTT-S Digest*.

[2] *MultiMatch RF and Microwave Impedance-Matching, Amplifier and Oscillator Synthesis Software*, Stellenbosch, South Africa: Ampsa (Pty) Ltd, http://www.ampsa.com, 2009.

[3] Edwards, M. L., and S. Cheng, "A Deterministic Approach for Designing Conditionally Stable Amplifiers," *IEEE Trans. Microwave Theory Tech.*, Vol. MTT-43, No. 7, July 1995.

[4] Babak, L. I., "Comments on A Deterministic Approach for Designing Conditionally Stable Amplifiers," *IEEE Trans. Microwave Theory Tech.*, Vol. MTT-47, No. 2, February 1999.

[5] Edwards, M. L., and S. Cheng, "Conditionally Stable Amplifier Design Using Constant μ-Contours," *IEEE Trans. Microwave Theory Tech.*, Vol. MTT-44, No. 12, December 1996.

[6] Vendelin, G. D., A. M. Pavio, and U. L. Rohde, *Microwave Circuit Design Using Linear and Nonlinear Techniques*, New York: John Wiley and Sons, 1990.

[7] Rohde, L. R., A. K. Poddar, and G. Böck, *The Design of Modern Microwave Oscillators for Wireless Applications*, New Jersey: John Wiley and Sons, 2005.

[8] Pozar, D. M., *Microwave Engineering*, Reading, MA: Addison-Wesley Publishing Company, 1990.

[9] Rollett, J. M., "Stability and Power Gain Invariants of Linear Two-Ports," *IRE Trans. Circuit Theory*, Vol. CT-9, March 1962, pp. 29–32.

[10] Edwards, M. L., and J. H. Sinsky, "A New Criterion for Linear 2-Port Stability Using a Single Geometrically Derived Parameter," *IEEE Trans. Microwave Theory Tech.*, Vol. MTT-40, No. 12, December 1992.

[11] Hauri, E. R., "Overall Stability Factor of Linear Two-Port in Terms of Scattering Parameters," *IEEE J. Solid-State Circuits*, Vol. 6, No. 12, pp. 413–415, December 1971.

[12] Boyles, J. W., "The Oscillator as a Reflection Amplifier: An Intuitive Approach to Oscillator Design," *Microwave Journal*, June 1986.

[13] Carson, R. S., *High Frequency Amplifiers*, New York: John Wiley and Sons, 1979.

[14] Abrie, P. L. D., and P. Rademeyer, "A Method for Evaluating and the Evaluation of the Influence of the Reverse Transfer Gain on the Transducer Power Gain of Some Microwave Transistors," *IEEE Trans. Microwave Theory Tech.*, Vol. MTT-33, No. 8, August 1985, pp. 711–713.

[15] Abrie, P. L. D., "A Series of CAD Techniques for Designing Microwave Feedback Amplifiers and Simplifying the Design of Reactively Matched Single-Ended Amplifiers," *IEEE MTT-S Digest,* 1990.

[16] Norton, D., U.S. Patent No. 3,426,298, 1969; "High Dynamic Range Amplifier," U.S. Patent No. 3,624,536, November 1971, Anzac Corporation.

[17] Mead, H. B., and G. R. Callaway: "Broadband Amplifier," U.S. Patent No. 4,042,887, August 1977, Q-bit Corporation.

[18] Norton, D. E., and A. F. Podell, "Transistor Amplifier with Impedance-Matching Transformer," U.S. Patent No. 3,891,934, June 1975, Anzac Corporation.

[19] Rohde, U. L., "Wideband Amplifier Summary," *Ham Radio*, November 1979.

[20] Rohde, U. L., "The Design of a Wide-Band Amplifier with Large Dynamic Range and Low Noise Figure Using CAD Tools," *IEEE Long Island MTT Symposium Digest*, April 28, 1987, pp. 47–55.

[21] Russel, K. J., "Microwave Power Combining Techniques," *IEEE Trans. Microwave Theory Tech.*, MTT-27, No. 5, 1979, pp. 472–478.

[22] Rauscher, C., "Large-Signal Technique for Designing Single-Frequency and Voltage-Controlled GaAs FET Oscillators," *IEEE Trans. Microwave Theory Tech.*, Vol. MTT-29, No. 4, April 1981.

[23] Johnson, K. M., "Large Signal GaAs MESFET Oscillator Design," *IEEE Trans. Microwave Theory Tech.*, Vol. MTT-27, No. 3, March 1979.

[24] Prigent, M., and J. Obregon, "Phase Noise Reduction in FET Oscillators by Low-Frequency Loading and Feedback Circuitry Optimization," *IEEE Trans. Microwave Theory Tech.*, Vol. MTT-35, No. 3, March 1987.

[25] Clarke, K. K., and D. T. Hess, *Communication Circuits: Analysis and Design*, Reading, MA: Addison-Wesley Publishing Company, 1978.

SELECTED BIBLIOGRAPHY

Bodway, G. E., "Two Port Power Flow Analysis Using Generalized Scattering Parameters," *Microwave Journal*, May 1967, pp. 61–69.

Ha, T. T., *Solid-State Microwave Amplifier Design*, New York: John Wiley and Sons, 1981.

Krauss, H. L., W. B. Bostian, and F. H. Raab, *Solid-State Radio Engineering*, New York: John Wiley and Sons, 1980.

Kurokawa, K., "Design Theory of Balanced Transistor Amplifiers," *Bell Syst. Tech. J.*, Vol. 44, No. 10, 1965, pp. 1675–1698.

Roddy, D., and J. Coolen, *Electronic Communications*, Reston, VA: Reston Publishing Company, 1981.

Tajima, Y., and P. D. Miller, "Design of Broad-Band Power GaAs Fet Amplifiers," *IEEE Trans. Microwave Theory Tech.*, Vol. MTT-32, No. 3, March 1984.

APPENDIX

THE UNBALANCED TRANSMISSION LINE

The basic equations associated with a transmission line when the currents are unbalanced [1] will be derived here. For the sake of simplicity, it will be assumed that there is no magnetic coupling between the two conductors of the line. The equivalent circuit shown in Figure A.1 applies.

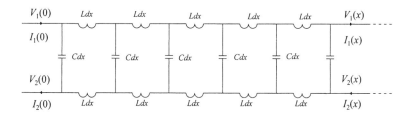

Figure A.1 The equivalent circuit of an unbalanced transmission line.

The current at position x on the line will be considered first. The current can be expressed in terms of balanced and unbalanced components as follows:

$$I_1(x) = I_b(x) - I_u(x) \tag{A.1}$$

and

$$I_2(x) = I_b(x) + I_u(x) \tag{A.2}$$

$I_b(x)$ in these equations is the balanced component of the current, while $I_u(x)$ is the unbalanced component.

It follows from these two equations that

$$I_2(x) = [I_b(x) - I_u(x)] + 2 I_u(x) = I_1(x) + 2 I_u(x) \tag{A.3}$$

$$= I_1(x) + I_0(x) \tag{A.4}$$

where $I_0(x)$ is the difference between the two currents.

Because there is no ground path on the line itself, the difference current $[I_0(x)]$ must be the same at all positions along the line (refer to Figure 6.17, if necessary), and (A.4) can therefore be simplified to

$$I_2(x) = I_1(x) + I_0 \tag{A.5}$$

With the currents defined, it follows that

$$D_x \cdot I_1(x) = -sCV_{12}(x) \tag{A.6}$$

$$D_x \cdot I_2(x) = -sCV_{12}(x) \tag{A.7}$$

$$D_x \cdot V_1(x) = -sL\,I_1(x) \tag{A.8}$$

$$D_x \cdot V_2(x) = sL\,I_2(x) = sL\,I_1(x) + sL\,I_0 \tag{A.9}$$

$$V_{12}(x) = V_1(x) - V_2(x) \tag{A.10}$$

Differentiation of (A.6) yields that

$$D_x^2 \cdot I_1(x) = -sC\,[D_x \cdot V_{12}(x)]$$

$$= -sC\,[D_x \cdot V_1(x) - D_x \cdot V_2(x)]$$

$$= -sC\,[-sL\,I_1(x) - sL\,I_1(x) - sL\,I_0]$$

$$= 2s^2\,LC\,I_1(x) + s^2\,LC\,I_0$$

that is,

$$(D_x^2 - 2s^2\,LC) \cdot I_1(x) = s^2\,LC\,I_0 \tag{A.11}$$

The solution to the equation is

$$I_1(x) = A\,e^{-\sqrt{2LC} \cdot sx} + B\,e^{\sqrt{2LC} \cdot sx} - I_0/2 \tag{A.12}$$

$$= I_0/2 + A\,e^{-\Gamma x} + B\,e^{+\Gamma x} \tag{A.13}$$

where

$$\Gamma = \sqrt{2LC} \cdot s = j\omega \cdot \sqrt{2LC} \tag{A.14}$$

The equation for $I_2(x)$ can be derived similarly and is given by

$$I_2(x) = I_0/2 + Ae^{-\Gamma x} + Be^{+\Gamma x} \tag{A.15}$$

An expression for $V_1(x)$ can now be obtained by integrating (A.8) after substitution of $I_1(x)$ and by using (A.13):

$$V_1(x) = V_1(0) - Z_0/2 \cdot (A - B) + Z_0/2 \cdot [Ae^{-\Gamma x} - Be^{\Gamma x}] \\ + sLx \cdot I_0/2 \tag{A.16}$$

where

$$Z_0 = \sqrt{\frac{2L}{C}} \tag{A.17}$$

The required expression for $V_2(x)$ follows similarly:

$$V_2(x) = V_2(0) + Z_0/2 \cdot (A - B) - Z_0/2 \cdot [Ae^{-\Gamma x} - Be^{\Gamma x}] \\ + sLx \cdot I_0/2 \tag{A.18}$$

An expression for $V_{12}(x)$ can be obtained by using (A.6):

$$V_{12}(x) = V_1(x) - V_2(x) = Z_0[Ae^{-\Gamma x} - Be^{\Gamma x}] \tag{A.19}$$

The equations derived are summarized below:

$$I_1(x) = -I_0/2 + Ae^{-\Gamma x} + Be^{+\Gamma x}$$

$$I_2(x) = I_0/2 + Ae^{-\Gamma x} + Be^{+\Gamma x}$$

$$V_1(x) = V_1(0) - Z_0/2 \cdot (A - B) + Z_0/2 \cdot [Ae^{-\Gamma x} - Be^{\Gamma x}] \\ + sLx \cdot I_0/2$$

$$V_2(x) = V_2(0) + Z_0/2 \cdot (A - B) - Z_0/2 \cdot [Ae^{-\Gamma x} - Be^{\Gamma x}] \\ + sLx \cdot I_0/2$$

$$V_{12}(x) = V_1(x) - V_2(x) = Z_0[Ae^{-\Gamma x} - Be^{\Gamma x}]$$

REFERENCE

[1]   Abrie, P. L. D., "Impedance-Matching Networks and Bandwidth Limitations of Class B Power Amplifiers in the HF and VHF Ranges," Master's Thesis, University of Pretoria, 1982.

ABOUT THE AUTHOR

Pieter Abrie received a B.Eng. in electrical engineering from the University of Pretoria (South Africa), in 1975, and his master's degree and engineering doctorate in electronics in 1982 and 1986, respectively.

In 1977 he joined the Department of Electronic Engineering of the University of Pretoria as a lecturer and specialized in the design of analog and radio-frequency circuits. While lecturing at the university, he presented many short courses on the design of radio-frequency circuits to the local industry through the Laboratory of Advanced Engineering (LGI), and developed the first version of his MultiMatch Impedance-Matching software.

Dr. Abrie spent a one-year sabbatical from May 1985 to April 1986 at the GEC Hirst Research Centre in Wembley (United Kingdom) in the compound semiconductor laboratory, where he was involved in the design of distributed amplifiers. He also developed a program for determining the constructional features of the optimum MESFET for a distributed amplifier with prescribed gain performance during this period.

His first book, *The Design of Impedance-Matching Networks for Radio-Frequency and Microwave Amplifiers*, was published by Artech House in 1985. An expanded version of this book, *Design of RF and Microwave Amplifiers and Oscillators*, was published in 1996, also by Artech House.

Dr. Abrie founded Ampsa CC in May 1986 to market his MultiMatch Synthesis software. He left the University of Pretoria in 1989 as an associate professor to pursue his software development and marketing activities on a full-time basis. Since then, most of his time was invested in development of the MultiMatch Software, which includes the MultiMatch Impedance-Matching Wizard and the MultiMatch Amplifier Design Wizard.

Dr. Abrie moved to the United States in July 2002 to promote his MultiMatch Synthesis software. He returned to South Africa in February 2004 to join Stratex Networks (Cape Town Operations) as a principal engineer (RF), where he was responsible for the design and development of transmit and receive modules. He joined Hittite Microwave Corporation in April 2006 as a principal engineer and there he designed various broadband and narrowband millimeter-wave MMIC LNAs until the end of March 2007. He joined SAAB Avitronics (Pretoria) in February 2009 as a principal microwave engineer. His current responsibilities include designing broadband limiting amplifiers. His MultiMatch software development activities continue through Ampsa (Pty) Ltd.

Dr. Abrie is also a member of the IEEE Microwave Theory and Techniques Society.

INDEX

1:4 transmission line transformer, 194
1-dB compression point, 64, 77, 83, 87

ACPR, 64, 79

active biasing, 433
admittance plane, 57, 374, 376, 392
Al-product, 126, 128
American wire gauge, 112
amplifier chain, 82
attenuation, 128–129, 239
attenuation constant, 132
auto-transformer, 193
auxiliary amplifier, 74
available power, 5, 20
available power gain, 391, 401
available power gain circle, 413

balanced amplifier, 436–438
balun, 124–125, 226
balun core, 125, 165
bandpass identities, 304
bandwidth, 139, 141, 152, 153, 160, 162,
        170, 194, 210, 220
bias point, 404
bipolar transistor, 88, 404–406, 408–410
bond wire, 237, 254
bond wire inductors, 252
boundary, 67
boundary lines, 50, 86
branch, 16, 31
branch multipliers, 31
building block, 201, 205

capacitance, 111, 193, 203, 248
capacitance ratio, 457
capacitor, 105–106, 231, 248
cascade, 82, 87, 91
cascade amplifiers, 424
cascade network, 62, 92–94
cascade representation, 58, 60
cascaded networks, 11
center frequency, 141
ceramics, 108
chamfer, 262
characteristic impedance, 128–129, 132, 134,
        193, 200, 209–210, 236, 240–241
Chebyshev function, 290

chip capacitors, 107
chip coils, 111
circle, 57, 374, 381, 392, 395
circuit Q, 149
circulator, 428, 434
circular loop, 252
class A, 64–65, 69, 87, 224
class AB, 64, 404
class B, 64, 67–71, 76, 87–88, 200, 224, 226
class C, 224
class E, 51
class F, 51
class-A load line, 65
class-B load line, 67
class-C load line, 70
class-E load line, 73
class-F stage, 71
class-F load line, 70
clipping, 67, 87, 89
coaxial cable, 128–129, 193, 210
coaxial lines, 129
coil, 111, 118, 121–122
combiner, 200
common-base, 7
common-collector, 8
common-drain, 98–100
common-emitter, 8
common-gate, 97–99
common-source, 97–98, 100
compensation, 216–219, 222
complex impedance, 443
complex terminations, 158
compression, 439
compression point, 83
conditional stability, 370
conductor, 113
conduction angle, 64, 70
configuration, 147, 196–198, 200, 430, 439, 455
conjugate match, 390
constant operating power gain, 394, 396
constant operating power gain circle, 396
constant noise figure circle, 400
contours, 69
control network, 424
copper losses, 133, 167
core, 124–125, 193, 203, 209
core dimensions, 116
core size, 126–127

correlation matrix, 52, 58, 61
coupled coils, 188
coupling, 193, 203, 241, 246, 453
coupling factor, 168, 170–171, 174, 180,
        186, 188
Cripps approach, 64, 84–85, 89
current, 193–194, 202
current-representation, 58–59
curving radius, 263
cutoff frequency, 170, 209, 217

Darlington synthesis, 282, 291
dc power, 66, 68
dc dissipation, 68
decoupling, 105, 235
definitions, 4, 14, 16
device-modification, 415
diagram, 16
dielectric constant, 129, 132
dielectric puck, 454–455
dielectric resonator oscillator, 455
dielectric resonators, 452
diode, 434
discontinuity effects, 259
dispersion, 132
dissipation factor, 107–108, 133
distortion, 77, 80, 82
distributed model, 232
Doherty amplifiers, 51, 64, 75
Doherty topology, 76
double-tuned transformer, 179
dualism, 144
dynamic range, 77–79, 391, 431

edge capacitance, 237
effective output power, 404
efficiency, 50, 68, 73, 224
electrical field, 113
equivalent circuit, 2–4, 9, 14, 85, 106, 109, 165,
        169–170, 176, 181, 211, 237, 240, 242,
        246, 273, 392, 455–456
equivalent noise sources, 52
equivalent passive problem, 370, 391, 413
error function, 405
external load, 49, 88

feedback, 85, 439
feedback network, 383
ferrite materials, 115
FET, 88
FETs, 404–406, 408–410
field strength, 113
fields, 113
film resistor, 231, 233–234
fixed-valued component, 443
flicker noise, 459

flow diagram, 441
flow graph, 32
flux, 196
flux coupling, 165
flux density, 117, 125, 128, 213, 221
Fourier series, 67
free-space, 240, 242
frequency response, 140
Friss' formula, 62
fundamental tone, 67–68, 70–72, 76, 82

gain compression, 79–80, 439
gain circle, 380, 395, 397, 410–411
gain leveling, 429
gain slope, 415
gain tapering, 225
gain-bandwidth constraints, 271, 416
gap capacitor, 231, 235, 249
gradient vector, 347
ground plane, 105, 237, 252–253

half-sinusoid, 67
harmonic, 50–51, 71
harmonic components, 79
harmonic currents, 87
highpass identities, 303
hybrid combiner, 436
hybrid couplers, 428
hybrid divider, 220, 436
hybrid transformer, 201

I/V-plane, 67, 86
I/V-constraints, 70–71, 76
ideal transformer, 165
image reflection, 253
impedance function, 272, 283
impedance matrix, 15
incident current, 18
increment frequency, 308, 310
indefinite admittance matrix, 7
indefinite S-matrix, 35
inductance, 106, 116, 118, 125, 128
inductors, 105, 109, 124, 231
infinite chain, 62, 83
inherently stable, 395
input admittance, 6
insertion loss, 63, 139, 160, 162, 185, 238–239
intercept point, 80–83, 87
intrinsic load, 49, 66, 87–89
intrinsic load termination, 69
intrinsic load line, 84
intrinsic power, 85
isolation, 201, 429

L-section, 139, 146–147, 149–150, 226
L_1 error, 405

large-signal, 447
LC transformers, 286–287
leakage, 169
leakage flux, 167
leakage inductance, 165
line segments, 67, 305
linearity, 64, 77
Linville stability factor, 375

load impedance, 168
load line, 66, 69, 75, 438, 442
load stability factor, 379
load-line area, 86
load-pull, 50
load-pull contours, 90, 405
loaded Q, 385
loop gain, 372, 383–385, 438, 442,
     444–445, 451
loop product, 31
loops, 33, 34
loss tangent, 107, 133
losses, 85, 106, 109, 112, 116–117,
     132, 166–167
lossless feedback, 429, 431–432
lossless feedback amplifier, 429
lossless network, 11
lossless two-port, 43
lowpass identities, 303
LSM, 274–275, 279, 307, 314–319, 321–324,
     328–329

MAG, 5, 391
magnetic core, 125, 194
magnetic flux, 165
magnetic field, 112, 202
magnetic material, 105, 109, 124, 202, 209
magnetizing inductance, 165, 167, 195, 209,
     211, 213, 222
mapping parameters, 86
materials, 235
matrix, 2
maximum controlled mismatch stable gain, 381
maximum double-sided mismatched gain, 370
maximum relative deviation, 347
maximum single-sided-matched stable gain, 381
maximum tunable gain, 389
microstrip, 232–235, 246, 248–250, 259
microstrip bend, 261
microstrip lines, 128–129, 132
MIM capacitor, 249–250
minimum detectable signal, 77
minimum-admittance functions, 273
minimum-impedance, 272
minimum-impedance function, 311
miter, 263

model, 50, 88, 167, 232, 248–250, 404, 440
modification circuit, 422
modification network, 416, 418, 420
modification sections, 391
MSG, 5
mu, 378
mu-prime, 379
multiplate capacitors, 233
multistage, 424
multistage power amplifiers, 50

node, 30, 31
noise circles, 402
noise figure, 6, 52–53, 56, 62–63, 391,
     400, 414
noise figure circle, 57–58, 401
noise floor, 77
noise match, 6
noise matching, 403
noise measure, 62, 83
noise parameters, 49, 52
noise power, 54
noise sources, 52, 60
normalizing impedance, 16, 25–26, 29, 38
Norton coupler amplifier, 432

odd-mode stability factors, 371
odd-mode oscillation, 371
oneport, 25–26
open-ended stub, 259
operating power gain, 390–391, 446–447
optimum terminations, 224
optimum power terminations , 421
oscillation frequency, 439
oscillations, 371, 382–383
oscillator, 385, 437–438, 451, 455
oscillator configurations, 438
oscillator design, 441
oscillator topologies, 440
output current , 68, 70–71, 76
output power, 49, 64, 66 , 404, 438–439, 446
output voltage, 70–71, 76
overlay capacitor, 233, 251

package parasitics, 405
parallel double-tuned transformer, 180
parallel feedback, 451
parallel-plate capacitor, 231–232, 237–240, 242,
     249, 250
parasitic absorption, 281–282
parasitic capacitance, 109, 119
parasitic inductance, 457
parasitics, 105
passive cascade, 63
passive network, 62–63, 91

path, 33–34
peak envelope power, 356
PEP, 356
phase noise, 440, 459
phase shift, 438
phase velocity, 131
PI-section, 139, 151–155, 157, 343–345, 438,
    442, 451
planar transmission lines, 106
point junction, 106
polarity, 196
poles, 38
potentially unstable, 396
power, 19, 40, 49, 64, 139, 160, 166, 404, 442
power amplifier, 223–224, 226–227, 230, 356
power contours, 69, 83–85, 91
power gain, 4, 390
power parameter approach, 64, 404
power parameters, 49–50, 64, 86, 90–91, 93–95,
    97–100, 438
power splitter, 200
power termination, 421, 423
primary current, 166
primary inductance, 169
propagation constant, 113
proximity effect, 115, 134
puck, 453
push-pull, 200

Q, 119
Q-factor, 108, 110, 122, 141–142
Q-values, 112
quality factor, 107

ratings, 223
ratios, 194
reactive loading, 371, 457
reactive load line, 69
real-frequency, 305
reference temperature, 53
reflected component, 18
reflected power, 20
reflection amplifier, 435
reflection coefficients, 12, 22, 292, 376,
    437, 438
reflection gain, 372, 382
reflection parameter, 20, 23, 25, 27, 29, 435
rejection, 142, 180
representations, 52
residues, 273
resistance, 110, 114, 195
resistive load line, 69
resonant circuit, 87, 140, 144, 146
resonant frequency, 107, 110–111,
    121, 235–236, 239–240
resonating frequencies, 105

resonator, 443, 456
resonator circuit, 452, 456
ribbon, 237
ribbon inductor, 252
Richards' transformation, 299
ripple, 179, 180
RLC networks, 354
RLC matching network, 225
rod core, 125
rods, 124
Rollette stability factor, 378, 384
route, 33

S-parameters, 11, 14, 22, 31, 42, 44, 88, 188,
    189, 320, 437
saturation, 446
saturation resistance, 65, 223
saturation voltage, 65, 223
scattering matrix, 20, 40
scattering parameters, 27

secondary inductance, 169
secondary winding, 188
self-capacitance, 111, 122–123
semi-infinite functions, 309–310
sensitivity, 424–427
sensitivity factor, 426
series double-tuned transformer, 185
series feedback, 95, 448, 451
signal flow graphs, 30
single-layer capacitors, 234
single-tuned transformer, 172
size, 125
skin depth, 113–114
skin-effect, 129, 134, 234
small-signal model, 88, 90, 404–410
Smith chart, 58, 376, 378, 380, 392–393,
    396, 398, 401, 413, 444
solenoidal coil, 111, 118, 255
solenoidal inductors, 255
source stability factor, 379
spiral inductor, 233, 255, 258, 259
spurious free dynamic range, 78–79
square spiral inductor, 257
square wave, 72
stability, 370, 415
stability circle, 376–377, 380
stability factors, 371
stabilizing resistance, 372
stable area, 375, 377
stacked cores, 124, 214
stacked toroids, 125
standard wire gauge, 112
step junction, 260
Sterne stability factor, 375
strip inductor, 252

stub, 238
superposition, 54

T-junction, 264
T-section, 151, 155, 157, 158–159, 344,
      444, 448
tapped coil, 173, 176
temperature, 118
thermal runaway, 108
thin-film resistors, 105
third-order intercept point, 77
TOI, 77
topologies, 440
toroid, 124–125
toroidal core, 125, 127, 170, 213–214
toroids, 165
transconductance, 440
transducer power gain, 77, 161, 382, 392–393,
      434–435, 438
transducer power gain circles, 400
transformation, 146, 150, 152, 154
transformation diagram, 151, 158
transformation factor, 177
transformation matrices, 60
transformation Q, 147, 149
transformation ratios, 196
transformation step, 147
transformer, 165, 193, 205–206, 208–209, 282
transistor, 404
transmission line, 13, 128, 133, 193, 201–203,
      211, 233, 342–343
transmission matrix, 10, 61, 234
transmission parameter, 10, 23
transmission zeros, 320
transmission-line models, 249
transmission-line transformer, 193, 195, 225
tunability factor, 388
turns ratio, 168
twisted pairs, 128, 133
two-tone intercept point, 64
two-tone products, 77
two-tone signal, 80
twoport, 3, 9, 11, 25–26, 31–32, 42, 52, 57

unbalanced current, 194
unit element, 300, 301, 304
unitary matrix, 40
unloaded Q, 125, 146, 160, 178, 454

varactor, 443, 457
varactor network, 457
varactor type, 442, 457
via holes, 265
voltage, 145

voltage gain, 189
voltage representation, 59
voltage-controlled oscillator, 443

waveforms, 65
winding, 193
wire thickness, 121

Y-parameter matrix, 373
Y-parameters, 1, 44, 244, 246, 394

Z-parameters, 8, 168, 188–189
zeros, 38

Recent Titles in the Artech House Microwave Library

Active Filters for Integrated-Circuit Applications, Fred H. Irons

Advanced Techniques in RF Power Amplifier Design, Steve C. Cripps

Automated Smith Chart, Version 4.0: Software and User's Manual, Leonard M. Schwab

Behavioral Modeling of Nonlinear RF and Microwave Devices, Thomas R. Turlington

Broadband Microwave Amplifiers, Bal S. Virdee, Avtar S. Virdee, and Ben Y. Banyamin

Computer-Aided Analysis of Nonlinear Microwave Circuits, Paulo J. C. Rodrigues

Designing Bipolar Transistor Radio Frequency Integrated Circuits, Allen A. Sweet

Design of FET Frequency Multipliers and Harmonic Oscillators, Edmar Camargo

Design of Linear RF Outphasing Power Amplifiers, Xuejun Zhang, Lawrence E. Larson, and Peter M. Asbeck

Design Methodology for RF CMOS Phase Locked Loops, Carlos Quemada, Guillermo Bistué, and Iñigo Adin

Design of RF and Microwave Amplifiers and Oscillators, Second Edition, Pieter L. D. Abrie

Digital Filter Design Solutions, Jolyon M. De Freitas

Distortion in RF Power Amplifiers, Joel Vuolevi and Timo Rahkonen

EMPLAN: Electromagnetic Analysis of Printed Structures in Planarly Layered Media, Software and User's Manual, Noyan Kinayman and M. I. Aksun

Essentials of RF and Microwave Grounding, Eric Holzman

FAST: Fast Amplifier Synthesis Tool—Software and User's Guide, Dale D. Henkes

Feedforward Linear Power Amplifiers, Nick Pothecary

Foundations of Oscillator Circuit Design, Guillermo Gonzalez

Fundamentals of Nonlinear Behavioral Modeling for RF and Microwave Design, John Wood and David E. Root, editors

Generalized Filter Design by Computer Optimization, Djuradj Budimir

High-Linearity RF Amplifier Design, Peter B. Kenington

High-Speed Circuit Board Signal Integrity, Stephen C. Thierauf

Intermodulation Distortion in Microwave and Wireless Circuits, José Carlos Pedro and Nuno Borges Carvalho

Introduction to Modeling HBTs, Matthias Rudolph

Lumped Elements for RF and Microwave Circuits, Inder Bahl

Lumped Element Quadrature Hybrids, David Andrews

Microwave Circuit Modeling Using Electromagnetic Field Simulation, Daniel G. Swanson, Jr. and Wolfgang J. R. Hoefer

Microwave Component Mechanics, Harri Eskelinen and Pekka Eskelinen

Microwave Differential Circuit Design Using Mixed-Mode S-Parameters, William R. Eisenstadt, Robert Stengel, and Bruce M. Thompson

Microwave Engineers' Handbook, Two Volumes, Theodore Saad, editor

Microwave Filters, Impedance-Matching Networks, and Coupling Structures, George L. Matthaei, Leo Young, and E.M.T. Jones

Microwave Materials and Fabrication Techniques, Second Edition, Thomas S. Laverghetta

Microwave Mixers, Second Edition, Stephen A. Maas

Microwave Radio Transmission Design Guide, Trevor Manning

Microwaves and Wireless Simplified, Third Edition, Thomas S. Laverghetta

Modern Microwave Circuits, Noyan Kinayman and M. I. Aksun

Modern Microwave Measurements and Techniques, Second Edition, Thomas S. Laverghetta

Neural Networks for RF and Microwave Design, Q. J. Zhang and K. C. Gupta

Noise in Linear and Nonlinear Circuits, Stephen A. Maas

Nonlinear Microwave and RF Circuits, Second Edition, Stephen A. Maas

QMATCH: Lumped-Element Impedance Matching, Software and User's Guide, Pieter L. D. Abrie

Practical Analog and Digital Filter Design, Les Thede

Practical Microstrip Design and Applications, Günter Kompa

Practical RF Circuit Design for Modern Wireless Systems, Volume I: Passive Circuits and Systems, Les Besser and Rowan Gilmore

Practical RF Circuit Design for Modern Wireless Systems, Volume II: Active Circuits and Systems, Rowan Gilmore and Les Besser

Production Testing of RF and System-on-a-Chip Devices for Wireless Communications, Keith B. Schaub and Joe Kelly

Radio Frequency Integrated Circuit Design, John Rogers and Calvin Plett

RF Design Guide: Systems, Circuits, and Equations, Peter Vizmuller

RF Measurements of Die and Packages, Scott A. Wartenberg

The RF and Microwave Circuit Design Handbook, Stephen A. Maas

RF and Microwave Coupled-Line Circuits, Rajesh Mongia, Inder Bahl, and Prakash Bhartia

RF and Microwave Oscillator Design, Michal Odyniec, editor

RF Power Amplifiers for Wireless Communications, Second Edition, Steve C. Cripps

RF Systems, Components, and Circuits Handbook, Ferril A. Losee

Stability Analysis of Nonlinear Microwave Circuits, Almudena Suárez
and Raymond Quéré

System-in-Package RF Design and Applications, Michael P. Gaynor

*TRAVIS 2.0: Transmission Line Visualization Software and User's
Guide, Version 2.0,* Robert G. Kaires and Barton T. Hickman

Understanding Microwave Heating Cavities, Tse V. Chow Ting Chan
and Howard C. Reader

For further information on these and other Artech House titles,
including previously considered out-of-print books now available
through our In-Print-Forever® (IPF®) program, contact:

Artech House Publishers	Artech House Books
685 Canton Street	46 Gillingham Street
Norwood, MA 02062	London SW1V 1AH UK
Phone: 781-769-9750	Phone: +44 (0)20 7596 8750
Fax: 781-769-6334	Fax: +44 (0)20 7630 0166
e-mail: artech@artechhouse.com	e-mail: artech-uk@artechhouse.com

Find us on the World Wide Web at: www.artechhouse.com